The *Natuurkundige Commissie* in the Netherlands Indies (1820–1850)

Emergence of Natural History

VOLUME 8

The titles published in this series are listed at *brill.com/enh*

The *Natuurkundige Commissie* in the Netherlands Indies (1820–1850)

By

Pieter van Wingerden

BRILL

LEIDEN | BOSTON

Cover illustration: 'Onze legerplaats in de vlakte Wienoto' ('Our camp on the plain of Winoto'), by Pieter van Oort. Plate 29 (Timor) from the volume *Land- en volkenkunde* of the *Verhandelingen*, NCAO, 871_029. Public domain. An indigenous man shows a snake to two members of the *Natuurkundige Commissie* that can possibly be identified as Salomon Müller on the left and Pieter van Oort on the right. In Gasso, E., Stork, L., Weber, A. et al. ed., *Natuurkundige Commissie Archives Online*. Leiden: Brill 2020. Naturalis Biodiversity Center Leiden; The Netherlands. https://dh.brill.com/nco/view/nco_NNM001000871_029/makingsense

The Library of Congress Cataloging-in-Publication Data is available online at https://catalog.loc.gov
LC record available at https://lccn.loc.gov/2025002962

Typeface for the Latin, Greek, and Cyrillic scripts: "Brill". See and download: brill.com/brill-typeface.

ISSN 2452-3283
ISBN 978-90-04-69686-0 (hardback)
ISBN 978-90-04-69687-7 (e-book)
DOI 10.1163/9789004696877

This book is printed on acid-free paper and produced in a sustainable manner.

Contents

Acknowledgements

This book was born from the chance reading of an English translation of *Histoire de l'ornithology* by Valérie Chansigaud, a beautifully illustrated account of humanity's fascination with birds.[1] In her chapter on the nineteenth century, she mentioned something called the Natuurkundige Commissie. When I went looking for a book on the Commissie and could not find a recent or comprehensive one, I decided to have a go at it and write the book that I wanted to read. The rest is history.

I would like to express my heartfelt thanks to Frans van Lunteren and Andreas Weber. Both, independent of each other, responded positively to my request for a meeting to discuss the possibility of a PhD on the Natuurkundige Commissie. We first met on August 24, 2017, in Frans' office in Leiden. During the next five years, Frans and Andreas helped me to navigate the pitfalls of writing a PhD. After having successfully defended it on October 10, 2023, Andreas guided me through the process of turning my thesis into a monograph. A heartfelt thanks also to both anonymous reviewers who provided me with many helpful comments.

I am grateful for my colleagues in the library world, both national and international, for their continued support and the occasional scan. I am also indebted to the countless unnamed and invisible people who have diligently worked on filling the Internet Archive, the Biodiversity Heritage Library, Delpher, and Google Books with just the publications that I needed. Thanks to Karien Lahaise, archivist at Naturalis Biodiversity Center, who has made my life easier by arranging access to both physical and digital archival resources. Thanks to my fellow historians of science (mostly through Gewina) for your interest, conversations, encouragement, and occasional critical comments.[2]

Last but not least, thank you, Hanna-Ruth, for listening to my endless stories of Forsten and Diard. You have not faltered in your support for me, even when we were hit by a pandemic and all sorts of other things. It is to you that I owe an unspeakable and immeasurable amount of gratitude. I'm blessed beyond measure with you in my life.

1 Chansigaud, *History of Ornithology*, published in North-America as *All About Birds*.

2 I must mention Marie-Odette Scalliet by name, who kindly shared her research notes on Diard with me.

Figures, Maps, and Tables

Figures

Maps

Tables

Archives Used

Erfgoed Leiden en Omstreken, Leiden

NCAO	Natuurkundige Commissie Archives Online, https://dh.brill.com/nco/
NL-ArnGA	Gelders Archief, Arnhem
NL-HaaNHA	Noord-Hollands Archief, Haarlem
NL-HaNA	Nationaal Archief, The Hague
2.02.01	Algemene Staatssecretarie
2.02.04	Kabinet des Konings, Kabinet der Koningin, 1841–1897
2.10.01	Ministerie van Koloniën, 1814–1849
2.10.30	Ministerie van Koloniën, 1850–1900
2.10.36.22	Ministerie van Koloniën: Stamboeken Burgerlijke Ambtenaren, Serie A
2.10.50	Ministerie van Koloniën: Stamboeken en pensioensregisters Militairen KNIL Oost-Indië en West-Indië
2.13.04	Stamboeken Officieren Landmacht en koloniale troepen
2.21.018	Collectie 197 Familiearchief Van Beresteyn
4.MIKO	Ministerie van Koloniën – Kaarten en Tekeningen
NL-LdnNAT	Naturalis Archief, Leiden
NL-LdnSAL	Stadsarchief Leiden, Leiden
NL-MidZA	Zeeuws Archief, Middelburg
NL-TieRAR	Regionaal Archief Rivierenland, Tiel
NL-RotSAR	Stadsarchief Rotterdam, Rotterdam
NL-UtrUA	Het Utrechts Archief, Utrecht

Abbreviations used in archival references:

f.	folio
inv.no.	inventory number
KB	Koninklijk Besluit = Royal Decree
Res.CG	Resolution of the commissioner-general
Res.GG	Resolution of the governor-general

Note on the Use of Historical Terms and Names

This monograph studies historical events in the colonial world of the nineteenth century. Our attitude to and our language about colonialism has changed in recent decades. However, due to the historical nature of this study, I have had no choice but to occasionally use historical terminology and quote historical passages that are nowadays considered to be derogatory, racist and Eurocentric. I only use these words and phrases because to do otherwise would be to deny the reality of history.

In addition, I have chosen to use historical names for some of the more well-known locations in the Indonesian Archipelago that feature in this study, like Batavia (Jakarta), Buitenzorg (Bogor), Borneo (Kalimantan), or Celebes (Sulawesi). For other place names I have tried to use the current name and spelling as much as possible, like Bengkulu, Dejima, or Banten. If the current name of a place could not be established, I have followed the spelling in the original (mostly Dutch) sources.

Introduction

> [One is tempted] to ask whether Kuhl and van Hasselt, Boie and Macklot, Horner and Forsten did not die in vain for their idol, science.[1]

This is how Ernst Stresemann, writing in 1951, concludes his chapter on the Natuurkundige Commissie.[2] The Natuurkundige Commissie was the most prominent, long-lasting, and expensive government-sponsored expression of the Dutch desire to understand what could be gained from the Netherlands Indies during the first half (and possibly the whole of) the nineteenth century.[3] The Commissie was active from 1820 to 1850 and was founded by the Dutch government in order 'to promote the knowledge of the natural condition and the products [of nature] of our possessions in the East Indies.'[4] When Stresemann evaluates the work of this Commissie, he expresses himself rather hesitantly, but it seems that by asking the question he already tells us in which direction his answer can be found. While Stresemann may have truly believed that the members of the Natuurkundige Commissie all died in vain, this statement was made in a very specific context.

Stresemann's perspective is that of a European naturalist, who considered the fact that much of the collections of the Natuurkundige Commissie remained unpublished at the time as a serious waste of opportunity. Promoting the knowledge of nature in the Netherlands Indies could only be done by publishing the collections. Not doing this sufficiently was seen as a serious shortfall and led to the suggestion that the young men who did not return from their

---

1 '[Man ist] versucht, sich zu fragen, ob nicht die Kuhl und van Hasselt, die Boie und Macklot, die Horner und Forsten vergebens für die vergötterte Wissenschaft in den Tod gegangen sind.' Stresemann, *Entwicklung der Ornithologie*, 154. Translation from Stresemann, *Ornithology*, 152.

2 The Dutch word *natuurkunde*, literally means 'knowledge of nature.' Nowadays, the meaning of *natuurkunde* is limited to physics, but in the early nineteenth century it carried a meaning equivalent to 'natural history.' The word *commissie* can be translated as 'commission,' and carries the meaning of a group of people appointed to perform a certain task. The most accurate English translation would be Natural History Commission, although in secondary English literature variations occur. In order to avoid confusion, I have chosen to use the Dutch term throughout this book.

3 When I use 'Netherlands Indies' or 'Indies' in this book I refer to the Netherlands East Indies.

4 'Bevorderen de kennis van de natuurlijke gesteldheid en van de voortbrengsels onzer bezittingen in de Oost-Indiën.' KB, May 2, 1820, no. 10. NL-HaNA, Staatssecretarie,2.02.01, inv.no. 988.

 | DOI:10.1163/9789004696877_002

collecting mission died in vain. In Stresemann's view, the publication of the *Verhandelingen* was too little too late.[5]

Over the years, Stresemann's work has become a classic text in the history of ornithology. As a result, his view has been repeatedly picked up and reiterated without questioning the validity of his statement. In 2003, Michael Walters went even further in *A Concise History of Ornithology* and outrightly blamed Temminck, the driving force behind the Commissie, for its failure:

> Temminck had failed to understand the excellent work done by all these men in surveying the natural history of the area, for he was really only interested in spectacular specimens for his museum.[6]

Even in the introduction of the standard field guide for the birds of Indonesia, Bas van Balen stayed in the tradition of Stresemann and Walters and wrote in 2016 that 'the Commission had proved to be disappointingly ineffective, aggravated by untimely deaths of four of its naturalists.'[7] To add insult to injury, in an early published version of a chapter for a popular history book celebrating 200 years of Naturalis Biodiversity Center, Menno Schilthuizen described the Natuurkundige Commissie as 'Temminck's cannon fodder,' a rather negative qualification that was scrapped for the final version.[8]

Stresemann's authoritative hesitation is clearly a factor that has played a role in the study of the Natuurkundige Commissie. His tentative dismissal of the value of the work of the Commissie may have slowed down or even prevented any further research into the nature of the Commissie. As a result, the Natuurkundige Commissie is seriously understudied, not only in the narrow sense of studying its collections, but also in the wider sense of studying the context in which these collections were gathered. No one has picked up the gauntlet that Stresemann has thrown down, and a scholarly monograph on

5 The first instalment of the major publication associated with the Natuurkundige Commisie (*Verhandelingen over de natuurlijke geschiedenis der Nederlandsche overzeesche bezittingen door de Leden der Natuurkundige commissie in Indië en andere schrijvers, uitgegeven op last van den koning door C. J. Temminck*, edited by J. A. Susanna, hereafter referred to as the *Verhandelingen*) was published on December 18, 1839. Twenty-eight further instalments were published between 1840 and 1847, divided in three unnumbered parts: *Zoologie* (Zoology), *Botanie* (Botany), and *Land- en volkenkunde* (Geography and Ethnography). Traditionally, *Zoologie* is considered to be volume one, *Botanie* volume two and *Land- en volkenkunde* volume three. See Husson and Holthuis, "Dates of Publication," 22.

6 Walters, *Concise History*, 93.

7 Van Balen, "Ornithological History," 17.

8 'Temminck's kanonnenvoer.' Schilthuizen, "Terabytes aan tropenjaren," 8. A chapter of the same name was later published in Schilthuizen and Vonk, *Wie wat bewaart*.

the Natuurkundige Commissie has been lacking for decades. With this study, I hope to fill this lacuna and invite further research.

As Nicholas Jardine, James A. Secord, Emma Spary, and Helen Anne Curry have helped us to understand, natural history is a vibrant and far-reaching complex that cannot be understood in glorious isolation. It cannot be presented as knowledge 'generated by isolated individuals working wholly within the domain of the mind.' Rather, it is 'the product of conglomerates of people, natural objects, institutions, collections, finances, all linked by a range of practices of different kinds.'[9] The Natuurkundige Commissie is one such conglomerate and in order to understand it properly, it needs to be treated as such and has to be considered as a complex reality with a vast range of participants (members, assistants, indigenous employees) and stakeholders (the king, the museum, the colonial government).

The main goal of this study is to find out how the Natuurkundige Commissie functioned across space (between the Netherlands and the Indies) and time (over a period of thirty years). This profoundly simple question is not easy to answer. In order to arrive at an answer, I will focus on the following two subquestions: How did the participants in the Natuurkundige Commissie relate to the stakeholders and to each other? How did these relationships influence the practice of the participants? In what follows, I take a tour through recent literature in relevant fields that has led me to using an underlying methodology of geography and a complex of layered loyalties as the best way to answer my question.

But first, I need to briefly stress that this is not a study of the natural historical collections of the Natuurkundige Commissie. Of course, objects from these collections will feature occasionally, but it is the context in which these collections were gathered that interests me. By studying the context of the creation of the natural historical collections of the Natuurkundige Commissie, this study contributes to the necessary decolonisation of this vast body of material colonial heritage, but it does not study the collections themselves. In a similar way, the (context of the) ethnographical collections of the Natuurkundige Commissie fall outside of the scope of this study as well.[10]

9 Jardine, Secord, and Spary, *Cultures of Natural History*, 8.

10 For recent scholarship on the ethnographical aspects of the work of the Commissie, see Ter Keurs, "Introduction"; Barnes, "Recording Cultures"; Ter Keurs, "Collecting in the Colony."

1 From Natural History to Travel and Exploration

Studying the practice of natural history has long been a standard ingredient of the study of the history of natural history. A classic example of this type of study is David Elliston Allen's *The Naturalist in Britain*, in which he studied the social historical component of being a naturalist in Great Britain from the seventeenth to the twentieth century.[11] However, there is more to studying natural history than just the social aspects. Jardine, Secord and Spary distinguish the following types of practices: material, social, literary, bodily, and reproductive:

> As a rough and ready guide we may distinguish the following types of practices. (1) *Material* practices are ways of making, handling and transforming things; [...] (2) *Social* practices cover the whole range of associations, recruitments, delegations and negotiations; [...] (3) *Literary* practices are conventions of genre, representation and persuasion; [...] (4) *Bodily* practices are forms of bodily, sartorial or gestural self-representation and normative accounts of physical and emotional experience in response to particular situations; [...] (5) Last, but by no means least, there are *reproductive* practices, that is, the means by which skills and knowledge are handed on from generation to generation.[12]

The authors of this passage argue that a balance is required in studying these practices. Too much or too little emphasis on one of these sets of practices can seriously distort our perception of natural history as a whole. Their *Cultures of Natural History* is a bird's eye view on the wide variety of ways of understanding natural history that emerges when such a balance is observed. Due to the complexity of the Natuurkundige Commissie, it is impossible to study the full extent of these sets of practices in the context of the Natuurkundige Commissie within the scope of one dissertation.

In the 2018 companion volume *Worlds of Natural History*, Curry, Jardine, Secord and Spary pick up on the variety in the study of natural history and identify three common criteria that unite the previously mentioned practices of natural history: materiality, order, and spatiality. Through the trends and turns between *Cultures* (1996) and *Worlds* (2018), these have been the stable factors in the study of natural history. While *Cultures* (1996) was steeped in the 'decentring' and 'defamiliarisation' of the cultural turn, *Worlds* (2018)

11 Allen, *Naturalist in Britain*. See also Förschler and Mariss, *Akteure, Tiere, Dinge*.

12 Jardine, Secord, and Spary, *Cultures of Natural History*, 8–9. Italics in the original, enumeration added.

shows the impacts of the material turn and the spatial turn. In the study of the history of natural history, these turns resulted in a strong interest in 'the intimate links between global communication and exchange of knowledge on the one hand, and exploration, empire and commerce on the other.'[13] Curry, Jardine, Secord and Spary identify two important trends that have grown out of this: stronger emphasis on the role of indigenous agency and knowledge, and 'greatly increased attention to networks as a model of natural historical practice.'[14]

While this study certainly owes a debt to both these trends, it is important to recognise a third trend that has influenced this study in equal measure. In 1996, incidentally the year when *Cultures* was published, an equally influential volume *Science in the Field* was published by Henrika Kuklick and Robert Kohler. This volume did not focus on the single discipline of natural history but saw it as part of a wider group of scientific disciplines that engage in field work. Field work was seen as the practice of science in the field in 'uncontrolled settings' as opposed to laboratory science.[15] The validity of this approach for natural history had already been signified by Dorinda Outram in her contribution to *Cultures*.[16]

When Kohler looked back on *Science in the Field* in 2011, he identified that out of the study of field sciences three trends could be seen to emerge. The first theme deals with 'the connection between expert and popular or [...] "vernacular" science,' of which *Cultures* is a prime example according to Kohler.[17] Kohler's second theme deals with 'large-scale scientific practices.'[18] But it is Kohler's third theme that this study is most indebted to. This is the theme of travel and exploration. Kohler asserts, and rightly so, that 'exploration for commercial or strategic purposes was often combined with scientific reconnaissance,'[19] which is also recognised throughout in *Cultures* and *Worlds*. While the study of the history of natural history can shine when it concentrates on the field sciences, such as in the 2018 volume *Naturalists in the Field*,[20] it cannot be denied that Kohler's wider dimension of field history holds great merit and that engagement with the developing field of the study of travel and exploration will further enlighten the practices of natural history.

---

13 Curry et al., *Worlds of Natural History*, 7.
14 Curry et al., *Worlds of Natural History*, 7.
15 Kuklick and Kohler, *Science in the Field*, 1.
16 Outram, "New Spaces."
17 Kohler, "History of Field Science," 214.
18 Kohler, "History of Field Science," 216. For an example, see Kohler, *All Creatures*.
19 Kohler, "History of Field Science," 219.
20 MacGregor, *Naturalists in the Field*.

A seminal text in the literature on travel and exploration is Felix Driver's *Geography Militant*, a study of cultures of exploration using the work of the Royal Geographical Society. Driver intentionally latches onto the approach of Jardine, Secord and Spary in *Cultures of Natural History*, and understands exploration 'as a set of cultural practices which involve the mobilization of people and resources.'[21] Just as Kohler widened the study of natural history to the study of the field sciences in general, Driver widens the study of natural history to the study of exploration and looks for material, social, literary, bodily and reproductive practices, creating a tapestry of geographical exploration just like Jardine, Secord and Spary created a tapestry of natural history.

Following on from Driver, Dane Kennedy, in *The Last Blank Spaces*, has written an inspiring comparative study on British exploration in Africa and Australia.[22] His goal is to elucidate how knowledge was produced in the context of British exploration. Kennedy explores the tension that exists between 'exploration as experience and exploration as epistemology.'[23] Even though exploration was about generating knowledge, the explorers were faced by a life in the field that made this process of gathering knowledge complicated and cumbersome beyond anticipation. Africa and Australia were vastly different from the British social club, just as, in the case of the Natuurkundige Commissie, the tropical forests and volcanoes of the Malay Archipelago were vastly different from the natural history museum in Leiden.

A further publication by Dane Kennedy, *Reinterpreting Exploration*, throws some necessary doubt on the direct association between expedition and imperialism, a feature that is central to Martin Thomas' *Expedition into Empire*. Thomas' volume is motivated by the existence of 'a considerable gulf between the effects of expeditions and what we know about them as cultural entities.'[24] While Kennedy argues that we need to 'decenter our understanding of exploration as an imperial enterprise,'[25] Thomas writes that 'a new and more critical dialogue around the concept of expedition [...] will see it recognized as a cultural and [...] as an imperial formation.'[26] There is no denying that the Natuurkundige Commissie has a strong imperial component, even though it was first and foremost a natural history project. As an expedition in a

21 Driver, *Geography Militant*, 8.
22 Kennedy, *Last Blank Spaces*.
23 Kennedy, *Last Blank Spaces*, 2.
24 Thomas, *Expedition into Empire*, 1.
25 Kennedy, *Last Blank Spaces*, 100. As quoted by Thomas, *Expedition into Empire*, 5.
26 Thomas, *Expedition into Empire*, 7.

colonial context, the Natuurkundige Commissie intrinsically harbours a tension between natural history and imperialism.

2 Geography of Science

A comparable tension existed between how the work of the Natuurkundige Commissie was conceived in the Netherlands and how it actually played out in the Netherlands Indies. Like many expeditions, the Commissie simultaneously existed in two different geographical worlds. On the one hand, the members had received their education, their assignment and their inspiration in a European context, while, on the other hand, the realisation of their plans took place in a colonial context in the Netherlands Indies. What happened there was far removed from the world of the museum and laboratory naturalists in the Netherlands. The social practice of the members of the Natuurkundige Commissie 'was grounded in concrete historical and geographical circumstances.'[27] If the Natuurkundige Commissie is to be properly understood, this geographical dichotomy in which it functioned must be taken into account.

The term 'geography of science' is now firmly embedded in academic research,[28] and 'covers the [...] questions regarding the local conditions and shaping of scientific knowledge and spatial diffusion of scientific practices and institutions.'[29] An important text in the development of the concept of geography of science is David Livingstone's *Putting Science in its Place*.[30] While in earlier discussions place and space were recognised as influential elements in the development of science, Livingstone's book gave rise to an extensive body of work dealing with the topic of how science is influenced by where it is created.

In the introduction to a special issue of the *British Journal for the History of Science*, Simon Naylor reviewed the field and identified three central directions in geography of science: places, contexts, and cartographies.[31] While the first concentrates on the actual locations where science takes place

27 Livingstone, *Putting Science in Its Place*, 180.

28 Livingstone and Withers, *Geographies of Nineteenth-Century Science*, ix.

29 Frenken, "Geography of Scientific Knowledge," 1008.

30 Livingstone, *Putting Science in Its Place*. For other work on this topic by Livingstone: Livingstone, "Text, Talk and Testimony"; Meusburger, Livingstone, and Jöns, *Geographies of Science*; Livingstone and Withers, *Geographies of Nineteenth-Century Science*. See also the work of Diarmid Finnegan, e.g. Finnegan, "Spatial Turn"; Finnegan and Wright, *Spaces of Global Knowledge*.

31 Naylor, "Introduction: Historical Geographies of Science."

('laboratories, field sites, museums, caves and pubs'),[32] and the third focuses on the spatial conceptions as produced by science (with Von Humboldt's isolines and Wallace's biogeographical line as examples), the second takes all forms of 'contexts, including the city, the region, and the nation' into account.[33] While the field notes of the Commissie participants contain ample material to study 'places,' and Commissie member Salomon Müller prefigured Wallace's line in his work on biogeography, it is the 'contexts' of the Natuurkundige Commissie that interest me most for this study.

An example of how these contexts matter is Sujit Sivasundaram's analysis of knowledge transmission on Ceylon. The topography of the island, and especially the way in which the British used these sites, was responsible for which types of natural historical knowledge were mediated to the British.[34] In much the same way, the Natuurkundige Commissie operated in a variety of topographical units that were used in different ways by the Dutch. The most important distinction, of course, is the one between the Netherlands and the Netherlands Indies. In addition to this, the activities of the Commissie in the Netherlands Indies often took place in areas that were only nominally under Dutch control and each required different approaches in terms of the shape an expedition would take and the specific tasks that were extended to the participants by the government in Batavia.[35] All in all, the Commissie operated in a thoroughly colonial and imperial context in the Netherlands Indies, which opens up the field of science and empire as an important conversation partner.

3 Science and Empire

In the past, our way of thinking about science and empire was heavily influenced by George Basalla's simple model of the spread of western science from 1967.[36] Science was seen as something that was produced and directed in the European centre or metropole, while the colonial periphery did not generate any science by itself and was only useful insofar as it produced data with which science could be generated. For a long time, Basalla's model was a departure point for many fruitful discussions on colonial science. One only has to call

32 Naylor, "Introduction: Historical Geographies of Science," 6.
33 Naylor, "Introduction: Historical Geographies of Science," 3.
34 Sivasundaram, "Islanded."
35 For an analysis of the 1836 expedition to Borneo in the context of Dutch imperialism, see Van Wingerden, "Op reis vanuit een drijvende stad." For the Natuurkundige Commissie in a context of war on Sumatra, see Van Zanen and Weber, "Schreeuwende apen."
36 Basalla, "Spread of Western Science."

to mind MacLeod's volume *Science and Empire* and Stuchtey's volume *Science Across the European Empires* in order to see the long-lasting significance of Basalla's working model.[37] Over the years, we have come to the conclusion that Basalla's theory displays a fundamental flaw.

The language of centre and periphery was born out of Eurocentrism, and the view that the European centre directed the science of the colonial periphery is no longer accepted.[38] As James Secord puts it, 'divisions between center and periphery [have been] replaced by patterns of mutual interdependence.'[39] No longer was the situation in the centre the benchmark, but the 'periphery' was moved to equal footing. No longer was the colony a faraway place in the margins. In the words of Frederick Cooper and Ann Laura Stoler, 'metropole and colony' needed to be treated 'in a single analytic field' because the resulting transformations were 'a product of both global patterns and local struggles.'[40] This realisation changed the direction of scholarship and required new ways of understanding the relationship between science and colonialism. Flowing from the realisation that even the word colonialism creates inequal categories, the study of science in a colonial setting is now much more often referred to in terms of science and empire, even though the words empire and imperial in and of itself are also not free from a sense of inequality.

It has long been recognised that the relationship between science and empire is an intimate and complex one. Studies like James E. McClellan III's *Colonialism and Science* and Richard Drayton's *Empire of Nature* focus our attention on the fact that the practice of science in colonial climes was heavily influenced and up to a certain point directed by institutions in the coloniser's heartland.[41] An institution like Kew Gardens with it widely dispersed network of colonial correspondents and colonial botanical gardens deliberately invested in becoming and remaining the centre of botanical research in the British empire.[42] At the same time, some of Kew's colonial partners had an agenda of their own as testified to by William Kelleher Storey's *Science and Power in Colonial Mauritius*.[43] Volumes like David Mackay's *In the Wake of Cook*, David Philip Miller and Peter Hanns Reill's *Visions of Empire* and Londa

37 MacLeod, *Nature and Empire*; Benedikt Stuchtey, *Science Across the European Empires*.

38 Sivasundaram, "Sciences and the Global," 146. See Raina, "From West to Non-West?"; Anderson, "Remembering the Spread of Western Science."

39 Secord, "Knowledge in Transit," 669. See also MacLeod, "On Visiting the 'Moving Metropolis'."

40 Cooper and Stoler, *Tensions of Empire*, 4.

41 McClellan, *Colonialism and Science*; Drayton, *Nature's Government*.

42 See e.g. Brockway, *Science and Colonial Expansion*.

43 Storey, *Science and Power in Colonial Mauritius*.

Schiebinger and Claudia Swan's *Colonial Botany* inform us about the direct ways in which science and imperialism interacted with each other.[44] It was sometimes hard to distinguish whether the goal of an expedition was primarily imperialistic or scientific, leading to the suggestion that oftentimes they were one and the same.[45] The field of the study of science and empire has developed to such maturity that 2021 even saw the publication of Andrew Goss' *The Routledge Handbook of Science and Empire*, the first such reference work that 'brings together the findings from a generation of scholarship examining imperial networks of scientists, imperial ways of knowing, and knowledge entangled with colonialism.'[46]

There are many ways to try to understand these existing 'imperial networks of scientists' and the 'patterns of mutual interdependence' that were the hallmarks of these networks. Bruno Latour's model of 'cycles of accumulation' already denoted that the field experience of the naturalist-explorer was the main driving force in the accumulation of knowledge, not only in an individual naturalist, but also between naturalists.[47]

The equivalent of the move from thinking in centre and periphery to thinking in terms of mutual interdependence can also be seen in imperial history at large. In what is termed New Imperial Histories, there has been a strong move towards seeing empire as a web of connections.[48] Some of these webs will have their focus in an 'old-fashioned' imperial centre, but the webbed nature of empire allows for many peripheral centres to be recognised. The shift of focus from the centre to the web of connectivity has led to a vibrant world of imperial networks, some very localised, some spreading far and wide.[49]

The nodal nature of such networks has also opened our eyes towards the many actors that were previously hidden in the margins of the periphery that now turn out to have played essential roles in keeping the wheels of networks turning. Simon Schaffer, Lissa Roberts, Kapil Raj and James Delbourgo in *The*

44 Mackay, *In the Wake of Cook*; Miller and Reill, *Visions of Empire*; Schiebinger and Swan, *Colonial Botany*.

45 Many overviews of these types of expeditions exist. See e.g. Zobel, "Naturalistes voyageurs français"; Drouin, "De Linné à Darwin"; Rice, *Voyages of Discovery*; Williams, *Naturalists at Sea*; Riffenburgh, *Journeys of Discovery*. An admirable bibliography of the travel narratives of these expeditions was published in 2016, with a supplement in 2018: Troelstra, *Bibliography of Natural History Travel Narratives*; Troelstra, *Supplement to Bibliography of Natural History Travel Narratives*.

46 Goss, "Introduction," 3.

47 Latour, *Science in Action*, 215–257.

48 Raben, "New Dutch Imperial History?," 6. For an overview of the field of New Imperial history, see Howe, *New Imperial Histories Reader*.

49 For the widespread nature of these networks, see Lambert and Lester, *Colonial Lives Across the British Empire*.

Brokered World have drawn our attention to those previously mostly disregarded go-betweens that played an essential role in the networks of knowledge creation,[50] while Kapil Raj in *Relocating Modern Science* and Felix Driver and Lowri Jones in *Hidden Histories of Exploration* have uncovered how essential the role of indigenous actors was in the creation of knowledge.[51] At the same time, Clare Anderson in *Subaltern Lives* has shown how even the most marginal people about whom we can only find tiny shreds of information can teach us much about the workings of imperial networks.[52]

4 Dutch Imperial Studies

An additional word needs to be said about Dutch imperial studies. As Henk Wesseling already noted in 1988, 'the absence of the Dutch in the international debate on imperialism is striking.'[53] Despite Susan Legêne's *De bagage van Blomhoff en Van Breugel* (1998), and Wesseling's own *The European Colonial Empires* (2004),[54] which incorporated Dutch imperial developments, Remco Raben rightly concluded in 2013 that the field of Dutch imperial studies had still not really developed, mainly because there has been little internal pressure from the Netherlands to move to a new direction.[55] Not much has changed since Raben had to conclude this. One of the more progressive works of recent years in this regard is Martin Bossenbroek's *De wraak van Diponegoro*.[56] Using both Dutch and Indonesian perspectives, Bossenbroek paints a picture of both the start and end of formal Dutch colonisation. By connecting both the beginning and the end of the Netherlands Indies to each other, Bossenbroek contributes to our understanding of the rationale of Dutch imperialism.

This theme was undoubtedly inspired by the current (and still unresolved) discussion in Dutch society regarding the 'status' of the atrocities committed in the Indonesian war of independence from 1945 to 1949. As long as the answer to these questions is still heavily debated, the trauma of losing an empire will clearly not have been processed in the Netherlands and decolonisation of our colonial past is still in its infancy.[57] This continuing struggle with our colonial

50 Schaffer et al., *Brokered World*.
51 Raj, *Relocating Modern Science*; Driver and Jones, *Hidden Histories*.
52 Anderson, *Subaltern Lives*. See also her more recent publication: Anderson, *Convicts*.
53 Wesseling, "Giant That Was a Dwarf," 59.
54 Legêne, *Bagage van Blomhoff en Van Breugel*; Wesseling, *European Colonial Empires*.
55 Raben, "New Dutch Imperial History?."
56 Bossenbroek, *Wraak van Diponegoro*.
57 See Honings, Van 't Veer, and Bel, *Postkoloniale spiegel*, 9–12; Kuitert, *Met een drukpers de oceaan over*, 17.

past has turned much of our discussion of the period of formal colonisation of the Netherlands Indies (1816–1949) inward.

The literature on the history of Dutch endeavours in the Netherlands Indies in this period is mostly written in Dutch, and therefore difficult to access for the wider international academic community. This also means that many of the innovations in the wider historiographical community have mostly passed by this vast body of Dutch literature on the Indies, except in the area of economic studies where the Dutch cultivation system on Java and subsequent developments in e.g. sugar cultivation have attracted ample attention.[58] In other areas, examples like Catia Antunes and Jos Gommans' *Exploring the Dutch Empire* (2015), Matthijs Kuipers' *A Metropolitan History of the Dutch Empire* (2022), and the growing body of publications by Bernhard Schär on the intersection of imperial history and history of science are the rare positive exceptions.[59]

The study of the history of science in the Netherlands Indies is also a field that is still very much in development. In the past, much literature was published in Dutch, but in recent years several scholars have taken to publishing in English. Notable examples are Andreas Weber's *Hybrid Ambitions* (2012), Peter Boomgaard's *Empire and Science in the Making* (2013), Fenneke Sysling's *Racial Science and Human Diversity in Colonial Indonesia* (2016), and Marieke Bloembergen and Martijn Eickhoff's *The Politics of Heritage in Indonesia* (2020).[60] Chapter 1 contains an overview of recent literature on the Natuurkundige Commissie and the history of science in the Netherlands Indies in general.

5 From Geography of Science to Layered Loyalties

With all this in mind, how then do we proceed to find out how the Natuurkundige Commissie functioned? How does this help to establish how the participants in the Natuurkundige Commissie related to the stakeholders and to each other,

58 See e.g. Fasseur, *Politics of Colonial Exploitation*; Dick et al., *Emergence of a National Economy*; Bosma, *Sugar Plantation in India and Indonesia*.

59 Antunes and Gommans, *Exploring the Dutch Empire*; Kuipers, *Metropolitan History of the Dutch Empire*. Schär's output is of such quality and quantity that it promises to provide a platform towards the development of Dutch imperial studies into a proper field. See Schär, "Earth Scientists as Time Travelers"; Schär, *Tropenliebe*; Schär, "From Batticaloa via Basel to Berlin"; Schär, "Switzerland, Borneo and the Dutch Indies."

60 Weber, *Hybrid Ambitions*; Boomgaard, *Empire and Science in the Making*; Sysling, *Racial Science and Human Diversity*; Bloembergen and Eickhoff, *Politics of Heritage*. See also Pyenson, *Empire of Reason*; Boomgaard, "Making and Unmaking of Tropical Science."

and how these relationships influenced the practice of the participants? What has become clear is that the underlying geographical dichotomy is a factor that cannot be dismissed. In fact, geography is the central feature on which my analysis of the functioning of the Natuurkundige Commissie is based. It is impossible to understand the Commissie without taking into account the fact that it was a thoroughly Dutch project operating in the vast realms of a faraway colonial world that was unknown to them (see Map 1).

With the notion of this geographical dichotomy as a solid base, I use a complex of layered loyalties, very much coloured and influenced by the geographical undercurrent. As the objects of these layered loyalties, I take the three stakeholders in the Natuurkundige Commissie. As I wrote in an earlier article, each member 'was appointed by (a) the Dutch government to collect specimens of natural historical value for (b) 's Rijks Museum van Natuurlijke Historie in Leiden, under the orders of (c) the governor-general of the Dutch East Indies.'[61] Since no one can serve three masters and keep them all happy, this construction created tensions for the members in the field. What would happen if the interests of the stakeholders did not align in a specific situation?

Each of the members had some form of relationship with each of the three stakeholders. It was the strength of this relationship, or, in other words, the strength of the loyalty of each member to these stakeholders, that governed the practice of the members in the field. The complicating factor was that two of the stakeholders were located in the faraway Netherlands, while, during an expedition to any island other than Java, the governor-general in Buitenzorg was far away as well. It was up to the members to make the practical decisions regarding itineraries and everyday activities without direct input from the stakeholders.

These decisions were informed by the loyalty that each member felt towards each stakeholder. Given the fact that proximity was a big issue, it was the layering of loyalties towards each stakeholder at any given moment that decided the course of action. Not only did these layered loyalties affect individual decisions of individual members, it also was a process that affected the bigger picture of the development of the Natuurkundige Commissie. It mattered when and by whom a member had been appointed in the Natuurkundige Commissie. It mattered which of the stakeholders would mean most for the advancement of the career of a member.

However, the Commissie did not only exist of members, but also included assistants and indigenous employees. During the thirty years of its existence,

61 Van Wingerden, "Science on the Edge of Empire," 804. Segments of this article are incorporated in chapters 1, 2, 3, 6 and 7 without any further reference.

MAP 1 'Overzichtskaart van de Nederlandsch Oost Indischen Archipel.' From Stemfoort and Ter Siethoff, *Atlas der Nederlandsche bezittingen in Oost-Indië*. Public domain. Universiteitsbibliotheek, Rijkuniversiteit Groningen, uklu TC---- 19

thirty different individuals were appointed as members or assistants of the Natuurkundige Commissie. Fourteen of these people were appointed in the Netherlands, sixteen in the Netherlands Indies. Traces of these people can be found in governmental archives in both the Dutch and the Netherlands Indies administrations. In addition to this, a multitude of indigenous actors was permanently or temporarily employed in various capacities in the service of the members and assistants of the Commissie. We know some detail for a few of them, little detail for some of them and no detail for most of them.

Since it is impossible, if only because of lacking sources, to analyse the layered loyalties of each individual participant, this study rests on a categorisation of these participants. Two valid categorisations can be made, each consisting of three groups. On the one hand, we can categorise the participants in members, assistants, and indigenous employees. On the other hand, we can categorise the participants in members and assistants appointed in the Netherlands, members and assistants appointed in the Indies, and indigenous employees.

I will not use a strict categorisation throughout this study, but as need demands will refer to either set of categories.

Categorisation 1	**Categorisation 2**
Members	Members and assistants appointed in the Netherlands
Assistants	Members and assistants appointed in the Indies
Indigenous employees	Indigenous employees

To reflect the grounding in geography of the complex of layered loyalties the second categorisation serves as the basis for the division of this study in three main parts. Part 1 (chapter 1, 2 and 3) begins by taking a look at the Natuurkundige Commissie as a whole and then focuses on the relationships of the European members and assistants with the three stakeholders, using the career of Commissie member Eltio Alegondus Forsten (1811–1843) as a model. Part 2 (chapters 4 and 5) takes Commissie member Pierre-Médard Diard (1794–1863) as its subject. Diard is a prime example of a member who was appointed in the Indies and whose first loyalty was to the Indies government rather than to any of the Dutch stakeholders. Part 3 (chapters 6 and 7) investigates the roles played by a myriad of indigenous actors in the work of the Natuurkundige Commissie.

6 A Chapter-by-Chapter Overview

Chapter 1 treats the Natuurkundige Commissie as a whole. It begins by setting out the political circumstances in both the Netherlands and what was to become the Netherlands Indies. Following from this, it places the Natuurkundige Commissie in the context of science in the Netherlands Indies and sets out how the Natuurkundige Commissie relates to earlier natural historical work. After an evaluation of previous work on (members of) the Natuurkundige Commissie, a new analytical historical framework, based on a geography of science, is proposed and the Natuurkundige Commissie is interpreted in the light of this geographical framework. This includes a complete analysis of all European members and assistants, both those appointed in the Netherlands and those appointed in the Indies.

Chapters 2 and 3 investigate the complex of layered loyalties in the careers of the members of the Natuurkundige Commissie. The life and work of Commissie member Forsten is taken as its central line. Based on his experience, the relationships with the three stakeholders of the Natuurkundige Commissie are investigated. Chapter 2 looks at the two Dutch stakeholders (the Dutch government and 's Rijks Museum) and elucidates how Forsten and the other members related to them, also in light of their personal motivations.

Chapter 3 follows the developments in the loyalties of the members to the governors-general. Both mapmaking and mining inspections are identified as tasks defined by the governors-general rather than by either of the Dutch stakeholders. In the final section of chapter 3, the relationships of the members to the three stakeholders are drawn together and interpreted in the light of the complex of layered loyalties, leading to the conclusion that a gradual shift of loyalties took place in the Natuurkundige Commissie, weakening the ties to the two Dutch stakeholders (each for reasons of their own) and resulting in a much stronger loyalty towards the governors-general. The changing of this balance also explains why the Natuurkundige Commissie was dissolved when it was.

Chapters 4 and 5 look at the versatile career of Pierre-Médard Diard, one of only two members appointed in the Indies. Chapter 4 introduces him as a transimperial naturalist who crisscrossed Southeast Asia, giving rise to rather divergent perceptions of his life and work, resulting in a cheating British Diard, a cheated French Diard, and finally, a relatively obscure Dutch Diard. On his travels, he was followed by conflict over ownership of his collections, making him a contested 'mobile subject' almost everywhere he came. In an arena of interimperial competition, Diard was not allowed to render services to different empires simultaneously and finally settled in the Netherlands Indies in 1824, where, as member of the Natuurkundige Commissie (1828–1841), he became subject to the same complex of layered loyalties as the other members, but from a different starting point.

Chapter 5 follows Diard's career in agricultural experimentation in the Netherlands Indies from 1828 to 1841, the period in which he served as (directing) member of the Natuurkundige Commissie as well. Diard's involvement with sericulture, tea cultivation, and sugar manufacturing is studied and placed in the context of his view of natural history. Diard's involvement with the Colonial Museum in Batavia and the speech he held at the Museum in the presence of prince Hendrik illustrate how his view was markedly different from Temminck's, and therefore in direct conflict with the goals that 's Rijks Museum had with the Natuurkundige Commissie. Diard's whole career is considered in the light of the complex of layered loyalties, showing how his much stronger loyalty to the only Netherlands Indies stakeholder put him in a very different position from the other members.

Chapters 6 and 7 investigate the role of indigenous agency and indigenous knowledge in the expeditions of the Natuurkundige Commissie based on the available textual evidence. In chapter 6, the category of permanent employees of the Natuurkundige Commissie is analysed. Not only their practical contributions and their specific roles are revealed, but also the perception of their work in the Netherlands. By looking at the identity of these indigenous employees, an impression of the longevity of their connection to the Natuurkundige Commissie is gained, but also of their limitations when travelling to the Outer Islands.

Chapter 7 reviews the indigenous people that were only temporarily hired. The largest group of these people were the porters, who did not always have a choice regarding their involvement. The second group consists of knowledge brokers in the realms of language and natural history: translators and hunters. It shows how the ability to communicate allowed contact with indigenous people on an equal level. In addition, it shows how mediation of knowledge of the natural world was required to create collections in the Outer Islands, while the process of the mediation of this knowledge into the European world of zoology shows how gradually sight was lost of the fact that this originally was indigenous knowledge. A brief word is said about how the indigenous actors from both chapters 6 and 7 fit into the complex of layered loyalties.

PART 1

A Tale of Geographies

CHAPTER 1

The Natuurkundige Commissie between the Netherlands and the Indies

The Natuurkundige Commissie existed for a period of thirty years. It was founded by royal decree on May 2, 1820,[1] and disbanded, also by royal decree, on April 17, 1850.[2] In this period, eight naturalists were appointed as members of the Commissie by royal decree and six assistants, both artists and preparators, were attached to them. The appointments took place in six instalments, in 1820, 1823, 1830, 1835, 1838 and 1842.[3] Strictly speaking, the assistants were not formal members of the Commissie, but two of them would be promoted and received member status in the Indies.[4] Of the eight naturalists only one returned, with six dying of disease and one being violently killed during a local uprising.[5] Of the six assistants, only one returned while still in service, with three dying of disease and two resigning their position and staying on in the Indies.[6] The fact that only two of the fourteen appointees returned to the Netherlands is a gruesome reminder of the tough conditions that they were faced with in the Netherlands Indies and the general state of health care in the early nineteenth century.[7] This extremely low survival rate is also a major contributing factor to the relative obscurity that the Natuurkundige Commissie fell into after its demise. After all, the tragic deaths of a number of relatively unknown, although promising, naturalists are hardly fuel for fame.

The Commissie was founded in a thoroughly Dutch context to operate in a profoundly different colonial world in the Malay Archipelago. Throughout the thirty years of its existence the members of the Natuurkundige Commissie

1 KB, May 2, 1820, no. 10. NL-HaNA, Staatssecretarie, 2.02.01, inv.no. 988.

2 KB, April 17, 1850, no. 14. NL-HaNA, KdK 1841–1897, 2.02.04, inv.no. 609.

3 The naturalists were Kuhl and Van Hasselt (1820); Boie and Macklot (1823); Korthals (1830); Horner (1835); Forsten (1838); Schwaner (1842). The assistants were Keultjes and Van Raalten (1820); Müller and Van Oort (1823); Van Gelder and Overdijk (1830). See below for a more detailed outline of these appointments.

4 Müller and Van Oort.

5 Korthals returned; Kuhl, Van Hasselt, Boie, Horner, Forsten and Schwaner died of disease; Macklot was murdered.

6 Müller returned; Keultjes, Van Raalten and Van Oort died of disease; Van Gelder and Overdijk resigned and stayed on in the Indies. Van Gelder would eventually make his way back to the Netherlands.

7 Pols, "Notes from Batavia"; Pols, *Nurturing Indonesia*, 6.

 | DOI:10.1163/9789004696877_003

found themselves caught between these two radically different geographical realities. They were selected and appointed by academics and politicians in the Netherlands to work for colonial officials in the Netherlands Indies. The expectations of these academics, politicians and colonial officials were not aligned and created a great deal of tension. Balancing these diverging and sometimes contradictory expectations with each other and with their personal ambitions was a complex daily task for the members of the Commissie that created continuous conflicts of loyalties.

1 The Historical Context in the Netherlands and the Netherlands Indies

Even though the members and their assistants were appointed by king Willem I (1772–1843, r. 1813–1840, Figure 1), he was not the driving force behind the founding of the Commissie.[8] That honour goes to Coenraad Jacob Temminck (1778–1858, Figure 2),[9] gentleman-naturalist and collector, who was to become director of 's Rijks Museum van Natuurlijke Historie [National Museum of Natural History] in Leiden, the precursor of what is now Naturalis Biodiversity Center.[10] Temminck made very clever use of the political developments and brought his ambitions to the attention of the king at exactly the right time. The previous decades had seen a major political and social upheaval in the Netherlands and across Europe. The Republic had been considerably weakened by the Fourth Anglo-Dutch War (1780–1784), which

8 Only one member was appointed by Willem's successor and son Willem II (1792–1849, r. 1840–1849).

9 For Temminck, see Gassó Miracle, *Coenraad Jacob Temminck*, which incorporates both Gassó Miracle, "Significance of Temminck's Work on Biogeography" and Gassó Miracle, "On Whose Authority?" Older work on Temminck includes Raat, "Coenraad Jacob Temminck" and Susanna, "Levensschets." For a recent reissue of the images from one Temminck's scientific works, see Reeuwijk and Wijnhoven, *Duiven van Temminck en Knip*.

10 The most recent publication on the history of the museum is Schilthuizen and Vonk, *Wie wat bewaart*. Older publications include Visser, "Rijksmuseum van Natuurlijke Historie in de 19e eeuw"; Holthuis, *Rijksmuseum van Natuurlijke Historie*; Raat, "Rijksmuseum van Natuurlijke Historie in Leiden"; Gijzen, *'s Rijks Museum van Natuurlijke Historie*. For the museum as part of the Leiden natural historical landscape, see Van der Land, *History of Natural History in Leiden*. For two publications celebrating the current museum and its collection, see Gassó Miracle and Monquil-Broersen, *Van onschatbare waarde* and Van Impelen, *Nieuwe Naturalis*. For Naturalis in the context of Dutch natural history museums, see Visser, "Ontwikkeling van het natuurhistorische museum" and Fred de Ruiter, *Natuurschatten*.

FIGURE 1 Portrait of king Willem I, king of the Netherlands, by Joseph Paelinck, oil on canvas, 1819, 227 × 155.5 cm, SK-C-1460, Rijksmuseum, Amsterdam. Public domain HTTP://HDL.HANDLE.NET/10934/RM0001.COLLECT.11886

was essentially a naval war that was mostly played out in the colonial arena. Stadtholder Willem V (1748–1806, r. 1751–1795), the father of Willem I, was widely considered responsible for losing the war and for its economic consequences. Resistance to his reign became bolder and bolder, to such an extent that many members of the Patriot party, the banner under which all opponents of Willem V had gathered themselves, were violently ejected from the country in 1787 with the help of Friedrich Wilhelm II of Prussia (1744–1797,

FIGURE 2 Portrait of Coenraad Jacob Temminck, by Jan Adam Kruseman, 1836, Naturalis Biodiversity Center, Leiden. Public domain

r. 1786–1797), the stadtholders brother-in-law. In the unstable aftermath of the French Revolution of 1789, the Patriots returned to the Netherlands in 1795 with the French and ousted stadtholder Willem V from his position, founding the Batavian Republic. Willem V withdrew to Britain and lived out his days in exile.[11]

Meanwhile, European developments tumbled over each other in quick succession. French army general Napoleon Bonaparte (1769–1821) took control of

11 Israel, *Dutch Republic*, 1090–1130; Schama, *Patriots and Liberators*. See Moore and Van Nierop, *Colonial Empires Compared*, for a comparative evaluation of Anglo-Dutch relations in the colonial theatre.

the French Republic as first consul in 1799 and had himself crowned emperor in 1804. Napoleon's imperialistic politics plunged all of Europe in a protracted war that was only concluded with the famous battle of Waterloo in June 1815. The Batavian Republic did not survive the Napoleonic wars. From its infancy the Republic was already not much more than a vassal state of the French Republic. The Batavian Republic was eliminated in 1806 and turned into a kingdom with Napoleon's brother Louis Napoleon at the helm as king of Holland. This too lasted for only a short while. When Louis Napoleon incurred the wrath of his brother, he was removed from his office by Napoleon in 1810 and the kingdom of Holland was annexed by France. Meanwhile, prince Willem Frederik, son of Willem V and future king Willem I, had been biding his time. When the anti-French coalition finally succeeded in beating Napoleon decisively at the Battle of Nations in Leipzig in October 1813, opportunity presented itself to Willem Frederik when he was invited by a delegation from the Dutch provisional government to return to the Netherlands. When Willem Frederik landed in the Netherlands on November 30, 1813, he assumed the title of Sovereign Prince. As a result of the European peace negotiations in Vienna in the following year, the Netherlands were turned into a kingdom, absorbed the Southern Netherlands, and Willem Frederik became king Willem I on March 16, 1815.[12]

The Dutch colonial empire did not emerge from this period of international upheaval unscathed. During the Fourth Anglo-Dutch War both the British and the French had occupied several Dutch possessions in both the West and East Indies. The Treaty of Paris of 1784 returned most of these to the control of the Dutch VOC (Verenigde Oost-Indische Compagnie, United East Indies Company).[13] The VOC, which effectively governed the Dutch colonies in the east, was already not in good shape when the war broke out, but the economic repercussions of the Fourth Anglo-Dutch War sent it into a downwards spiral it would not recover from.[14] When Willem V fled to Britain in 1795, he took up residence in Kew and issued what has become known as the Kew Letters, a series of letters to Dutch colonial officials to urge them to throw in their lot with the British and effectively relinquish control of the Dutch colonies to the British in order to counteract the French.[15] In combination with military threat and action, this led to a substantial reduction of the Dutch colonial empire (the

12 See Koch, *Koning Willem I*, for a recent full biography of Willem I. Unfortunately, little attention is paid to (Willem's attitude to) the colonial world in this study. An in-depth study of Willem I's colonial policy is sadly lacking.

13 See Tarling, *Anglo-Dutch Rivalry*, for a detailed history of Anglo-Dutch relations in the Malay Archipelago.

14 See Gaastra, *Geschiedenis van de VOC*, translated as Gaastra, *Dutch East India Company*.

15 See Bossenbroek, *Wraak van Diponegoro*, 56.

FIGURE 3 Portrait of Sir Thomas Stamford Bingley Raffles, by George Francis Joseph, oil on canvas, 1817, 139.7 × 109.2 cm, NPG 84, National Portrait Gallery, London

Cape was lost in 1795 and Ceylon in 1796). Later during the Napoleonic wars, the British invaded and occupied the Dutch possessions on Java in 1811, even though this was a somewhat unsanctioned ambitious project of colonial official and British governor of Java Thomas Stamford Raffles (1781–1826, g. 1811–1816, Figure 3), whom we will encounter again in more detail in chapter 4.[16]

16 Raffles has been a popular subject in both academic and popular writing. For a contemporary account, see Raffles, *Memoir of the Life*. John Bastin has devoted over fifty years of study to Raffles and his family. His most recent effort is Bastin, *Raffles and Some of His*

After Napoleon's defeat, a series of peace talks were held, and a number of treaties were put in place to restore some clarity and order to the war-weary world. European borders were redefined, and colonial empires were reshuffled. Due to their geographical location in Europe, the Netherlands were seen by the British as a potential buffer state. The Kingdom of the Netherlands was founded with Willem I at the helm and its geographical territory was considerably enhanced in order to safeguard against any future French ambitions. In addition, the prize colony of Java was restored to allow the Netherlands to develop themselves into the required strong buffer state.[17] Article 60 of the constitution of 1815 placed full control of the Dutch overseas possessions exclusively in the hands of the king himself,[18] allowing him to develop Java into a source of much-needed revenue in order to finance his limitless ambitions of restoration and development after decades of dissent, war, and chaos. The constitution placed Java in the hands of the king as a tool to be used to bolster the international reputation of a fledgling kingdom and its ambitious king. Natural history publications could be used to enhance international prestige. In addition, more intimate knowledge of the natural world of the colonies could result in new ways to increase revenue. The time was ripe for an ambitious project like the Natuurkundige Commissie.

2 Science in the Netherlands Indies

In previous centuries, most natural history in the Dutch East Indies had been undertaken by (not necessarily Dutch) VOC company officials and travellers, mainly because the East Indies fell under direct control of the VOC. As such, it was the VOC (and not the Dutch government in the Netherlands) that was involved in directing science to a certain extent in the east.[19] From previous

Friends. For Raffles' affinity for natural history, see Bastin, "Raffles and the Study of Natural History." For a more popular recent biographical account of Raffles, see Glendinning, *Raffles and the Golden Opportunity*.

17 Conveniently, the British did not return the Cape and Ceylon.

18 'De koning heeft bij uitsluiting het opperbestuur over de volkplantingen en bezittingen van het Rijk in andere werelddeelen.' Constitution of the Kingdom of the Netherlands (1815), article 60: https://www.denederlandsegrondwet.nl/id/vi6jejadg5su/zesde_afdeeling_van_de_magt_des_konings.

19 See Bethlehem and Meijer, *VOC en cultuur*, and Blussé and Ooms, *Kennis en Compagnie*, for two sets of case studies on the relationship between the VOC and science. See Van Berkel, Van Helden, and Palm, *History of Science in the Netherlands*, 210–228; Friedrich, Brendecke, and Ehrenpreis, *Transformations of Knowledge*, and Boomgaard, *Empire and Science in the Making*, for three recent overviews of science in the Dutch colonial world. A monograph on the exact sciences in the Netherlands Indies is Pyenson, *Empire of Reason*.

scholarly work 'two conflicting interpretations of the VOC's role in promoting scientific research' emerge. Harold Cook argued that 'Dutch commercial mentality enhanced interest in empirical enquiries,' while Klaas van Berkel argued that the VOC functioned as an unwilling benefactor of science. '[The VOC] created the infrastructure that enabled research [but] it had no serious regard for scientists.'[20]

One of the naturalists who embodies the role of science for the VOC is Georg Eberhard Rumpf, known as Rumphius (1627–1702).[21] Susanne Friedrich has recently used the case of Rumphius to argue that 'a more nuanced interpretation of the relationship between the VOC and scholarship is needed, one that distinguishes between the goals of the Company and those of the men connected to it, as well as between the kinds of knowledge that interested the different parties.'[22] Friedrich concludes that Rumphius, 'while he was free to research nature and to exchange his findings with fellow researchers in Asia and Europe as long as he did not spread "secrets," [...] lost the power to disseminate his works as he saw fit.'

Working as a naturalist alongside the VOC had both advantages and limitations. This was the case for much of the seventeenth and eighteenth centuries. Naturalists in VOC-service such as Rumphius, Hendrik van Rheede tot Drakenstein (1636–1691), Paul Hermann (1646–1695), François Valentijn (1666–1727), Josua van Iperen (1726–1780), Jacob Radermacher (1741–1783), and Friedrich von Wurmb (1742–1781) were first and foremost company officials.[23] As Friedrich's case study of Rumphius has shown, a good relationship with their superiors was a necessary condition for any company official with natural historical ambitions and it helped enormously if the results of their research were considered useful in the context of the VOC's commercial goals.

Already in 1796, very soon after the founding of the Batavian Republic, the VOC was nationalised. On December 31, 1799, the VOC was finally dissolved, ushering in a period of instability and upheaval in what now effectively had become the Netherlands Indies and officially fell under control of the Dutch

Two useful but dated surveys of scientific endeavours in the Indies and their history are Rutten, *Science in the Netherlands East Indies*, and Honig and Verdoorn, *Science and Scientists in the Netherlands Indies*.

20 Cook, *Matters of Exchange*, and Van Berkel, "Onwillige mecenas?" as summarised in Friedrich, "Importance of Being a Good Employee."

21 A classic work on Rumphius is Greshoff, *Rumphius gedenkboek*. For recent scholarship on Rumphius, see Yoo, "Wars and Wonders"; Friedrich, "Importance of Being a Good Employee," and Kießling, *Wissenspoetik und koloniale Naturgeschichte*.

22 Quotes all from Friedrich, "Importance of Being a Good Employee," 185–186.

23 The last three played a prominent role in the early years of the Bataviaasch Genootschap.

government in the Netherlands. This created a different climate for naturalists, who lost both a patron and an inhibitor when the VOC was dissolved. The Bataviaasch Genootschap van Kunsten en de Wetenschappen (Batavian Society for Arts and Sciences), founded in 1778, was initially very active, but had, already in the early 1790s, fallen into a slumber that it would only begin to wake up from during the period of British control of Java (1811–1816).[24]

British governor Thomas Stamford Raffles was a keen naturalist himself and became well-known as the author of *History of Java* (1817)[25] and for his patronage of the sciences, which included support for the work of American naturalist Thomas Horsfield (1773–1859), who had been present in the Netherlands Indies from 1799 to 1819 and would publish his *Zoological Researches in Java and the Neighbouring Islands* in 1824.[26]

The true revival of Dutch science in the Netherlands Indies only took shape when the Dutch committee that was tasked with taking over the Netherlands Indies from the British arrived in 1816. This *commissie-generaal* consisted of three commissioners-general, including the future governor-general Godert Alexander Gerard Philip van der Capellen (1778–1848, g. 1819–1826, Figure 4).[27] They were accompanied by the 'director of agriculture, arts, and sciences of Java and the neighbouring islands,' as was the full job title of Carl Georg Caspar Reinwardt (1773–1854, Figure 5),[28] who was directly responsible for the founding of 's Lands Plantentuin (National Botanical Garden) in Buitenzorg (Bogor). which is still in full operation today as a centre of scientific botanical research in Indonesia.[29] Reinwardt was succeeded as its director in 1822 by his

24 See The and Van der Veur, *Verhandelingen van het Bataviaasch Genootschap*, for a bibliography of the scientific output of the Genootschap. See Groot, *Van Batavia naar Weltevreden*, for a detailed history of the Genootschap.

25 Raffles, *The History of Java*, republished in 2010 as Raffles, *A History of Java*.

26 Horsfield, *Zoological Researches in Java*, republished in 1990 with an introduction by John Bastin. For more on Horsfield, see Bastin and Moore, "Geological Researches" and Bastin, "Natural History Researches."

27 See Jeurgens, "Op zoek naar betrouwbare informatie," for the work of this *commissie-generaal*.

28 Reinwardt's travel narrative was posthumously published as De Vriese and Pijnappel, *Reis naar het oostelijk gedeelte van den Indischen Archipel*. For recent work on Reinwardt, see Weber, *Hybrid Ambitions*; Weber, "Encountering the Netherlands Indies"; Weber, "Bitter Fruits of Accumulation." For summaries of Reinwardt's extant letters, see Van Heiningen, *Correspondence of Caspar Georg Carl Reinwardt*.

29 See Treub, *'s Lands Plantentuin te Buitenzorg*, and Rijnberg, *'s Lands plantentuin, Buitenzorg*, for an overview of the history of the garden. See Goss, *Floracrats*, and Weber, "Garden as a Niche," for diverging evaluations of the scientific output of the garden in the early nineteenth century. See Blais, *Empire de la nature*, for the garden in an international colonial context. For other work on the garden, see Badenhuizen, "Botanical

FIGURE 4 Portrait of Godert Alexander Gerard Philip, Baron van der Capellen, governor-general of the Netherlands Indies, by Cornelis Kruseman, oil on canvas on hardboard, 1816–1857, 103.5 × 89.5 cm, SK-A-3795, Rijksmuseum, Amsterdam. Public domain HTTPS://ID.RIJKSMUSEUM.NL/20028654

assistant and botanist Carl Ludwig Blume (1796–1862, Figure 6), who would become director of 's Rijks Herbarium (the Dutch national botanical museum) in 1830.[30]

Garden at Bogor"; Wille, "From Laboratory Lichens to Colonial Symbiosis"; Weber and Wille, "Laborious Transformations"; Goss, "Reinventing the Kebun Raya"; Wille, *Mannen van de microscoop*.

30 Blume worked in the Indies from 1818 to 1826, became director of 's Lands Plantentuin in 1822, director of 's Rijks Herbarium (National Botanical Museum) in the Netherlands in 1830, and had a distinguished career as a botanist. For more on Blume, see Den Ouden, "C. L. Blume"; Van Steenis, "Dedication"; Fransen, Holthuis, and Adema, "Type-catalogue

FIGURE 5 Portrait of Caspar Georg Carl Reinwardt, lithograph by Reinier Vinkeles, after a painting by Mattheus Ignatius van Bree, before 1816, RP-P-OB-64.069, Rijksmuseum, Amsterdam. Public domain HTTP://HDL.HANDLE.NET/10934/RM0001.COLLECT.189869

Reinwardt's appointment on January 11, 1815, was a direct result of the stabilising political situation in Europe. King Willem I was an ambitious and industrious leader and determined to make himself count, especially in the Netherlands Indies, whose governance had become his special privilege. Eager to find out what the riches of the Indies were and how a profit could be made from it to fund his plans for the Netherlands, he had appointed Reinwardt as the scientific arm of the *commissie-generaal*. Science in Europe had always benefited from royal patronage. During the previous fifty years, the British and

of the Decapod Crustacea," 64–66; Weber, "Garden as a Niche"; Honings, "Naar het 'Verlokkend Oost'."

FIGURE 6 Portrait of Carl Ludwig Blume, lithograph by Jean Baptiste Madou, 1828–1835, RP-P-OB-40.229, Rijksmuseum, Amsterdam. Public domain
HTTP://HDL.HANDLE.NET/10934/RM0001.COLLECT.148357

the French had published monumental works with scientific descriptions of the exotic peoples and faraway lands of their colonial realms.[31] However, the lack of royal patronage and the difficult circumstances for the VOC had not

31 See Mackay, *In the Wake of Cook*, and Boomgaard, *Empire and Science in the Making*.

left much space for similar publications from explorations of the Dutch colonial world.[32] The time was ripe for an ambitious monarch to extend his royal patronage to naturalists who were tasked with discovering the riches of the natural realm of the newly restored Dutch colonial possessions. Publishing the results of this research would bolster the prestige of the small Netherlands with a large colonial empire and allow Willem I to take his rightful place in the European constellation of monarchs.

As Andreas Weber has convincingly shown, Reinwardt was a 'hybrid traveler' in the sense that his instructions and duties were extremely varied.[33] Weber identifies three main strands in the instructions issued to Reinwardt:

> First, he was supposed to survey the colony's economy and agriculture [...]. Second, he was tasked with improving the education of European civil servants and the medical care in hospitals and apothecaries. And third, Reinwardt was ordered to collect plants, animals and minerals.[34]

During his six-year stay in the Indies, from April 27, 1816, to June 19, 1822, Reinwardt tirelessly worked towards meeting his goals.[35] His work in the Indies included a long inspection tour that led him all the way to the Moluccas and back (February 27, 1821–March 18, 1822).[36] However, the results of Reinwardt's labour to fulfil the third strand of his instructions never fully reached the Netherlands. Four shipments with Reinwardt's collections of objects of natural history were lost at sea, which prevented his achievements in the colony from becoming part of European scientific discourse and significantly, and through no fault of his own, diminished the fame that Reinwardt could have had as a travelling naturalist.[37] Also, the fact that part of his travel history was only published posthumously did not advance his renown.

In the end, the manifold nature of his instructions left Reinwardt limited space to remedy the loss of his collections, as he was constantly occupied with

32 Van Berkel, "Empire without Science?"

33 Weber, *Hybrid Ambitions*, chapter 4.

34 Weber, *Hybrid Ambitions*, 122.

35 For Reinwardt's arrival date, see Vriese and Pijnappel, *Reis naar het oostelijk gedeelte van den Indischen Archipel*, 52. For his departure date, see *Bataviasche Courant*, June 22, 1822, no. 25, p. 2.

36 Departure date from Batavia and arrival date in Buitenzorg. Both dates from Vriese and Pijnappel, *Reis naar het oostelijk gedeelte van den Indischen Archipel*, 86. De Vriese notes that Reinwardt had already set foot on Javanese soil in Besuki on December 16, 1821, and continued the journey back to Batavia overland.

37 Gassó Miracle, *Coenraad Jacob Temminck*, 104.

other matters. It is this hybridity that sets Reinwardt apart from the members of the Natuurkundige Commissie. They did not have the three competing strands in their instructions but were simply tasked with collecting plants, animals, and minerals. As such, they were theoretically able to devote their undivided attention to their single, primary task, a luxury that Reinwardt was never afforded. This single focus was the result of the tireless lobby work of Temminck, who correctly gauged the mood of the time and was able to steer the ambitions of the king and his personal ambitions to a close alignment.

3 The Founding of the Natuurkundige Commissie

The history of the Natuurkundige Commissie is inextricably entwined with that of 's Rijks Museum van Natuurlijke Historie and its first director Coenraad Jacob Temminck. Most of the Commissie members had ties with Temminck or worked at 's Rijks Museum before they were selected. A recommendation by Temminck was an almost certain path to membership. This history of the founding of 's Rijks Museum is well-researched,[38] especially since Eulàlia Gassó Miracle's recent study of Coenraad Jacob Temminck.[39] In chapters two and three she sketches the developments that led to the founding of 's Rijks Museum. Gassó Miracle shows that the idea of a national museum of natural history was already suggested in 1814 by Temminck to Anton Reinhard Falck (1777–1843, Figure 7), then personal secretary to the king, and that it took some years of slowly boiling and simmering before Falck wrote a final proposal to the king that resulted in the founding of 's Rijks Museum van Natuurlijke Historie on August 9, 1820. In those six years, several ideas were launched and lobbied.

It was the death of Leiden professor Sebald Justinus Brugmans (1763–1819) that swayed the process definitively towards Temminck's proposals and ideas. According to Falck's proposal, the country did not have any cabinet that was comparable to the museums in Paris and Vienna. The two collections that were the main candidates to become such a museum in the Netherlands were not complete enough (the cabinet of Leiden University headed up by Brugmans) or important enough ('s Lands Kabinet van Natuurlijke Historie in Amsterdam with Reinwardt as its paper director). The only solution was to take Temminck up on his offer and combine the two existing collections with Temminck's

38 Gijzen, *'s Rijks Museum van Natuurlijke Historie*, focuses mostly on the origin of the collection of the museum while Holthuis, *Rijksmuseum van Natuurlijke Historie*, devotes ample attention to the people who made the museum.

39 Gassó Miracle, *Coenraad Jacob Temminck*.

FIGURE 7 Portrait of Anton Reinhard Falck, lithograph by Charles Baugniet, 1842, RP-P-OB-39.950, Rijksmuseum, Amsterdam. Public domain HTTP://HDL.HANDLE.NET/10934/RM0001.COLLECT.76002

impressive private collection, which would result in a collection that would be worthy to serve as a national museum. Of course, Temminck became the director of the new 's Rijks Museum van Natuurlijke Historie.[40]

Temminck had not been idle while the plans for the founding of a national museum slowly gravitated towards his proposal. The collections that Reinwardt was supposed to have collected were meant for 's Lands Kabinet van Natuurlijke

40 Falck's proposal and the final royal decree are published in full in Gijzen, *'s Rijks Museum van Natuurlijke Historie*, 9–14.

Historie in Amsterdam. Due to the unfortunate shipwrecks and Reinwardt's lack of opportunity to replace the lost collections, the results of his collecting mission were a deep disappointment. Temminck, who served as co-director of 's Lands Kabinet in Reinwardt's absence witnessed this first-hand. Again, he correctly gauged the mood of the time and started lobbying for a second collecting mission to the Netherlands Indies, while Reinwardt was even still in the Indies. Again, Temminck's friendship with Falck proved to be of value.

On January 18, 1819, Falck sent a proposal to the king in which he argued that another person needed to be prepared to take over Reinwardt's work. The British and the French had sent naturalists to all their colonies, and even the Prussians, who did not have any colonies, financially supported Von Humboldt in his expedition to Asia.[41] The Dutch had become an object of 'taunt and contempt'[42] because of their lack of scientific attention to their colonies. Although this was not the fault of the current administration, the fact that the British had published more comprehensively on the Dutch colonies than they had ever done themselves should serve as an example to the Dutch. While Reinwardt's expedition was a good start, it was necessary to find a person to assist and eventually succeed him.

Falck proposed to support the development of the talented Heinrich Kuhl, a German student from Groningen, who came well-recommended by Temminck, and was about to be snatched up by the Prussian government. Kuhl should be given a grant in order to visit the natural historical collections in Paris and London and prepare himself for an expedition to the Indies. The king approved Falck's proposal and awarded Kuhl f1,200 to travel to Paris and London.[43] Kuhl not only travelled to Paris and London, but all across Europe.[44] On May 2, 1820, Kuhl was considered prepared enough to be appointed as the first member of the Natuurkundige Commissie, together with his close friend Johan Conrad van Hasselt.[45]

---

41 Falck may refer to Von Humboldt's attempts to set up an expedition to India, which did not materialise because the British East India Company withheld their permission. See Wulf, *Invention of Nature*, 162–165.

42 'Schimp en minachting.'

43 KB, January 23, 1819, no. 123. NL-HaNA, Staatssecretarie, 2.02.01, inv.no. 757. The KB includes Falck's original proposal from January 18, 1819. Falck's proposal carries the incorrect date of January 18, 1818.

44 See Klaver, *Inseparable Friends*, 13–17, for Kuhl's activities in this period.

45 KB, May 2, 1820, no. 10. NL-HaNA, Staatssecretarie, 2.02.01, inv.no. 988.

4 Sources and Scholarship on the Natuurkundige Commissie

As Eulàlia Gassó Miracle rightly states, 'a more comprehensive history of the Natuurkundige Commissie has yet to be written.'[46] This does not mean to say that there is not a firm base to stand on, but it does indicate that the study of the Natuurkundige Commissie is not as straightforward as it may look. In general, there have been two approaches. Some historians have tried to treat the Natuurkundige Commissie as a whole, mostly in the wider context of the history of science of the Netherlands Indies, even though more recent work has taken the history of colonialism and imperialism into account.[47] Other historians have concentrated their efforts on the work of a single member or a couple of members that worked alongside each other.[48] In addition to this, there is a large group of publications on the history of science or the history of natural history that have studied the Commissie or its members as part of a larger project. Some of these contain original research on the Commissie of various depths, while others do no more than use existing work on the Commissie as their main source.[49]

What most of these studies have in common is that their approach to the archival sources that are available has been somewhat cursory and haphazard at best and virtually non-existent at worst. This does by no means reflect upon the scholarly quality of these studies, but it illustrates the fact that there is such a vast amount of archival material available that a limited selection

46 This sentence is from Gassó Miracle's original PhD thesis from 2019 (Gassó Miracle, "Temminck's Order," 101) but was cut on the road to publication and does not appear in the published version from 2021. To my mind, this sentence stills holds true, even though Weber, "Collecting Colonial Nature," and Van Wingerden, "Science on the Edge of Empire," have made welcome contributions.

47 Veth, *Overzicht van hetgeen gedaan is*, and Sirks, *Indisch natuuronderzoek*, are the most significant studies that place the Natuurkundige Commissie in the context of history of science, while Weber, "Collecting Colonial Nature," and Van Wingerden, "Science on the Edge of Empire," are the first attempts to address issues of imperialism and colonialism in the history of the Natuurkundige Commissie. Other notable or recent works that include historical narratives of the work of the Commissie are Smit, "International Influences"; Mees, "Vogelkundig onderzoek"; Mörzer Bruyns, *Met de Triton en Iris*; Van Balen, "Gevleugeld eremetaal"; Jacobs and Shea, "Eight Weeks in Lobo Bay."

48 Recent examples are Klaver, *Inseparable Friends*; Manse, "Kennis is macht"; Van Zanen and Weber, "Schreeuwende apen"; Van Wingerden, "Op reis vanuit een drijvende stad"; Jacobs, "Salomon Müller"; Jacobs, "Heinrich Christian Macklot." Extensive references to work on individual members of the Natuurkundige Commissie can be found in Appendix 2.

49 Goss, *Floracrats*, is an example that contains original research on a Commissie member, while Stresemann, *Entwicklung der Ornithologie*, relies on previous publications which is an approach that is justified by the scope of his project.

will already yield much that is of interest. The publications that are steeped in archival material are of such accuracy and quality that they can easily serve as reliable sources for further study, doing away with the need to engage with the archives themselves for those who study the Commissie in the context of a larger project.

There are two main destinations for the researcher who wishes to directly engage with the archival material on the Natuurkundige Commissie. The first destination are the archives of the Natuurkundige Commissie themselves, held at Naturalis Biodiversity Center in Leiden. These archives have recently been made available digitally in open access.[50] The more than 20,000 scans contain correspondence, diaries, drawings, field notes (consisting of general notes and specimen/object lists) and publications. The second destination are the archives of the ministry of colonial affairs, held at the Nationaal Archief in The Hague. Relevant parts of this collection have also been made available digitally in open access in recent years.[51] This vast archive contains a multitude of material, including reports, correspondence, decrees, personnel files, and maps. The archives of the ministry of colonial affairs are more unwieldy for the researcher of the Natuurkundige Commissie, since these archives contain material on virtually anything that happened in the Dutch colonies and only a tiny percentage of this vast collection concerns scientific affairs.

In addition to these two major collections, there are many other archive collections at the Nationaal Archief that contain substantial selections of material that touch upon the Natuurkundige Commissie and its members, while snippets of information can be found all over the nineteenth-century archival collections. The creation of a complete index of all archival material that pertains to the Natuurkundige Commissie in the Nationaal Archief would require more than a lifetime of research. In addition to this, a third, mostly unexplored resource are the archives of the colonial administration that are held at the

50 Gassó Miracle et al., "Natuurkundige Commissie Archives Online." These archives are available online in open access at https://dh.brill.com/nco/. An example of a full URL for a reference to e.g. inventory number 1075 is: https://dh.brill.com/nco/view/nco_NNM001001075/makingsense. Any specific scans from this inventory number can be accessed by adding _001 to 1075, leading to: https://dh.brill.com/nco/view/nco_NNM001001075_001/makingsense. For the sake of brevity in the references, I will refer throughout to these archival documents by only using NCAO (Natuurkundige Commissie Archives Online) followed by the inventory number and scan number, e.g. 1075_001. Access to these documents can be gained by replacing 1075_001 in the URL above by the reference given in the notes. For the advanced text and image recognition tool (MONK) used on NCAO, see Weber et al., "Towards a Digital Infrastructure."

51 NL-HaNA, Koloniën, 1814–1849, 2.10.01 and NL-HaNA, Koloniën, 1850–1900, 2.10.30.

Arsip Nasional Republik Indonesia. I have not been in a position to visit these archives, but it is only to be expected that the archival collections at the Arsip Nasional include material that pertains to the history of the Natuurkundige Commissie that is not available at either Naturalis or the Nationaal Archief.

Another important source for the history of the Natuurkundige Commissie are the published correspondence and travel reports of the individual members. In addition to this, a decent amount of the material collections of the Natuurkundige Commissie has been used as the basis for a variety of nineteenth-century taxonomical publications and has continued to be of use in twentieth- and twenty-first-century biological work.[52] A full overview of these two types of publications is not available. Although easier to create than a full archival index, this would still be a substantial task.[53] The reference section in Appendix 3 contains a basic, albeit undoubtedly not complete selection of material from the first category, while only a very small number of the second category are represented in the references in this study, simply because this study deals mostly with the practice in the field in the Netherlands Indies and not with the afterlife of the collections of the Natuurkundige Commissie.

A thorough evaluation of the staggering amounts of objects that were collected by the members of the Natuurkundige Commissie and their (lasting) significance in the areas of taxonomy, biology, ecology, archaeology, anthropology and ethnography would be a series of full studies in itself that have yet to be written. One has to remember that the history of natural history collections is not a topic that is high on the agenda for most of the guardians of these collections.[54] Until Naturalis Biodiversity Center develops a strong interest in the history and provenance of its own collections and makes serious work of decolonising its collections, a study of this nature will most likely remain unwritten.[55]

The overview of types of publications and sources above serves as an indication that the study of the history of the Natuurkundige Commissie is not without its own specific problems. No one in recent times has written a monograph on the Natuurkundige Commissie, and the publication that comes

---

52 For recent examples of the continued relevance of the collections of the Natuurkundige Commissie, see Klein Douwel, "Dus je bent er nog wel"; Jacobs, "Frühe Beschreibungen und Darstellungen"; Jacobs and Koch, "On the Discovery of the Emerald Tree Monitor."

53 See the bibliography of Klaver, *Inseparable Friends*, for the extent of material that is available for just the first two Commissie members Kuhl and Van Hasselt.

54 See Das and Lowe, "Nature Read in Black and White."

55 The two anniversary publications are a step in the right direction but are more anecdotal than systematic in nature: Schilthuizen and Vonk, *Wie wat bewaart*; Gassó Miracle and Monquil-Broersen, *Van onschatbare waarde*.

closest to a monograph dates from 1879 and only devotes half its pages to the Natuurkundige Commissie.[56] It seems to me that it is the vastness of the resources available that has prevented such a monograph from being written. In a decade of dedicated research, a single researcher may still only scratch at the surface if they do not make the difficult choice to limit themselves to a subset of the resources.

Approaching the Natuurkundige Commissie as a monolithic enterprise is unrealistic at best. An extensive treatment of all aspects of the work of the Natuurkundige Commissie in one volume is simply not possible without losing the interpretative edge that is required. The most obvious solution to this problem has been for researchers to shift focus from the entire Commissie to an individual member, or sometimes an individual expedition. This approach has revealed much of significance,[57] but is insufficient to reveal the intricacies of a project of such complexity and longevity as the Natuurkundige Commissie. This is a case where the sum of the study of individual members of the Commissie is actually much less than the study of the full Natuurkundige Commissie, which I have just argued is unrealistic. This leads to the paradoxical situation where we cannot see the forest for the trees, but when we step back and focus too much on the forest, we lose sight of the individual trees without which there would not be a forest.

In the next section of this chapter, I propose an analytical historical framework that will help resolve this apparent paradox in the study of the Natuurkundige Commissie. This framework allows the student of the history of the Natuurkundige Commissie (including myself) to approach individual members in such a way that the view of the full Commissie is not obscured and vice-versa. This framework opens up the space to look beyond the eye-catching individual hero-explorers and reveals that the Commissie needed many more people than the naturalists who were the public (but skewed) face of the Natuurkundige Commissie in the Netherlands.

5 Exploration and Expeditions

In the nineteenth century, 'exploration was conceived of as a particular kind of travel associated with the sight of new landscapes, peoples, plants and animals.

56 Half (pp. 20–121) of the 204 pages of Veth, *Overzicht van hetgeen gedaan is*, deal with the Natuurkundige Commissie.

57 Glaubrecht, "'Experience' of Nature," and the introductions in Weber and Van Zanen, *Kunstenaar op Java*, and Weber and Van Zanen, *Kunstenaar op Sumatra*, are some recent examples of this.

To lay eyes upon a lake, a peak, a species, for the first time, was the dream of every aspiring explorer.'[58] One of the most important vehicles for exploration was the expedition or, to be more precise, the geographical or scientific expedition. The goal of expeditions generally was to gather previously unknown information. Of course, this information was only unknown to the Europeans and was fully part of well-functioning indigenous knowledge systems. When expeditions are presented as voyages into the unknown, it is implied that the unknown was only unknown to the Europeans. Much of the knowledge that was gathered in this fashion was taken out of context and incorporated into a European construct of knowledge. I will say more about this process in chapter 7. For now, I have no choice but to use the historically relevant language of exploring the unknown, with the implied knowledge that this is an inaccurate and heavily Eurocentric category.

While explorers were in search of knowledge and information, it was not just random information that expeditions were looking for. James Cook's first voyage, for example, had the concrete goal of observing the 1769 transit of Venus as part of an international effort to calculate the distance from the earth to the sun.[59] Cook's famous endeavours ushered in a period of marine exploration that caused a plethora of scientific measurements and observations to flood back into Europe.[60] All these efforts required specialised personnel and many of the oceanic expeditions included resident naturalists.

The table of contents of Glyn Williams' *Naturalists at Sea* reads like a who is who of these naturalists during the maritime exploration of the eighteenth and early nineteenth centuries. Williams devotes chapters to William Dampier, Georg Wilhelm Steller (travelling with Vitus Bering), Philibert Commerson (travelling with Louis-Antoine de Bougainville), Joseph Banks and Daniel Solander (travelling with James Cook), Johann Reinhold Forster (travelling with James Cook), James Cook and George Vancouver, Jean-François de Lapérouse and Antoine Bruni d'Entrecasteaux, Alejandro Malaspina, Nicolas Baudin and Matthew Flinders, and finally, Charles Darwin. All these travelling naturalists brought back boatloads of specimens and data and much of this material was published in Europe. In fact, many of these voyages would not have been as famous as they are if they would not have spawned the publications that they did. The publications consisted of both narrative travel reports with descriptions of 'landscapes, peoples, plants and animals,' and specialised

58 Driver, *Geography Militant*, 9.

59 See Wulf, *Chasing Venus*.

60 See Mackay, *In the Wake of Cook*.

overviews of what was collected. It is very clear that 'systemized collection and dissemination of data lie at [the] essence [of expeditions].'[61]

Just as the oceans, vast and unexplored as they appeared to the European travellers, were once perfect destinations for exploratory expeditions, so too the inland areas behind colonial trade posts were mostly unexplored and begged European adventurers and naturalists for further exploration. According to Dane Kennedy, the unknown interiors of the continents of Africa and Australia 'were the conceptual equivalents of oceans, vast and empty spaces that could become truly knowable only through the application of scientific methods that seaborne explorers had pioneered.'[62]

Many explorers 'emulated [the] approach' of the oceanic expeditions 'by keeping journals and other written records of their observations and experiences; by using precision instruments to map their geographical locations and measure meteorological conditions; and by collecting and preserving large quantities of botanical, geological, and other specimens and artifacts.'[63] The resulting data would then be sent back to Europe to be published in some form that would contribute to the growing body of European knowledge and would definitely also bring fame to the explorers.[64] So how does the work of the Natuurkundige Commissie measure up against these expeditions? And how can the convergence or divergence from what an expedition is and does help us to understand the Natuurkundige Commissie?

Dane Kennedy defines the shared characteristics of an expedition as follows:

> They were supported by institutional sponsors, such as governmental agencies and learned societies; they were supplied with instructions that outlined their operations and objectives; and they were structured with clear lines of command and defined duties.[65]

The Natuurkundige Commissie neatly matches Kennedy's characteristics. It has already become clear that the Commissie would not have come into being without the spiritual sponsorship of Temminck and his 's Rijks Museum and

---

61 Thomas, *Expedition into Empire*, 1.

62 Kennedy, *Last Blank Spaces*, 2.

63 Kennedy, *Last Blank Spaces*, 2.

64 It must be noted that the areas that seemed ripe for exploration by Europeans were, of course, well-known to the people already living there. European explorers had a tendency to obscure local support that was in almost every aspect indispensable to them. When I use the language of exploring, it is with the knowledge and the recognition that the areas that were 'explored' were not a vacuum.

65 Kennedy, *Last Blank Spaces*, 3.

the financial sponsorship of the Dutch government. Clear instructions were also given to the members, either as an integral part of a royal decree, as an additional document from the ministry, or packaged in a decree from the governor-general in the Indies. Lines of command were clarified by giving the members a status as civil servants within the hierarchy of the colonial government. In the next two chapters, these characteristics are an integral ingredient in my analysis of the layered loyalties that the members were subject to.

Dane Kennedy is very hesitant to automatically connect science and empire,[66] while the subservience of expeditions to the goals of empire was taken as the departure point for *Expedition into Empire*, edited by Martin Thomas. When measured against the four traits of expeditions in service of empire as defined by Thomas, the Natuurkundige Commissie comes out as somewhat unusual.[67] 'Physical hardship and preparedness to sacrifice oneself in the service of empire' (trait 1: man versus nature) is something that applies to the Natuurkundige Commissie, so much so that the loss of so many members eventually had an inversely proportional effect on the legacy of the Commissie. Instead of propelling them into fame as heroic servants of empire, they sank further into obscurity.

Also, 'the deployment of advanced technology, especially precision-navigational instruments' (trait 2: technology) seems largely absent from the expeditions of the Natuurkundige Commissie. Hardly anything is known about the methods and techniques used for land surveying or any other modern technologies. As we will see in chapter 5, Commissie member Diard, who was not even one of the members appointed in the Netherlands, was heavily involved in technical improvements, but not in an expeditionary context.

A further failure, according to Thomas' list of traits, is the lack of any significant follow-up expeditions that 'buil[t] on the successes' of the Commissie, or 'use[d] the disappearance or failure of their predecessors as the stimulus for new journeys.' (trait 3: machines for producing discourse). Undoubtedly, some travellers will have been inspired by the Natuurkundige Commissie, but that remained limited to the circle of naturalists and then only for a short while.

The final trait is one that the expeditions of the Natuurkundige Commissie did not lack. All members made 'diaries, maps, sketches, and other types of representation that would be published or in other ways disseminated,' (trait 4: media)[68] but the first significant output that was directly identified with the

---

66 Kennedy, *Reinterpreting Exploration*, 1–18, especially 9–10.

67 See Thomas, *Expedition into Empire*, 12–17.

68 See Driver, *Geography Militant*, for an extensive analysis of this aspect of expeditionary science.

work of the Commissie, the *Verhandelingen*, only saw the light of day in 1839 in an expensive and mostly unaffordable format, while the mass-market edition was only published in 1857, years after the demise of the Commissie.[69] Even though media representation was there, it did not have enough popular appeal to pervade the public domain in the Netherlands.

While the Commissie perfectly fits Kennedy's characteristics of an expedition, the above shows that the Commissie hardly meets any of Thomas' traits of expeditions in the service of empire. The obvious conclusion would be that the Commissie did not serve imperial needs at all. However, from chapter 3 it will become clear that the Commissie was heavily involved in imperial reconnaissance by creating maps and inspecting mines for the colonial government, while in this chapter we have already seen that the goal of producing an eminent work on the riches of the Dutch colonial empire was indeed part of a grand Dutch scheme to gain international prestige as a strong colonial power.

Therefore, it cannot be denied that the Commissie had a strong imperialistic element to it, which was perhaps in some aspects more coincidental than intended. When we take into account that the original and clearly defined mission of the Commissie was to collect objects of natural history, and that this was the foremost task that the members dedicated themselves to in the field, Andreas Weber is slightly overemphasizing the importance of the service of the Commissie 'to the colonial state in managing its overseas possessions.'[70] As Weber argues and we will see in chapter 3, the colonial government did indeed manage to make use of the Commissie expeditions, but even the governor-general had to follow the Dutch royal decrees that expressly defined researching the natural world as the primary task of the naturalists, although it left enough room for creative solutions.

What sets the Natuurkundige Commissie apart from the expeditions analysed by Kennedy and Thomas is the perception of the project in the Netherlands. Traditional expeditions (with or without imperial connotation) are subject to a strong divergence between the perception at home and the reality in the field. The problem with the Natuurkundige Commissie was there were no popular publications in the public eye that could create such a public perception at home until after its demise. One of the main reasons, as explained above, is the high mortality rate of the naturalists and the modest profile of the only two members that did return. Salomon Müller was sent out as an assistant-preparator so never took centre-stage, while Pieter Willem Korthals was a systematic museum botanist, a profession that still does not

69 Müller, *Reizen en onderzoekingen*.

70 Weber, "Collecting Colonial Nature," 81.

appeal to the masses.[71] Because of this, what happened in the field mostly stayed in the field.

In the Netherlands, the Commissie was very much a vehicle for Temminck's ambitions for 's Rijks Museum, which found favour because of the element of imperial fame that the king stood to gain if the project turned out to be a success. The members were appointed in six instalments, each following and replacing one another, turning the Natuurkundige Commissie, from a Dutch administrative point of view, into one large 30-year expedition with a changing cast of characters. The common goal of all the members of the Commissie was to collect objects for 's Rijks Museum. It did not matter who the naturalist in the field was as long as they sent back crates of specimens. The death of a naturalist in the field caused grief and lament for those who had personally known them, but it also created new opportunities for eager young naturalists to advance their career. From a Dutch point of view, new appointments basically meant nothing more than the hiring of new staff to fill a job opening in a long-term project. New appointments were only made when a position had become vacant through death or return. While the Natuurkundige Commissie may initially not have been intended to last for thirty years, the steady renewal of the personnel pool gave it a long lifespan.

It was only after the publication of last instalment of the *Verhandelingen* on December 24, 1847,[72] which was the main means to foster imperial fame, that the Commissie lost its relevance and faded from view. From a political point of view, the Commissie had done what it was meant to have done and it was unclear why more money should be poured into what had now essentially become a purely scientific project. There were also no further economic benefits to be expected, since the contribution of the work of the Natuurkundige Commissie to the agricultural development of Java had been fairly minimal. King Willem II only appointed one new member in 1842 who, as a geologist, was responsible for investigating the potential of coal mining on Borneo.

However, there was no reason why such a task should take place in the context of an expensive Natuurkundige Commissie and already in 1844 sentences like 'not so much can be expected from its continuation' started appearing in official correspondence.[73] King Willem III, for whom the Commissie was of

71 Franz Wilhelm Junghuhn did make a splash when he returned to the Netherlands. However, he vocally distanced himself from what he considered the limiting constraints of the Natuurkundige Commissie. See Junghuhn, *Java*, volume 1 (2e uitgave), 3–4.

72 This twenty-ninth and final instalment was, in fact, already somewhat of an afterthought and followed two full years after the twenty-eighth instalment, published on December 23, 1845. See Husson and Holthuis, "Dates of Publication," 24. See Appendix 4 for a full table of contents of the *Verhandelingen*.

73 Quoted by Goss, *Floracrats*, 28.

much less importance than it had been to his grandfather Willem I, lost all interest in it. The loss of royal privilege when it came to dealing with the colonies may also have played an important role.[74] The fading and eventual loss of royal and political interest left the Commissie without any solid ground to stand on. Finally, this resulted in the dissolution of the Commissie on April 17, 1850, when there was only one member left in the field.[75]

The situation in the field was entirely different. While the expedition started for the Dutch supporters and policymakers as soon as the naturalists set foot on board the ship that would take them to the Indies, for the naturalists themselves this was only a first step towards their final goal of collecting specimens. Sometimes years would pass between leaving the Netherlands and setting out on an actual expedition in the Indies. Many of the members spent some time in Batavia and its immediate environment (including Buitenzorg, Figure 8) before setting out further afield. In the early years, this was not a problem, because Batavia and its surroundings offered enough nature to explore and plenty of specimens to collect. But for any trip that was going to lead them far afield, they needed administrative permission to travel since they were ultimate placed under orders of the governor-general.

In the course of the first decade, this slowly turned Batavia into a strange no-man's land for Commissie members on hold; a twilight zone between arrival and departure. This time was sometimes filled with collecting what their hands found to collect in an already explored area, but always with an eye on setting out what they were meant to do, namely travelling to scientifically unknown areas to discover and collect new specimens. In the 1820s several expeditions farther afield on West Java were organised, but it was not until 1828 that the first expedition left with a destination that lay outside of Java (New Guinea and Timor). This would be followed by expeditions that would take the Commissie to the main Outer Islands (Sumatra, Borneo, Celebes, Moluccas) with a dedicated (though sometimes small) crew of support staff.

In a sense, the members of the Commissie unwittingly undertook expeditions on two levels: in the eyes of the Dutch policymakers, they undertook a single expedition to the Indies, but in their own eyes and the eyes of the colonial authorities, they undertook various expeditions to the various districts of Java and the Outer Islands. What was not seen from the Netherlands

74 The revised 1840 Constitution repeats that the king has exclusively control over the colonies, but adds that he needs to provide financial statements to the Staten-Generaal, while the 1848 Constitution omits the phrase 'bij uitsluiting' ('exclusively') and adds that government regulations are to be established by law. See https://www.denederlandsegrondwet.nl/id/vi6jejadzrsv/ artikel_60_opperbestuur_volkplantingen.

75 The only other remaining member, Junghuhn, was in the Netherlands, while Müller was on long-term furlough.

FIGURE 8 'Vue du Palais de Buitenzorg.' ('View of the Palace of Buitenzorg'), lithograph by H. J. Backer, 1842, KITLV 47B25, Leiden University Libraries, Digital Collections. Original plate from Van Braam, *Vues de Java*. Public domain HTTP://HDL.HANDLE.NET/1887.1/ITEM:852507

was that through the constant personnel renewal each expedition consisted of a different group of Commissie members and assistants, sometimes even dividing forces to survey different territories independently and simultaneously. What remained entirely invisible is that the members of the different sets of appointments mixed in the Netherlands Indies. Sometimes expeditions were undertaken by members from one appointment, sometimes from several appointments. The view of the operating structure of the Natuurkundige Commissie was perceived very differently in both the Netherlands and the Netherlands Indies.

If we want to properly understand the Commissie, it is important to take this geographical dichotomy between Dutch appointments and expeditions in the Netherlands Indies into account. Focusing solely on the appointments and their achievements, runs the risk of seeing the Commissie as a monolithic whole and losing sight of the diversity and unique features of each individual expedition. Focusing on single Commissie members, on the other hand, runs the risk of losing sight of the perception that existed in the Netherlands and the strong focus on collecting that was an integral part of the makeup of the Commissie. I propose that we can only begin to understand the Commissie by recognising the fact that there are two equally valid, but very disparate pictures of the Commissie that each individually do not do full justice to the Commissie. Only by taking the two geographic realities of the Commissie into account can we begin to understand the significance of the Commissie in both a European and a global colonial context.

In what follows I sketch the series of appointments in which the Commissie members left for the Indies, contrasted with the additional appointments that were required as events unfolded in the Indies. I refer the reader who is unfamiliar with the narrative of the Natuurkundige Commissie and the biographies of the participants to the appendices, as my analysis of the identity of the members and assistants will not take place in the form of a chronological narrative and contains limited biographical detail.[76] My focus instead is to show that each member and assistant had their own unique story but was subject to several loyalties to the various stakeholders in the conglomerate of the Natuurkundige Commissie. In the next two chapters I use a case study of the work of Commissie member Forsten with parallels from the experience of other members to map out these layered loyalties to the various stakeholders who were divided over the Netherlands and the Netherlands Indies, and I show how the balancing act between those layered loyalties significantly influenced the practice of the Commissie in the field.

6 Appointments in the Netherlands

Between 1820 and 1842, in a series of six appointments, eight naturalists were appointed by royal decree to collect specimens for 's Rijks Museum in the Netherlands Indies. They were supported by six assistants, both preparators and draughtsmen (Table 1). It may seem odd to note that the term 'Natuurkundige Commissie' is not mentioned until the third appointment decree. Veth already noted this and explains that the name 'Natuurkundige Commissie' was attached to all of those doing natural history research in the Netherlands in hindsight.[77] As I have indicated above, there is a clear line of succession of members of the Natuurkundige Commissie who succeed in the footsteps of Kuhl and Van Hasselt. As a result, the use of the name 'Natuurkundige Commissie' for those appointed in the first two decrees is fully justified even

---

76 Veth, *Overzicht van hetgeen gedaan is*, has done an excellent job in composing a full historical narrative of the Commissie from the Dutch point of view. In Appendix 1 I make my own attempt at a basic chronological narrative of events in the Indies that is more extensive than what I can present here in this chapter. The narrative in Appendix 1 is supported by a set of biographical narratives of all participants in the Indies in Appendix 2.

77 Veth, *Overzicht van hetgeen gedaan is*, 20.

TABLE 1 Members, assistants, and associates of the Natuurkundige Commissie with years of their involvement indicated in grey. This overview excludes time spent in the Netherlands working on publications (Korthals, 1837–1842; Müller, 1837–1850; Junghuhn, from 1848), or any continued involvement after the formal dissolution (Schwaner, 1850–1851, Junghuhn from 1850)

	1820	1821	1822	1823	1824	1825	1826	1827	1828	1829	1830	1831	1832	1833	1834	1835	1836	1837	1838	1839	1840	1841	1842	1843	1844	1845	1846	1847	1848	1849	1850
Kuhl	■	■																													
Van Hasselt	■	■	■	■																											
Van Raalten	■	■	■	■	■	■	■	■	■	■																					
Keultjes	■	■																													
Maurevert		■	■	■																											
Bik				■																											
Boie				■	■	■	■	■																							
Macklot				■	■	■	■	■	■	■	■	■	■																		
Müller				■	■	■	■	■	■	■	■	■	■	■	■	■	■	■													
De Bruijn					■	■																									
Fievez					■								■																		
Kollmann					■																										
Van Oort						■	■	■	■	■	■	■	■	■	■																
Zipelius								■	■																						
Latour									■																						
Diard									■	■	■	■	■	■	■	■	■	■	■	■	■	■									

TABLE 1 Members, assistants, and associates of the Natuurkundige Commissie with years of their involvement indicated in grey (*cont.*)

	1820	1821	1822	1823	1824	1825	1826	1827	1828	1829	1830	1831	1832	1833	1834	1835	1836	1837	1838	1839	1840	1841	1842	1843	1844	1845	1846	1847	1848	1849	1850
Spanoghe											■																				
Korthals											■	■	■	■	■	■	■	■													
Van Gelder											■	■	■	■	■	■	■	■	■												
Overdijk											■	■	■	■	■	■	■	■	■	■	■								■	■	
Bürger														■	■																
Horner																■	■	■	■												
Junghuhn																			■	■						■	■	■	■		
Forsten																			■	■	■	■	■	■							
Von Gaffron																					■	■	■	■	■	■	■				
Schwaner																							■	■	■	■	■	■	■	■	■
Comblen																								■	■	■	■				
Van Pers																											■	■			
François																													■		
Strung																													■		
Brendonck																													■		
Oehler																													■	■	■

though during their lifetimes they may not have known that they were members of the Natuurkundige Commissie. Most of the eight members were medical doctors by education, which included a substantial amount of training in the natural sciences, and most of them had prior working experience at 's Rijks Museum or had shown a strong affinity for natural history.

6.1 *Appointment 1 (1820): Kuhl and Van Hasselt with Keultjes and Van Raalten*

The first appointment took place on May 2, 1820.[78] The king appointed Heinrich Kuhl (1797–1821, Figure 9) and Johan Conrad van Hasselt (1797–1823, Figure 10) to 'travel through the Dutch possessions in the East Indies with the express goal of expanding the scientific knowledge of the products of nature.'[79] As assistants were appointed Gerrit van Raalten (1793–1829), anatomist, and Gerrit Laurens Keultjes (1786–1821), artist.

FIGURE 9 Portrait of Heinrich Kuhl, lithograph by Friedrich Fleischmann, probably after a painting by Fleischmann, before 1820. Lithograph held at Naturalis Biodiversity Center, reproduced by Karien Lahaise. Public domain. Current whereabouts of the original painting unknown.

78 KB, May 2, 1820, no. 10. NL-HaNA, Staatssecretarie, 2.02.01, inv.no. 988.

79 'De Nederlandsche bezittingen in Oost Indie te bereizen, met het bepaalde doel, om de wetenschappelijke kennis van de voorbrengselen der Natuur in die landen, uit te breiden.'

FIGURE 10 Portrait of Johan Conrad van Hasselt, artist unknown, before 1820, Familievereniging Van Hasselt. Photograph by Evert van Hasselt. Public domain

6.2 *Appointment 2 (1823): Boie and Macklot with Müller and Van Oort*

The second appointment took place on December 5, 1823.[80] 'Taking into account that through the deaths of doctor Kuhl and of the artist Keultjes it is necessary to send a new mission of persons to continue the commenced research trip in our East Indies possessions,'[81] the king appointed Heinrich Boie (1794–1827, Figure 11) and Heinrich Christian Macklot (1799–1832) to continue the work of Kuhl.[82] As assistant was appointed Salomon Müller (1804–1863), preparator. When these three naturalists finally left for the Indies in December 1825, they

80 KB, December 5, 1823, no. 84. NL-HaNA, Staatssecretarie, 2.02.01, inv.no. 1831.

81 'In aanmerking genomen hebbende, dat, ten gevolge van het overlijden van Doctor Kühl en van den teekenaar Keultjes, het noodzakelijk is eene nieuwe uitzending van personen te doen ter voortzetting der aangevangene natuurkundige onderzoekings-reis in Onze Oost-Indische bezittingen.'

82 The news of the death of Van Hasselt in September 1823 had not yet reached the Netherlands.

FIGURE 11 Portrait of Heinrich Boie, by Jakob Wilhelm Roux, 1826, Dithmarscher Landesmuseum, Meldorf, photograph by Peter Geißler. Public domain. On October 12, 2024, Topaz Gigapixel ai was used to increase the resolution of this image.

were accompanied by Pieter van Oort (1804–1834, Figure 12), draughtsman. He had been appointed on April 16, 1825,[83] as a replacement for Jan Cornelis de Bruijn (1794–1862), who had been appointed on February 17, 1824, but was relieved from his position on February 13, 1825, at his own request.[84]

83 KB, April 16, 1825, no. 120. NL-HaNA, Staatssecretarie, 2.02.01, inv.no. 2209.

84 KB, February 17, 1824, no. 173. NL-HaNA, Staatssecretarie, 2.02.01, inv.no. 1891; KB, February 13, 1825, no. 86. NL-HaNA, Staatssecretarie, 2.02.01, inv.no. 2161.

FIGURE 12 Self-portrait, by Pieter van Oort, chalk drawing, ca. 1826–1834, 45.5 × 36 cm, 9324, Centraal Museum Utrecht. Public domain. The text in the lower left corner reads: 'Na mijn dood mijne ouders te zenden.' ('To be sent to my parents after my death.') HTTPS://HDL.HANDLE.NET/21.12130/COLLECT.32709171-61EE-4F3F-8666-90291D18FF59

6.3 *Appointment 3 (1830): Korthals with Van Gelder and Overdijk*

The third appointment took place on July 25, 1830.[85] 'Because of the deaths that have recently occurred in the persons who occupy themselves with natural history research in our East Indies possessions,'[86] the king appointed

85 KB, July 25, 1830, no. 40. NL-HaNA, Staatssecretarie, 2.02.01, inv.no. 3437.

86 'Wegens de sterfgevallen, welke in den laatsten tijd hebben plaats gehad onder de personen, welke die zich onledig houden met natuurkundige onderzoekingen in Onze Oost-Indische bezittingen.'

FIGURE 13 Portrait of Pieter Willem Korthals, photograph of a lithograph by J. M. Kierdorff after a painting by J. L. Cornet, unknown date. Photograph held at Naturalis Biodiversity Center, reproduced by Karien Lahaise. Public domain. Current whereabouts of the original painting or the lithograph unknown.

Pieter Willem Korthals (1807–1892, Figure 13) as botanist.[87] As 'preparators or assistants'[88] were appointed Dominicus Hermanus Righard van Gelder (1809–1879) and Bartholomeus Nicolaas Overdijk (1807–1851).

87 Boie (1827) and Van Raalten (1829) had recently died, as had Indies appointed botanical assistant Zipelius (1828).

88 'Preparateurs of assistenten.'

6.4 *Appointment 4 (1835): Horner*

The fourth appointment took place on February 23, 1835.[89] When Temminck suggested to the minister that Ludwig Horner (1811–1838, Figure 14) could be a suitable replacement for the unfortunate Macklot,[90] his recommendation was brought to the king and as a result Horner was appointed to take up the disciplines of geology and mineralogy.

FIGURE 14 Portrait of Ludwig Horner, by Johann Andreas Hirnschrot, oil on copper(?), 1834, 39 × 32.2 cm, Inv 50, Zentralbibliothek Zürich. Public domain
HTTPS://DOI.ORG/10.7891/E-MANUSCRIPTA-85725

89 KB, February 23, 1835, no. 71. NL-HaNA, Staatssecretarie, 2.02.01, inv.no. 4003.
90 Macklot was murdered in 1832.

6.5 *Appointment 5 (1838): Forsten*

The fifth appointment took place on February 4, 1838.[91] When the return of Müller and Korthals opened up the possibility to appoint a new member, the choice was made to appoint Eltio Alegondus Forsten (1811–1843, Figure 15) as zoologist who had already sent in an open application to become a member in the previous year.[92]

6.6 *Appointment 6 (1842): Schwaner*

The sixth and final appointment took place on January 9, 1842.[93] Carl Anton Ludwig Maria Schwaner (1817–1851) was appointed as member of the Natuurkundige Commissie. The same royal decree also stipulated that no new members would be appointed for the foreseeable future and that effort should be taken to both limit the costs of natural history research in the Indies and speed up the publications of the results in the Netherlands.

Taken together, these six appointments formed the most visible aspect of the Natuurkundige Commissie in the Netherlands. All these members and assistants were in their early careers and departed from the Netherlands in their twenties. Keultjes was the only one who was already in his thirties and married when he departed for the Netherlands Indies. All the other members and assistants were young and unattached, giving them the freedom to spend however long they needed or wanted to in the tropics. Just as their expeditionary travels served as an event that propelled them into fame for Joseph Banks, Alexander von Humboldt, and Charles Darwin, the members and assistants of the Natuurkundige Commissie were eager to use their own travels as a steppingstone towards a future career in natural history. A famous expedition with a magnificent publication would certainly be of great value towards this.

Of the eight members, four were German (Kuhl, Boie, Macklot, Schwaner), one was Swiss (Horner) and three were Dutch (Van Hasselt, Korthals, Forsten). Of the six assistants, only one was German (Müller) and five were Dutch (Van Raalten, Keultjes, Van Oort, Overdijk, Van Gelder). By education, four of the members were medical doctors (Van Hasselt, Macklot, Forsten, Schwaner), one was a trained surgeon (Horner), one had a law degree (Boie), one had an honorary degree in mathematics and physics (Kuhl). Boie had followed extensive courses in the natural sciences, like both Macklot and Schwaner who had a strong affinity for geology. Korthals did not have a university education prior to his departure to the Indies but was trained as an apothecary and had worked

91 KB, February 4, 1838, no. 98. NL-HaNA, Staatssecretarie, 2.02.01, inv.no. 4334.

92 KB, February 27, 1837, no. 51. NL-HaNA, Staatssecretarie, 2.02.01 inv.no. 4226.

93 KB, January 9, 1842, no. 66. NL-HaNA, KdK 1841–1897, 2.02.04, inv.no. 79.

FIGURE 15 Portrait of Eltio Alegondus Forsten, painter unknown, 1830–1838, measurements unknown, Naturalis Biodiversity Center. Public domain
PHOTOGRAPHY BY KARIEN LAHAISE

for several years as botanist for Reinwardt.[94] None of the assistants had a university education but all were well-trained in their profession. Van Raalten had prior working experience in the zoological cabinet of Theodoor Gerard van Lidth de Jeude (1788–1863) in Utrecht, Müller was a protégé of Boie and as such had worked with him in Heidelberg and at 's Rijks Museum in Leiden, while both Overdijk and Van Gelder worked at 's Rijks Museum as preparators before being appointed as assistants. Keultjes was said to have learnt his trade as an artist in the orphanage in Utrecht where he grew up[95] and Van Oort was educated as an artist by his painter-father and had worked with Van Lidth de Jeude as well.[96]

The most striking element from the overview above is perhaps the prevalence of non-Dutch nationals in an official colonial project of the Dutch government. Six out of fourteen appointees were German-speaking foreigners, while only three out of eight members were Dutch nationals. Of the German-speaking foreigners only Kuhl received his education in the Netherlands, in Groningen, while Boie, Macklot, Horner and Schwaner had a German education. So how did these foreigners end up being appointed to a national Dutch project? There are several reasons for this. As Philipp Teichfischer has shown, there was an 'increasing number of medical students' in the German-speaking world in the early nineteenth century. In combination with a 'stagnation in the number of practising doctors,' this led to a surplus in medical doctors.[97] As a result, many German medical doctors were in search of employ. Colonial service was an attractive prospect, but because there was no German colonial empire, many doctors enlisted in the Dutch colonial medical service. And while Germany was not a competing colonial power, the Dutch happily allowed German doctors to enlist, leading to the statistical fact that from 1816 to 1857 about one third of the medical doctors and pharmacists employed by the Dutch colonial health service were Germans.[98] Given the fact that five out of eight of the Commissie members had a degree in the medical realm, it is no surprise that at least some of them were from the German-speaking world.

Horner is actually exemplary of Teichfischer's findings. As a Swiss national, Horner also came from a country that had no colonial empire of their own. He had already enlisted with the Dutch colonial medical service and was set to travel to the Indies as a second-class surgeon. He was planning on occupying

94 Karstens and Kleibrink, *Leidse Hortus*, 66.
95 'Gerardus Laurentius Keultjes.' RKD database, https://rkd.nl/explore/artists/44173.
96 Weber and Van Zanen, *Kunstenaar op Java*, 21–22.
97 Teichfischer, "Transnational Entanglements," 73–74.
98 Teichfischer, "Transnational Entanglements," 66–67.

himself in his free time with collecting specimens. Because of his personal interest he visited 's Rijks Museum during his stay in Leiden in preparation of his departure. When Temminck found out that this young man from Switzerland intended to become a competitor of the Natuurkundige Commissie while he was already employed by the Dutch government, he stepped in and requested him to be appointed as member of the Natuurkundige Commissie in order to prevent any specimens from being sent to foreign museums. Temminck succeeded and Horner's prospective medical career turned into a naturalist one.

Horner's backstory illustrates another important reason for the prevalence of German-speaking members and assistants in the Natuurkundige Commissie. His appointment was a direct consequence of him becoming part of Temminck's international network. As an internationally renowned collector and zoologist, Temminck corresponded with many international naturalists and regularly visited colleagues in France, Britain and the German lands. He used these international networks to fill the more senior posts in 's Rijks Museum. Because German-speaking naturalists did not have their own colonial agenda, they could be more easily won over to exclusively serve Dutch colonial interests.

The involvement of Kuhl, Boie, Macklot and Müller was a direct result of Temminck's international network. During his studies, Kuhl travelled extensively in the Netherlands and Germany and in 1816/1817 he met Temminck in Amsterdam.[99] Boie's careful review of Temminck's *Manuel d'Ornithologie* in 1816[100] led to a correspondence between the two men and they met each other in Germany for the first time in 1817/1818.[101] Boie also happened to have been friends with Kuhl since 1814. After Boie's appointment as curator at 's Rijks Museum, he introduced two able naturalist-friends from his Heidelberg time, Macklot and Müller, to Temminck, who both also travelled to Leiden.

The fact that nationality was a potential issue for the Dutch government becomes obvious from the proposal included with the appointment decree for Boie and Macklot. Civil servant Daniël Jacob van Ewijck (1786–1858), who was heavily involved with the first appointment, writes to the king that

> Mister Temminck concludes a report to me by declaring that through his daily intercourse with the previously mentioned travellers [Boie, Macklot, Müller] he is convinced that they are imbued with a true sense of gratitude for the trust that they hope will be placed in them and that

99 Klaver, *Inseparable Friends*, 8.
100 Boie, "Manuel d'ornithologie."
101 Gassó Miracle, *Coenraad Jacob Temminck*, 68.

> they, although foreigners, will strive to emulate their Dutch colleague [Van Hasselt][102] and that it will definitely honour the motherland to have such foreigners working on Java.[103]

Clearly, Temminck felt the need to stress the fact that even though they were foreigners they would have such a sense of loyalty towards the Dutch government for honouring them with this position that they would work extra hard to outshine their Dutch colleagues. And it was not only the Dutch government to which they owed their position, but more directly so to Temminck, whose network served as a recruiting ground for members and assistants.

Of the eight members and six assistants at least five (Boie, Macklot, Müller, Van Gelder, Overdijk) were employed by 's Rijks Museum before leaving, at least two (Van Oort, Forsten) spent some preparatory training time at the Museum. Kuhl and Van Hasselt had gained experience by visiting and working at several (inter)national institutions of natural history research. They were never trained at 's Rijks Museum because it did not yet exist at the time of their departure. Of the remaining three members and two assistants, two had working relationships with Dutch naturalists from Temminck's network (Van Raalten with Van Lidth de Jeude, Korthals with Reinwardt and Blume). Horner's entry into Temminck's network has been explained above, while very little is known about the circumstances that led to Keultjes' appointment other than that he came well-recommended and hailed from Utrecht. It is not inconceivable that the positive recommendation he received came through the Van Lidth de Jeude/Van Raalten connection in Temminck's network. The little that is known of Schwaner's association with Temminck leaves room to speculate that his study period in Heidelberg may have made him part of Müller's network, who had returned to the Netherlands in 1837 and regularly spent time with his family in Heidelberg in the years leading up to Schwaner's appointment in 1842.

102 It was not yet known in the Netherlands that Van Hasselt had died almost two months before this recommendation was written.

103 'De Heer *Temminck* besluit een aan mij gedaan rapport met te verklaren, dat hij door een dagelijkschen omgang met de voormelde reizigers overtuigd is, dat zij met een waar gevoel van erkentelijkheid doordrongen zijn voor het vertrouwen, waarmede zij hopen vereerd te worden, en dat zij, alhoewel vreemdelingen, met hunnen Nederlanschen ambtsgenoot zullen wedijveren, en dat het gewis het moederland eer zal aandoen zoodanige vreemdelingen op Java te hebben doen uitkomen.' KB, December 5, 1823, no. 84. NL-HaNA, Staatssecretarie, 2.02.01, inv.no. 1831.

7 Appointments in the Netherlands Indies

As demonstrated above, from a Dutch perspective the Natuurkundige Commissie was a long-running continuous project where deceased members were replaced with 'fresh stock' from Temminck's network. However, the story in the Indies was an entirely different one. As soon as the newly arrived members and assistants arrived, they were slotted into the colonial hierarchy, where they did not have such a dedicated driving force like Temminck available to set them on their further path. The members and assistants had been given a rank in the colonial hierarchy: members were mostly third-class civil servants, assistants mostly fourth of fifth-class civil servants. They all became entirely dependent on colonial bureaucratic structures and the goodwill of the man in charge (the governor-general), not only for their finances, but also even for permission to travel the archipelago. Governor-general Van der Capellen introduced travel limitations for non-Javanese from May 20, 1823, which made the movements of the explorers even more dependent on bureaucratic goodwill.[104]

In practice, upon arrival in Batavia, the members first needed to make themselves seen and heard by the governor-general before they could even dream of setting out on an expedition. In addition, because the members and assistants were civil servants like any other, the governor-general also had the power to relieve them of their duties and appoint new people. An appointment from the Netherlands was not a guarantee for immunity against the governor-general, so much depended on the lobbying prowess of the members and their assistants, which mostly took place through the designated directing members.[105] Boie had been appointed directing member in the Netherlands. Upon his death Macklot took over, later joined and succeeded by Pierre-Médard Diard (1794–1863, Figure 16) who would play an essential role in the life of the Natuurkundige Commissie.[106]

When I started this chapter, I wrote that eight members and six assistants were appointed in the Netherlands. As has become obvious, these fourteen naturalists mostly dominated the narrative in the Netherlands. The sheer number of participants in the work of the Natuurkundige Commissie remained invisible and anonymous in the Netherlands. In chapters 6 and 7, I will discuss

104 Bossenbroek, *Wraak van Diponegoro*, 207.

105 The role of directing member had never been clearly defined. In practice, the directing member was responsible for most of the correspondence with the Dutch and the Indies governments and had the power to send in proposals to the ministry or the governor-general. He was often consulted and asked for advice by the governor-general in matters of natural historical relevance.

106 See chapters 4 and 5 on Diard.

FIGURE 16
Portrait of Pierre-Médard Diard as an older man, with the insignia of the Order of the Dutch Lion on the left and the French Ordre national de la Légion d'honneur on the right. Plate 1 (after page 10) from Peyssonnaux, "Vie, voyages et travaux." The original plate is reproduced from an original painting that was in the possession of Paul Viot from Amboise. Current whereabouts unknown. Public domain

the indigenous permanent and temporary employees of the Commissie. None of these were directly appointed by the governor-general and his administration, but employing them was the prerogative of the members, as long as they kept within the budgetary limits.

In this section, it is time to shed some light on the participants that were appointed by the governor-general in the Indies, mostly outside of the view of the Dutch stakeholders. Between 1820 and 1850 seventeen new participants were appointed by the governors-general. The governor-general did not have the authority to appoint new members. While this was never put in writing, the governors-general were very aware of the nature of the Natuurkundige Commissie as a royal project, which is confirmed when lieutenant-governor-general De Kock writes to commissioner-general Du Bus in an advice regarding a salary raise for Van Oort: 'The great interest that His Majesty apparently takes in the progress of this Commissie and the use and importance of the Commissie itself has induced the government to give this advice.'[107] The

107 'Het groote belang dat Zijne Majesteit blijkbaar in den goeden afloop van den Commissie stelt, in het nut en de belangrijkheid der Commissie zelve, heeft de Regering tot het uitbrengen van dit advies genoopt.' Res.GG, April 24, 1827, no. 6. NL-HaNA, Koloniën, 1814–1849, 2.10.01, inv.no. 2800.

FIGURE 17
Portrait of Franz Wilhelm Junghuhn, artist unknown, lithograph printed by Pieter Willem Marinus Trap (Leiden), published by F. Günst (Amsterdam), 1866, RP-P-OB-29.799, Rijksmuseum, Amsterdam. Public domain. Several other portraits of Junghuhn are known. HTTP://HDL.HANDLE.NET/10934/RM0001.COLLECT.185919

governors-general were aware of the fact that all resolutions would eventually end up at the department of colonial affairs in the Netherlands.

Given the fact that the rule of the colonies was the prerogative of the king himself, it is likely that many of these resolutions were read by king Willem I, or at least presented to him in summary. Willem was a very hard-working and conscientious ruler with a firm control on everything that happened in his government.[108] While the governor-general was given the freedom to set salaries and to direct the work of the Natuurkundige Commissie in the Indies, it was clear to all that the appointment of new members was not something that the governor-general was allowed to do without consulting the Netherlands. In the thirty years of the Commissie's existence, only two members were appointed in the Indies (Pierre-Médard Diard in 1828 and Franz Wilhelm Junghuhn (1809–1864, Figure 17) in 1845).

The initiative for both these appointments came from the Indies. Diard was proposed by Macklot as a replacement for the then recently deceased Boie, Junghuhn had previously worked as an assistant (1838–1839) and put himself forward. When Macklot proposed Diard as a replacement for Boie to

108 Koch, *Koning Willem I*, 370–381.

commissioner-general Du Bus, he referred the matter to lieutenant-governor-general De Kock and the Council of the Indies for advice. They responded on October 27, 1827, that they fully supported Diard for this position. However, recent objections from the side of the ministry of colonial affairs, even though Diard's response had, in the eyes of the Indies government, removed these objections, were still not fully resolved since the minister had not yet responded to the report from the Indies government.[109] With the impending departure of the entire Commissie to New Guinea and Timor, Du Bus could not wait for a response from the Netherlands and he appointed Diard on February 4, 1828, as successor to Boie.[110] It was only on July 14, 1828, that the matter was resolved in the Netherlands by royal decree and Diard was cleared to serve.[111]

Junghuhn's request to be appointed was treated with similar caution. At first, he was appointed as assistant for the limited period of one year from July 1, 1838, by governor-general De Eerens on Junghuhn's own request supported by a positive recommendation from directing member Diard.[112] Unfortunately for Junghuhn, his appointment was not renewed for financial reasons. De Eerens had given a raise to Horner and Van Gelder on December 1, 1837.[113] Because this news took some time reaching the Netherlands, the budget calculations in the Netherlands which concluded that there was space in the budget to appoint Forsten in February 1838 had not taken the increments in the salaries of Horner and Van Gelder into account. More than a year later, on December 13, 1838, De Eerens was kindly reminded that he was not allowed to add to the budget without prior approval from the Netherlands.[114] News of Junghuhn's appointment had apparently not reached the ministry yet, but De Eerens took the hint and quietly allowed Junghuhn's provisional appointment to lapse. When Diard allowed Junghuhn to continue his work (probably expecting his appointment to be prolonged), De Eerens was forced to explicitly cancel Junghuhn's appointment on September 29, 1839, three months after it had already expired.[115] Due to De Eerens' late decision, Diard eventually managed to squeeze a salary for

109 Res.GG, October 27, 1827, no. 3. NL-HaNA, Koloniën, 1814–1849, 2.10.01, inv.no. 2806.

110 Res.GG, February 4, 1828, no. 43. NL-HaNA, Koloniën, 1814–1849, 2.10.01, inv.no. 2513.

111 KB, July 14, 1828, no. 29. NL-HaNA, Staatssecretarie, 2.02.01, inv.no. 3013. Unfortunately, this specific KB was not present when I consulted inv.no. 3013, but the content of the decree is confirmed by Algemene Indices op de Koninklijke Besluiten. NL-HaNA, Staatssecretarie, 2.02.01, inv.no. 5099, f. 1751.

112 Res.GG, July 10, 1838, no. 13. NL-HaNA, Koloniën, 1814–1849, 2.10.01, inv.no. 2584.

113 Res.GG, December 1, 1837, no. 11. NL-HaNA, Koloniën, 1814–1849, 2.10.01, inv.no. 2577.

114 Res.GG, December 13, 1838, no. 7. NL-HaNA, Koloniën, 1814–1849, 2.10.01, inv.no. 2589.

115 Res.GG, September 29, 1839, no. 12. NL-HaNA, Koloniën, 1814–1849, 2.10.01, inv.no. 2598.

Junghuhn from July to October 1839 from the budget, but this was the end of Junghuhn's first placement with the Commissie.[116]

This did not deter Junghuhn from travelling and exploring, with a government commission to Sumatra in 1842 and extended research on Java in 1843 and 1844. When his reports met with approval, he decided to apply for membership again in 1844. This time, his request was forwarded to the ministry on October 21, 1844,[117] by governor-general ad interim Reijnst and after a lengthy discussion in the Netherlands, Junghuhn's appointment as a member was ratified in the Indies by a resolution on May 5, 1845. His appointment was backdated to April 1, 1845. It was Junghuhn's achievements, rather than the needs of the Natuurkundige Commissie that finally decided in favour of this appointment. King Willem II was in favour of letting the Natuurkundige Commissie run its course, as he had already made clear in Schwaner's appointment decree. Therefore, it required the combined lobbying power of Temminck, Reinwardt and Blume with minister of colonial affairs Baud to sway the king's mind and allow Junghuhn to be appointed.[118]

The cases of Diard and Junghuhn show that the appointment of new members in the Indies was a slow and cumbersome process. Du Bus, as commissioner-general in a higher position than the governors-general, felt confident enough to appoint Diard before a response from the Netherlands was received. De Eerens allowed himself a similar freedom of action, but was subtly reprimanded for this, which may have inspired Reijnst to dutifully await clearance from the Netherlands before appointing Junghuhn again.

An even more complicated case is the association of Heinrich Bürger (1804/6–1858, Figure 18) with the Natuurkundige Commissie. In response to the continuing unrest on Sumatra, it was decided that a military expedition was required. On May 18, 1833, general-major C. J. Riesz was appointed as the highest military and civil authority in the residency Sumatra's Westkust.[119] On the same day, governor-general Van den Bosch concluded that 'it must be considered effective to let several civil servants, destined for natural historical research, participate in the expedition to Sumatra's Westkust.'[120] Van den Bosch ordered Heinrich Bürger and 'the members of the Commissie of

116 Res.GG, April 5, 1840, no. 8. NL-HaNA, Koloniën, 1814–1849, 2.10.01, inv.no. 2605.

117 Res.GG, October 21, 1844, no. 2. NL-HaNA, Koloniën, 1814–1849, 2.10.01, inv.no. 2662.

118 Res.GG, May 5, 1845, no. 1. NL-HaNA, Koloniën, 1814–1849, 2.10.01, inv.no. 2669.

119 Res.GG, May 18, 1833, no. 1. NL-HaNA, Koloniën, 1814–1849, 2.10.01, inv.no. 2550.

120 'Dat het als doelmatig moet worden geacht, om eenige ambtenaren, bestemd voor het natuurkundig onderzoek, te doen deelnemen aan de ophanden zijnde expeditie naar Sumatra's Westkust.' Res.GG, May 18, 1833, no. 2. NL-HaNA, Koloniën, 1814–1849, 2.10.01, inv.no. 2550.

FIGURE 18
Photograph of part of a drawing with a portrait of Heinrich Bürger writing at his desk, by Kazan Watanabe. Photograph held at Naturalis Biodiversity Center, reproduced by Karien Lahaise. Full image available as plate 1 in Uéno, "Japanese Portrait," 94. Original drawing formerly owned by Nuiémon Kasuya. Current whereabouts unknown. Public domain

natural historical research'[121] Salomon Müller and Pieter van Oort to report to Riesz' expedition and await his orders. To all appearances, Bürger seemed to be a member of the Commissie, but in reality, he was never appointed as such. As a result, he must be considered an associate, rather than a member or an assistant.

Of the seventeen new participants appointed in the Netherlands Indies, only two were appointed full members after consultation with the Netherlands (Diard in 1828, Junghuhn in 1845). One of these had previously been temporarily appointed assistant for only a year (Junghuhn in 1838–1839). Of the remaining fifteen new participants, Bürger was only a member by appearance, while the other fourteen were assistants for specific tasks. These fourteen assistants were all directly appointed by the governors-general without consulting the ministry in the Netherlands, took on a variety of roles and sometimes served in two or three different roles, either simultaneously or at different times: expedition companion (Maurevert, Bik, Zipelius, Latour, Von Gaffron, Comblen, Van Pers), artist (Maurevert, Bik, Fievez dit de Malines, Latour, Von Gaffron,

121 'De leden der kommissie van het natuurkundige onderzoek.'

Van Pers), botanist (Zipelius), mining expert (Comblen), manuscript copier (Kollmann, Spanoghe), artist-copier (Fievez dit de Malines, Spanoghe), assistant to Schwaner in drawing up his reports and maps (François, Strung, Brendonck, Oehler). Finally, Overdijk, one of the assistants appointed in the Netherlands, was temporarily rehired in his old role of preparator.

Of the expedition companions, two were attached to Kuhl and/or Van Hasselt (Maurevert, Bik), two were attached to the 1828/1829 New Guinea/Timor expedition (Zipelius, Latour), one was attached to both Forsten and Schwaner (Von Gaffron), two were attached to Schwaner (Comblen, Van Pers). All of these assistants were already present in the Indies for some other reason prior to their appointment. Four had previous working experience at 's Lands Plantentuin (Bik, Fievez de Malines, Zipelius, Latour), two were involved in the military medical service (Kollmann, Junghuhn), five were in the army when they were appointed (Latour, Von Gaffron, Strung, Brendonck, Oehler). Of the seventeen new participants, five were German (Kollmann, Zipelius, Bürger, Von Gaffron, Junghuhn), four were Dutch (Bik, Fievez dit de Malines, François, Brendonck), four were Belgian (Latour, Spanoghe, Comblen, Van Pers), one was French (Diard), one was Austrian (Oehler), one was from India (Maurevert), and one is from unknown European provenance (Strung). Maurevert was the only non-European appointment.

Of these seventeen new participants, only Zipelius, Diard, Bürger and Junghuhn became visible in the Netherlands. Zipelius had been appointed as *hortulanus* at 's Lands Plantentuin in 1823, when Blume served as its director. Correspondence between the two men and Blume's special interest in Zipelius' botanical work secured his visibility in the Netherlands after Blume's return in 1826.[122] The conflict that surrounded Diard's expedition to Borneo in 1826 secured early interest in his actions from Temminck and led to an extensive discussion regarding his reliability. In addition, his appointment as a prospective member had to be approved in the Netherlands. His involvement with various agricultural projects kept him in the eye of governors-general Van den Bosch and Baud who would become ministers of colonial affairs after their return to the Netherlands. This prevented Diard from fading from view, even though it was his experimental agricultural work that kept him visible, more than his work for the Natuurkundige Commissie. Still, both Zipelius and Diard were quite unknown to the general public. Only Junghuhn would take centre-stage, but this would not happen until after the dissolution of the Natuurkundige Commissie in 1850. Junghuhn published his monumental

122 See Zipelius, "Uittreksel uit eenen brief."

four-volume work on Java between 1850 and 1854.[123] This work was based on his extensive expeditions, only some of which were done as assistant or member of the Natuurkundige Commissie. The preface to the second edition of the first volume show a rather spiteful side of Junghuhn:

> Many obstacles that prevented the publication of this work had to be cleared by me; I had to conquer the opposition of persons, who have had the habit in the previous two dozen years of considering all the results of the journeys, undertaken at the expense of the government in order to research nature in the Netherlands Indies, as their property, and who have attempted to thwart all my attempts to bring the fruits of my labour to the light myself.[124]

Without naming them, Junghuhn condemns Blume and Temminck's attitude towards his collections and publications, while he conveniently forgets the fact that his first appointment decree in the Indies clearly stated that all his collections (even those he had brought together before his appointment to the Natuurkundige Commissie) were to be considered government property,[125] while his instructions as member made no secret of the fact that it was the government that decided on when, where and how his findings would be published.[126]

Blume and Temminck, as the directors of 's Rijks Herbarium and 's Rijks Museum, rightfully considered Junghuhn's collections property of their museums and correctly believed they had the right to govern the way in which Junghuhn's findings were published. Nevertheless, Junghuhn publicly raged against this incessantly to such an extent that it severely tarnished Blume's reputation.[127] An additional, and perhaps even greater, claim to fame for Junghuhn was the publication of his *Licht- en schaduwbeelden uit de*

123 Junghuhn, *Java*.

124 'Vele hinderpalen, die de publicatie van dit werk beletteden, moesten door mij uit den weg worden geruimd; ik moest den tegenstand verwinnen van personen, die sedert een paar dozijn jaren gewoon zijn alle vruchten van reizen, ter onderzoeking der natuur in Neêrlandsch Indië op kosten van het Gouvernement ondernomen, als hun eigendom te bezigen, en die de pogingen, door mij aangewend om de vruchten van mijn arbeid zelf in het licht te doen verschijnen, trachtten te verijdelen.' Junghuhn, *Java*, volume 1 (2e uitgave), 3.

125 Res.GG, July 10, 1838, no. 13. NL-HaNA, Koloniën, 1814–1849, 2.10.01, inv.no. 2584.

126 Article 6 of the instructions that are included in Res.GG, April 5, 1846, no. 27. NL-HaNA, Koloniën, 1814–1849, 2.10.01, inv.no. 2681.

127 See Nieuwenhuys and Jacquet, *Java's onuitputtelijke natuur*, 127–133.

binnenlanden van Java (1853), a spiritual allegory that promoted a controversial agnostic worldview in a Javanese setting.[128] Both his tendency to fearlessly fight his fights in the public arena and his rather controversial views, propelled Junghuhn to greater fame than any of the other participants in the work of the Natuurkundige Commissie.

8 Conclusion

As has become clear from this chapter, the Natuurkundige Commissie is a much more complex conglomerate than it has yet been given credit for. The standard image of six members and eight assistants that were appointed by the king to research nature in the Netherlands Indies is, albeit true, sadly one-sided and lacking. It ignores the two additional members and fifteen additional assistants that were appointed in the Netherlands Indies, and the countless indigenous permanent and temporary employees. The Natuurkundige Commissie cannot be understood without taking the geographical dichotomy between the Netherlands and the Indies into account. Recognition of the fact that what was perceived in the Netherlands as one thirty-year long project, was in fact made up of a series of single expeditions in the Netherlands Indies, is the first step towards untangling the complexity of the Commissie. This geographical dichotomy obscured the view of the Commissie in the field, prevented the free flow of information between the two geographies and created two worlds: one world visible in the Netherlands that led to a distorted image of the Natuurkundige Commissie and would even tempt Stresemann to suggest that the members may have died in vain;[129] and another world visible only in the Netherlands Indies about which hardly anything was known in the Netherlands.

In addition, the public controversies created by Junghuhn drew far too much attention and took some of the space in the public arena that might otherwise have been taken by reports of the work of other members. As a result, Junghuhn is still widely considered to be the most important representative of the Natuurkundige Commissie, while he was only active for three years and much of his work did not even take place in the context of the Natuurkundige Commissie. Even the standard work on the history of science in the Netherlands

128 This book was so popular it was reprinted six times. See Mathijsen, *L*, 411.

129 Stresemann, *Entwicklung der Ornithologie*, 154, echoed by Schilthuizen, "Terabytes aan tropenjaren."

identifies Junghuhn as 'the most important investigator' of the Commissie, which leads to a one-dimensional and distorted image of the Commissie.[130]

As this chapter has shown, the conglomerate of the Natuurkundige Commissie was directed from the Netherlands with a whole unseen world of practice in the field in the Netherlands Indies. Communication between the Dutch government, the museum, and the government in the Indies was not always smooth, and much happened in the field that was not visible from the Netherlands. At the same time, the six members and eight assistants that were appointed in the Netherlands were very much aware of the Dutch dimension of the Commissie and displayed a remarkable tenacity in their commitment to the prime mission of the Commissie: collecting objects of natural history.

---

130 Van Berkel, Van Helden, and Palm, *History of Science in the Netherlands*. I do not blame Van Berkel for this view, since this was written in a time when little research had been done into the Commissie as a whole.

CHAPTER 2

Layered Loyalties: the Netherlands

On January 26, 1841, Dutch zoologist Eltio Alegondus Forsten (1811–1843, Figure 15) wrote to his confidante in Leiden from his lodge in Tondano in the Manado-region of Northeast Celebes: '[...] then I will arrive back on Java, and then, and then, and then as soon as possible with the whole bunch to Holland, to hopefully never set foot in the Indies again.'[1] Forsten's letter betrays his eagerness to leave his place of sojourn in the Netherlands Indies to return to his home in the Netherlands. These two geographical areas played a defining role in Forsten's scientific practice (see Map 1). As we have seen in the previous chapter, Forsten was one of the members of the Natuurkundige Commissie who had been appointed by the king in the Netherlands to travel to the Netherlands Indies in order to enhance the knowledge of the natural world in the Indies. Forsten had also been trained at 's Rijks Museum and was personally acquainted with the leading triumvirate of Temminck, Reinwardt, and Blume, and the returnees Müller and Korthals. Just like them, he would be sent to the Indies with a clear mission for 's Rijks Museum in mind, but he would also be placed under the orders of the governor-general.

1 Intrinsic Motivation of the Members

Forsten applied for membership of the Natuurkundige Commissie himself, an unusual move in the history of the Commissie. What was his motivation to do so? A closer look at his life story throws some light on this. Eltio Alegondus Forsten was born in Middelburg on July 2, 1811.[2] Young Eltio (Figure 19) attended the local Latin school. We can follow his progress there in the news bulletins of the Latin school in the *Middelburgsche Courant*, where he is reported to have started in April 1824 and finished on August 26, 1829.[3] The natural history content of Forsten's education was fairly high. We read of him successfully

1 'Dan kome ik op Java terug en dan, en dan, en dan zoo spoedig mogeljk met de geheele boel naar Holland, om zoo ik hoop nooit weder in Indie te komen.' Letter Forsten to Susanna, January 26, 1841, NCAO, 1230_003.

2 Different spellings of Forsten's first names appear in various modern publications. His birth certificate (Birth certificate, no. 229. NLMidZA, Burgerlijke Stand Zeeland (1796) 1811–1980, 1.1.078, inv.no. MDB-G-1811-1) clearly states Eltio Alegondus.

3 *Middelburgsche Courant*, no. 49 (April 22, 1824): 1.

 | DOI:10.1163/9789004696877_004

FIGURE 19
Portrait of Eltio Alegondus Forsten, unknown artist, miniature painting on ivory, ca. 1818–1824, 4.5 × 3.5 cm, private collection. Public domain. This miniature is part of a series of six. The other five portraits feature Forsten's parents, brother and two sisters.
HTTPS://RKD.NL/EXPLORE/IMAGES/165530

taking exams in Natural History, Anatomy, Physiology, Medicine and Chemistry (which included medical botany and apothecary skills).[4] The final act at the Latin school consisted of a public oration on a topic of the pupil's own choice. Forsten's topic was *De vita et meritis Caroli Linnaei* (On the life and works of Carolus Linnaeus),[5] which shows that Forsten had already been bitten by the natural history bug at a young age. The other four pupils spoke on entirely different topics relating to what we would now call the humanities, so the choice of an oration on Linnaeus illustrates Forsten's early affinity for natural history.

Just as his father did in Harderwijk, Forsten went on to study medicine at Leiden University.[6] A few events from his time in Leiden stand out. At the age of twenty, Forsten submitted a response to the question about chemistry in a prize contest organised by the university, which was opened on February 8, 1831.[7] His 54-page essay won the prize and was published in 1832 as *Responsio de aethere et naphthis*, answering the question: 'Expand on the chemical history, scientific knowledge, nature and use of ether and naphtha,

4 'Natuurlijke Geschiedenis, Ontleedkunde, Physiologie, Heelkunde, and Kruid-, Schei, en Artsenijmengkunde.' *Middelburgsche Courant*, no. 68 (June 5, 1828): 1.

5 *Middelburgsche Courant*, no. 104 (August 29, 1829): 1.

6 On the developments of the study of natural history and medicine at Leiden University, see Otterspeer, *Groepsportret met Dame III*, 267–300.

7 *Algemeene Konst- en Letterbode*, no. 10 (March 11, 1831): 146.

as well as of the other remaining material products of the action of putting acids in alcohol.'[8] Forsten concluded his studies in Leiden with a *dissertatio botanico-pharmaceutico-medica*, that discussed the tree *Cedrela febrifuga*, now *Toona ciliata* M.Roem, commonly known as the red cedar, and widespread throughout Southeast Asia, including Java. Rumphius already knew about the medicinal properties of the bark of this tree, while Blume published an extended paragraph on the wider family of the *Cedrela febrifuga* in his *Bijdragen tot de flora van Nederlandsch Indië* in 1825.[9] In his dissertation Forsten shows that he has read quite widely on the topic and quotes publications by Rumphius, Blume, and American naturalist Thomas Horsfield (1773–1859), who collected extensively in the Indies. Forsten successfully defended his work on December 16, 1836.[10]

This dissertation already points to Forsten's future, and the topic was chosen in line with his desires and hopes for the future. How better to qualify himself to become a member of the Natuurkundige Commissie of the Netherlands Indies than to write on one of the most useful trees from that area? It was this thesis that accompanied Forsten's application to the king for a post on the Natuurkundige Commissie.[11] Given his previous education and the choices he had made it is no surprise that Forsten stood first in line when there was an opportunity to go to the Indies to collect. An additional reason that will certainly have helped in drawing Forsten to the Indies was that both his grandfathers had held government positions in the Dutch East Indies. His maternal grandfather, Theodorus Knibbe (d. 1814), was Councillor of Justice in Batavia,[12] while his paternal grandfather, Jacob Balthasar Forsten (1757/8–1825), was a member of the supreme court in the Netherlands Indies.[13] Stories of the Indies will have been told often in the Forsten family home, painting pictures of the tropics to an impressionable boy.

Forsten's story shows a remarkable similarity to the stories of many of the members of the Commissie. His personal interests gravitated heavily towards natural history and the Netherlands Indies. This strong intrinsic motivation to study the natural world was present with all members of the Commissie. From a young age, Kuhl had shown 'a keen interest in natural history.'[14] At the

8 'Exponantur aetheris et naphtharum, tum reliquarum ex actione acidorum in alcoholem productarum materierum, historica chemica, ratio, natura et usus.'

9 Blume, *Bijdragen tot de flora*, 199–211.

10 Forsten, *De cedrela febrifuga*.

11 KB, February 27, 1837, no. 93. NL-HaNA, Staatssecretarie, 2.02.01, inv.no. 4226.

12 *Opregte Haarlemsche Courant*, no. 100 (August 22, 1815): 2.

13 *Middelburgsche Courant*, no. 46 (April 18, 1826): 2.

14 Klaver, *Inseparable Friends*, 5.

age of sixteen, he already met with Johann Wolfgang von Goethe (1749–1832).[15] Kuhl published his first monograph in 1817, at the age of nineteen.[16] When Theodorus van Swinderen offered him the opportunity to study the natural sciences in Groningen, along with the post of assistant at the museum of natural history at the university, Kuhl could not refuse.[17]

A similar pattern can be seen in Boie's life. While studying, Boie worked hard in his spare time to study natural history. Like Kuhl, he found a protector who offered him a post at a natural history cabinet. In Boie's case, the patron was Friedrich Tiedemann (1781–1861) and the cabinet was located in Heidelberg. When Boie and his friend Macklot, at some point, were discussing their hunting spoils at an inn, they spotted the interest of the son of the innkeeper. It turned out that this innkeeper's son, Salomon Müller, had taught himself to prepare specimens and had created a wonderful collection of mounted local wildlife for display in the inn. In the cases of Kuhl, Boie and Müller, it was their strong personal interest in natural history that launched them in a direction that would allow them to make a career out of it. The same pattern applies to Horner, who found a patron in Heinrich Rudolf Schinz (1777–1861) and spent all his free time during his studies travelling around Switzerland and Germany on mostly geological walking tours.[18] It is clear that all members of the Natuurkundige Commissie possessed a strong interest in natural history, which was nurtured by benevolent mentors. This path would lead them all to Temminck, who recognised their interest and talent and enlisted them in his collecting project.

Forsten and the other members come out of this narrative as keen naturalists, with an intrinsic motivation that served as a prime driver for their collection work in the Indies. No one needed to spur them on in their quest to find new and unknown species. However, they were in need of an infrastructure to pursue their personal interests. They found a stable platform for their work when they were appointed by the Dutch government to collect specimens of natural historical value for 's Rijks Museum van Natuurlijke Historie in Leiden under the orders of the governor-general of the Dutch East Indies. This pattern applies to all Commissie members appointed by the king. They all received an official appointment that put them under the orders of the governor-general, and they all had strong connections with 's Rijks Museum in Leiden.

15 See Phillips, *Acolytes of Nature*, for an account of the thriving scene of natural history in Germany during these years.

16 Kuhl, *Die deutschen Fledermäuse*.

17 Klaver, *Inseparable Friends*, 7.

18 Anonymus, "Untitled." In addition to what is available in the archives of the Natuurkundige Commissie (Horner's diary of his Sumatra expedition and several letters), a large amount of archival material is available in the Zentralbibliothek Zürich.

However, collecting specimens was not necessarily the goal for all people and institutions involved with the Commissie. In this chapter I look in more detail at how the two Dutch stakeholders sought to direct the practice of the members in the field and how the members were affected by this. I do this for the third stakeholder in the next chapter. It is a story of divergent interests and of how astute scientists negotiated and deftly manoeuvred between three different stakeholders, to all of which they owed a level of loyalty. They were eager to carve out their own career path that should have led to a respected position within the European scientific establishment after their return, while at the same time keeping those in positions of authority over them satisfied.[19]

2 Shaping the Commissie in the Netherlands

As we have seen before, the Natuurkundige Commissie was very much a royal project, with all matters governed by the king himself. His ministers would prepare a piece of advice that the king would then either approve, edit, or deny (Table 2). The ministers or other government officials corresponded with Temminck, supported from 1837 by Reinwardt and Blume. While the king may have endorsed the Natuurkundige Commissie, he did not come up with the conditions and instructions himself. That work was done for him by government officials. Throughout the thirty years of the Natuurkundige Commissie five of them stand out and deserve closer attention. They are Anton Reinhard Falck (1777–1843),[20] Daniël Jacob van Ewijck van Oostbroek en de Bilt (1786–1858),[21] and the trio of Johannes van den Bosch (1780–1844),[22] Hendrik Merkus de Kock (1779–1845),[23] and Jean Chrétien Baud (1789–1859).[24]

Falck was appointed general secretary by king Willem I, and he held this position from December 31, 1813, to March 19, 1818. This meant that Falck was present at the meeting between the king and his ministers that took place

19 See Gibson, "Careering Naturalists," for three examples of how individual the career path of a naturalist was in the British context.

20 See https://www.parlement.com/id/vg09llswffq6/a_r_anton_falck for career details.

21 See https://www.parlement.com/id/vg09llyzbmvd/d_j_van_ewijck_van_oost-broek_en_de_bilt for career details. The exact dates for Van Ewijck's career are taken from the expanded entry on parlement.com that is only available to researchers upon request.

22 See https://www.parlement.com/id/vg09llsmwiwb/j_graaf_van_den_bosch for career details. For a recent biography, see Sens, *Kolonieman*.

23 See https://www.parlement.com/id/vg09llsb8mzu/h_m_baron_de_kock for career details. See Bossenbroek, *Wraak van Diponegoro*, for a recent treatment of De Kock's role in the Java War (1825–1830).

24 See https://www.parlement.com/id/vg09lkxrbnwk/j_ch_baud for career details.

TABLE 2 Formal heads of all three stakeholders: Kings of the Netherlands, ministers of colonial affairs, directors of 's Rijks Museum, and governors-general and commissioners-general of the Netherlands from 1820 to 1850. Dates and job titles taken from profiles at http://www.parlement.com

Last name	First name	Dates	Position	From	To
			NETHERLANDS GOVERNMENT: KINGS		
Willem I	Willem Frederik	1772–1843	sovereign prince	November, 1813	March 16, 1815
			king	March 16, 1815	October 7, 1840
Willem II		1792–1849	king	October 7, 1840	March 17, 1849
Willem III		1817–1890	king	March 17, 1849	November 23, 1890
			NETHERLANDS GOVERNMENT: MINISTERS OF COLONIAL AFFAIRS		
Falck	Anton Reinhard	1777–1843	minister of Public Education, National Industry, and Colonies	March 19, 1818	March 30, 1824
Elout	Cornelis Theodorus	1767–1841	minister of National Industry and Colonies	March 30, 1824	April 5, 1825
			minister of Navy and Colonies	April 5, 1825	October 1, 1829
Quarles van Ufford	Jacques Jean	1788–1855	minister of Navy and Colonies ad interim	October 1, 1829	January 1, 1830
Gobbelschroy, van	Pierre Louis Joseph Servais	1784–1850	minister of Water Management, National Industry, and Colonies	January 1, 1830	October 4, 1830
Clifford	Gerard George	1779–1847	director-general of Water Management, National Industry, and Colonies ad interim	October 4, 1830	October 1, 1831

TABLE 2 Formal heads of all three stakeholders (*cont.*)

Last name	First name	Dates	Position	From	To
			minister of National Industry and Colonies ad interim	October 1, 1831	January 1, 1834
Brocx	Arnoldus	1779–1861	minister of Colonies ad interim	January 1, 1834	May 30, 1834
Bosch, van den	Johannes	1780–1844	minister of Colonies	May 30, 1834	January 1, 1840
Baud	Jean Chrétien	1789–1859	minister of Colonies ad interim	January 1, 1840	August 10, 1840
			minister of Navy and Colonies	August 10, 1840	January 1, 1842
			minister of Colonies	January 1, 1842	March 25, 1848
Rijk	Jules Constantijn	1787–1854	temporary minister of Colonies	March 25, 1848	November 21, 1848
Baud	Guillaume Louis	1801–1891	minister of Colonies	November 21, 1848	June 18, 1849
Bosch, van den	Engelbert Batavus	1789–1851	minister of Colonies	June 18, 1849	November 1, 1849
Pahud	Charles Ferdinand	1803–1873	minister of Colonies	November 1, 1849	January 1, 1856
MUSEUM: DIRECTORS					
Temminck	Coenraad Jacob	1778–1858	director 's Rijk Museum van Natuurlijke Historie	August 9, 1820	January 30, 1858

TABLE 2 Formal heads of all three stakeholders (*cont.*)

Last name	First name	Dates	Position	From	To
INDIES GOVERNMENT: GOVERNORS-GENERAL & COMMISSIONERS-GENERAL					
Capellen, van der	Godert Alexander Gerard Philip	1778–1848	commissioner-general	May 11, 1816	January 16, 1819
			governor-general	January 16, 1819	January 1, 1826
Kock, de	Hendrik Merkus	1779–1845	lieutenant-governor-general	May 8, 1822	June 7, 1830
			governor-general ad interim	January 1, 1826	February 4, 1826
Bus de Gisignies, du	Leonard Pierre Joseph	1780–1849	governor-general	February 4, 1826	January 16, 1830
Bosch, van den	Johannes	1780–1844	governor-general	January 16, 1830	July 2, 1833
			commissioner-general	January 17, 1832	February 1, 1834
Baud	Jean Chrétien	1789–1859	governor-general ad interim	July 2, 1833	February 29, 1836
Eerens, de	Dominique Jacques	1781–1840	lieutenant-governor-general	March 5, 1835	February 29, 1836
			governor-general	February 29, 1836	May 30, 1840
Hogendorp, van	Carel Sirardus Willem	1788–1856	governor-general ad interim	May 30, 1840	January 5, 1841
Merkus	Pieter	1787–1844	governor-general ad interim	January 6, 1841	February 14, 1843
			governor-general	February 14, 1843	August 2, 1844
Reijnst	Jan Cornelis	1798–1871	governor-general ad interim	August 2, 1844	September 30, 1845
Rochussen	Jan Jacob	1797–1871	governor-general	September 30, 1845	May 12, 1851

twice a week, which gave him both an intimate knowledge of the development within the government, and the ability to independently advise the king, since he was not formally subordinate to any of the ministers.[25] On March 19, 1818, Falck was appointed minister of public education, national industry, and colonial affairs. He would be at the head of this ministry until March 30, 1824. For Temminck, this meant that he had a friend in high places who had the ear of the king. Falck had already been instrumental in the establishment of 's Rijks Museum, even though the final event that swayed the project towards Temminck's proposal was the unexpected death of Leiden professor Sebald Justinus Brugmans (1763–1819).[26] And now Temminck could continue to make use of his influential friend.

However, Falck was in control of a sort of super-ministry with many different responsibilities. All practical matters regarding culture and science (including the Natuurkundige Commissie) would be dealt with on behalf of the minister by Van Ewijck (Figure 20). From 1820 to 1831, Van Ewijck would play a pivotal role in preparing proposals for further appointments and writing practical instructions for the members and assistants of the Natuurkundige Commissie. Van Ewijck served as secretary of Falck's ministry of public education, national industry, and colonial affairs from March 10, 1818, to March 17, 1824, with a special role as commissioner for the department of education, arts and sciences. During this period, both the appointments of Kuhl, Van Hasselt, Van Raalten and Keultjes, and of Boie, Macklot, and Müller reached the king in the form of a draft royal decree with a supporting rationale written and signed by Van Ewijck on behalf of the minister of colonial affairs (Falck). Van Ewijck corresponded with Temminck and based most of his proposals to the king on Temminck's proposals.

From March 17, 1824, to December 13, 1831, Van Ewijck was a department head within the ministry of internal affairs, which allowed him to continue to be involved with the instructions for the Commissie. When Van Ewijck intended to broaden Korthals' task beyond the systematic and taxonomic work that Temminck envisioned, Temminck disagreed, but Van Ewijck prevailed.[27] In the royal decree with the appointment of Korthals, Van Gelder, and Overdijk, instructions are not included, but it is clearly stated that instructions will be written by the minister of internal affairs, who most likely devolved this task to Van Ewijck. This shows that the Natuurkundige Commissie was not embedded in just one ministry. Both colonial affairs (and this department

25 See Steur, "Staatssecretarie en Kabinet des Konings," for the inner workings of the general secretariat during Willem I's reign.

26 See Gassó Miracle, *Coenraad Jacob Temminck*, chapter 2.

27 Gassó Miracle, *Coenraad Jacob Temminck*, 111.

FIGURE 20 Portrait of Daniël Jacob van Ewijck van Oostbroek en de Bilt, lithograph by Antonie Johannes Groeneveldt, 1858, RP-P-1909-774, Rijksmuseum, Amsterdam. Public domain HTTP://HDL.HANDLE.NET/10934/RM0001.COLLECT.388668

changed ministries frequently until 1834), and internal affairs were involved in the practical details surrounding the Natuurkundige Commissie, while, of course, financial affairs was also regularly consulted. Van Ewijck was the stabilising factor in the Dutch government during the first decade of existence of the Natuurkundige Commissie, and did much to guard continuity.

After van Ewijck's departure, there was no civil servant left with a vested interest in the Natuurkundige Commissie. Fortunately, shortly after Van Ewijck's departure, a trio of former (lieutenant-)governors-general ascended to the rank of minister. Both Van den Bosch and Baud became ministers of colonial affairs, Van den Bosch (Figures 21 and 22) from 1834 to 1840, Baud (Figures 23 and 24) from 1840 to 1848, while De Kock (Figures 25 and 26) became minister

FIGURE 21 Portrait of Johannes Graaf van den Bosch, governor-general of the Netherlands Indies, by Raden Saleh, oil on canvas, 1836, 115 × 97 cm, SK-A-3798, Rijksmuseum, Amsterdam. Public domain
HTTPS://ID.RIJKSMUSEUM.NL/20028657

FIGURE 22
Nepenthes boschiana, by Jan Gaykema Jacobsz. Plate 2 from the volume *Botanie* of the *Verhandelingen*, NCAO, 868_002. Public domain. This plant was named by Korthals after governor-general Johannes van den Bosch.
HTTPS://DH.BRILL.COM/NCO/VIEW/NCO_NNM001000868_002/MAKINGSENSE

FIGURE 23 Portrait of Jean Chrétien Baud, governor-general ad interim of the Netherlands Indies, by Raden Saleh, oil on canvas, 1835, 119.8 × 97.8 cm, SK-A-3799, Rijksmuseum, Amsterdam. Public domain
HTTPS://ID.RIJKSMUSEUM.NL/20028658

FIGURE 24
Dipterocarpus baudii, by Jan Gaykema Jacobsz. Plate 5 from the volume *Botanie* of the *Verhandelingen*, NCAO, 868_005. Public domain. This plant was named by Korthals after governor-general ad interim Jean Chrétien Baud.
HTTPS://DH.BRILL.COM/NCO/VIEW/NCO_NNM001000868_002/MAKINGSENSE

FIGURE 25 Portrait of Hendrik Merkus, Baron de Kock, army commander and after 1826 lieutenant-governor-general of the Netherlands Indies, by Cornelis Kruseman, oil on canvas, 1826–1857, 106 × 90 cm, SK-A-3796, Rijksmuseum, Amsterdam. Public domain
HTTP://HDL.HANDLE.NET/10934/RM0001.COLLECT.9969

FIGURE 26
Bauhinia kockiana, by Dominicus van Gelder and Jan Gaykema Jacobsz. Plate 10 from the volume *Botanie* of the *Verhandelingen*, NCAO, 868_010. Public domain. This plant was named by Korthals after lieutenant-governor-general Hendrik Merkus de Kock.
HTTPS://DH.BRILL.COM/NCO/VIEW/NCO_NNM001000868_010/MAKINGSENSE

of internal affairs from 1836 to 1840.[28] Their familiarity with the work of the Natuurkundige Commissie in the field helped immensely in thinking through the policy regarding the Commissie. However, their intimate knowledge of the Indies also gradually pushed the direction of the Natuurkundige Commissie away from its initial collecting mission towards a more utilitarian direction, as Van Ewijck had already initiated with Korthals' instructions. The Commissie slowly moved away from its original collecting mission and more economically useful elements of natural history, such as mineralogical research, were added to the mix.

Both Van den Bosch and De Kock were involved in writing proposals to the king regarding the appointment of Forsten and the decision regarding the publication of the *Verhandelingen*. Baud was also involved in an advisory capacity as council of state from 1836 to 1840 and would be the most important advocate for the Natuurkundige Commissie until his departure in 1848. King Willem II did not have a vested interest in the project and intended to let the Natuurkundige Commissie run its course after the appointment of Schwaner. It was Baud's lobbying that ensured royal approval for the appointment of Junghuhn in 1845, adding a couple of years to the intended lifespan of the Commissie. After Baud's departure, there was no one left within the government who saw the Commissie as an essential project. Temminck had long since realised that his influence was waning and the Natuurkundige Commissie was not the project anymore that he had intended it to be.[29] This led him to find other ways to obtain collections for 's Rijks Museum.

3 Government Directives for the Members

Forsten himself initiated contact with the Dutch government when he applied for a position on the Commissie in 1837. He received the response that there was no current vacancy, but that he would be considered for the next vacancy, if he would make sure to gain practical experience in natural history by working at 's Rijks Museum in Leiden. This was exactly according to the advice that Baud

28 De Kock had been lieutenant-governor-general from 1822 to 1830, and governor-general ad interim in 1826 to fill the short gap between Van der Capellen's departure from office and the arrival of Du Bus. Van den Bosch had been governor-general from 1830 to 1833 and commissioner-general from 1833 to 1834. Baud had been governor-general ad interim from 1833 to 1836. As such, they had seen fourteen years of the work of the Natuurkundige Commissie in practice.

29 Gassó Miracle, *Coenraad Jacob Temminck*, 112–114.

had given to minister Van den Bosch.[30] Dutifully, Forsten went on to undertake this task, until he was indeed appointed as member of the Natuurkundige Commissie on February 4, 1838.[31]

Direct contact between the members of the Commissie and the Dutch government was rare. Throughout the thirty years of the existence of the Commissie, Forsten is the only member who was appointed after having directly applied to the king himself, and even then, he was referred back to 's Rijks Museum. It was only the recommendation by Temminck, Reinwardt, and Blume that eventually secured his appointment. In previous years, Temminck's recommendation had been enough, which meant that only a path through 's Rijks Museum or Temminck's personal network would lead to an appointment. In the previous chapter, we have seen how this operated in practice for many of the members.

It needs to be noted that, from 1837 onwards, it was not just Temminck alone whose positive recommendation was required. A standing committee of the three resident powerhouses of natural history of the Indies, Reinwardt, Blume, and Temminck, had been established on July 28, 1837, as an advisory organ for the minister regarding the affairs of the Natuurkundige Commissie.[32] As Veth states, it is unclear how long this committee officially existed, but their advice was heeded at least until the mid-1840s.[33] Their advice, together with Baud's advocacy, ensured the final appointment of Junghuhn in 1845. So, while the members and their assistants were appointed by the king, there was no way around Temminck in gaining a spot in the Commissie. A positive recommendation by Temminck was almost a guarantee for an appointment. Of course, Temminck had to operate within the financial boundaries, but his relationships with Falck and Van Ewijck ensured a flying start, while the familiarity of De Kock, Van Den Bosch and Baud continued to carry the momentum until the project had finally run out of steam. It is undoubtedly true that the publication of the *Verhandelingen*, the first instalment of which was published in 1839, did much to convince the king that there was nothing more to be gained from

30 Baud's advice from February 10, 1837, is included with the royal decree of February 27, 1837, no. 51, that deals with Forsten's application, KB, February 27, 1837, no. 51. NL-HaNA, Staatssecretarie, 2.02.01, inv.no. 4226.

31 KB, February 4, 1838, no. 98. NL-HaNA, Staatssecretarie, 2.02.01, inv.no. 4334.

32 KB, July 28, 1837, no. 93. NL-HaNA, Staatssecretarie, 2.02.01, inv.no. 4273. According to Veth (Veth, *Overzicht van hetgeen gedaan is*, 90), news of the establishment of the committee only reached its members on September 11, 1837. Gassó Miracle places the establishment of the triumvirate in 1827 (Gassó Miracle, *Coenraad Jacob Temminck*, 108), which I believe to be a typo.

33 Veth, *Overzicht van hetgeen gedaan is*, 124.

keeping the Commissie alive. Inadvertently, rather than being the crowning glory of the Commissie, the *Verhandelingen* perhaps became the first nail in its coffin.

Upon Forsten's appointment, instructions were written by the ministry of internal affairs.[34] This document was to be the basis of and the rationale behind Forsten's practice of natural history in the colonies. The instructions cover eight pages and consist of thirteen articles. In these instructions, Forsten is charged with a zoological mission, but he is strongly encouraged not to pass up on any opportunity to collect botanical and geological specimens (article 1). He will be assisted by Van Gelder as artist,[35] and by a certain number of indigenous collectors and preparators (article 2).[36] He will concentrate his own work on zoology and make sure that his collected specimens will be transported to the Netherlands 'on a regular basis and at the earliest opportunity' (article 3).[37] He will follow the precepts of the directors of the two museums[38] when it comes to how to collect, prepare, and keep his specimens and will correspond with said directors to prevent him from sending too many copies of the same specimens (article 4). He will adopt as his main collection focus specimens not yet present at 's Rijks Museum, which is why he will visit parts of the Indies that have not been visited, or only casually so, by other naturalists (article 5). He will attach individual labels to each specimen with its scientific and local names and its collecting location (article 6). He will keep a diary with observations and discoveries, not just of a taxonomical nature, but also pertaining to way of life, feeding habits, habitat, etc. (article 7). He will send a semi-annual summary of his notes to the Indies government, accompanied by descriptions of hitherto unknown specimens, for publication in the Netherlands (article 8). He will have drawings made of new species in such a format that they can

34 Provisionele Instructie voor den Heer Med Doctor E. A. Forsten. NCAO, 1079_005 to 1079_009.

35 This actually never happened. Van Gelder left government service in 1838, before Forsten's arrival, to pursue a career as administrator of an estate. In 1840, he became co-owner of the rental contract, took over the full contract in 1849, and he eventually returned to the Netherlands a rich man in 1858. See Gols, "Rijke oud-Indië gangers," for more on Van Gelder's life after his return.

36 In chapters 6 and 7, I will look at the indigenous employees in more detail. For now, it will suffice to say that Forsten was accompanied to Celebes by at least seven Javanese assistants, but more names are mentioned in various places in Forsten's manuscripts. The returning names are Hendrik, Jeilar, Ammon, Droes, Modair, Jeilin, Jamin and Itjam.

37 'Geregeld en ten spoedigste.'

38 Both Temminck at 's Rijks Museum van Natuurlijke Historie and Blume at 's Rijks Herbarium.

be published without having to be redrawn (article 9).[39] He will not publish any of his descriptions in a foreign language or send them to a foreign society, unless after publication in the Netherlands (article 10). All Forsten's specimens and field notes will be the property of the government and, in case of his death, will be sent by the governor-general to the department of colonies (article 11). The length of his stay will be decided after the currently ongoing consultation on this topic will have delivered its results (article 12). He will be awarded the sum of fl,000 to purchase necessary equipment, which, upon his return, will be left in the Indies for the use of the remaining members of the Commissie (article 13).

Throughout the thirty years of the Commissie's existence the way in which the instructions were set up and communicated to the members changed substantially. Forsten's instructions were first sent by the minister of internal affairs to the triumvirate, Forsten's designation of the steering committee consisting of Temminck, Blume and Reinwardt.[40] In an accompanying letter, the minister tells the triumvirate to communicate these instructions to Forsten. After some financial remarks, the minister tells them that the department of colonial affairs has asked the governor-general to make Celebes Forsten's collecting locality. Undoubtedly, this was a response to an earlier request made by the three gentlemen.[41] Given the nature of the triumvirate, it is likely that the draft for the instructions and the request to send Forsten to Celebes had flowed from their pens. It is striking to note that the instructions were actually not directed to Forsten personally but were sent to him through the mediation of Reinwardt, Temminck and Blume. Direct contact between the Dutch government and Forsten was very limited. Forsten's only additional moment of direct contact with the government, which took place after his appointment, was an audience with minister of colonial affairs Baud, in the weeks before his departure.[42] It is not inconceivable that Müller and Korthals, who were on leave in the Netherlands and knew Baud from their time in the Indies, played a role in motivating Forsten to request this audience. I doubt whether this visit was anything more than a social networking call.

39 See Weber and Van Zanen, "Matters of Perspective," for a discussion on the requirements of natural historical drawings in the context of the Commissie.

40 Reisverhaal van Dr. Forsten naar Menado. NCAO, 1075_014.

41 This is confirmed by the governor-general in a later decree: Extract uit het Register der Besluiten van den Gouverneur Generaal van Nederlandsch Indië no. 4. (1839, Dec. 22). NCAO, 1079_001 to 1079_004; also available as Res.GG, December 22, 1839, no. 4. NL-HaNA, Koloniën, 1814–1849, 2.10.01, inv.no. 2601.

42 Letter Forsten to cousin Verstege, July 29, 1838. NL-HaNA, Van Beresteyn, 2.21.018, inv.no. 1814.

The instructions for the first two appointments of Kuhl, Van Hasselt, Van Raalten and Keultjes, and of Boie, Macklot and Müller were included in the royal decree, written by Van Ewijck. Both the appointment decrees for Korthals, Van Gelder and Overdijk, and Forsten mention explicitly that the minister of internal affairs is responsible for their instructions, while the appointment decrees for both Horner and Schwaner state that all practical affairs should be arranged between the ministers of internal and colonial affairs. That the instructions from the Netherlands were of relevance in the Indies is proven by the fact that governor-general Van den Bosch used article 11 from Korthals' instructions as rationale for Korthals' appointment as member of the Hoofdkommissie van Landbouw [Main Committee of Agriculture] on November 22, 1831.[43] Apparently, Korthals was instructed to investigate the inner workings of the Hoofdkommissie, that was responsible for agricultural development in the Indies.

I have not yet located copies of the actual text of the instructions for Korthals or Horner. It is possible that Horner did not receive instructions from the Netherlands and that he was expected to slot into the current instruction pattern of the other members of the Natuurkundige Commissie. After all, Horner departed for the Indies only three days after his appointment and any instruction forthcoming from the ministry would have to be sent on after him. This may also have been a foreshadowing of the future shift in the balance of power. Forsten was the last member to receive instructions for his mission from the Netherlands. Instructions for Schwaner and Junghuhn were written by the governor-general in the Indies and were of a very different nature than the instructions written in the Netherlands.[44] It is striking to note that the focus on collecting natural historical specimens, a task that featured heavily in the instructions that were generated in the Netherlands, is almost an afterthought in the instructions issued to Schwaner and Junghuhn.

All members, being keen naturalists, were intent on using their Indies adventure to create a career for themselves. Forsten, for this reason, was not always a faithful and obedient subject of the king and would let his own career interests prevail on occasion. This is clearly visible in his response to the news he received, when already in the field, about the intended publication of the *Verhandelingen*. How the findings of the Commissie needed to be published had always been a bone of contention and had never been

43 Res.GG, November 22, 1831, no. 15. NL-HaNA, Koloniën, 1814–1849, 2.10.01, inv.no., 1841.

44 Schwaner's instructions can be found in Res.GG, May 30, 1843, no. 10. NL-HaNA, Koloniën, 1814–1849, 2.10.01, inv.no., 2645. Junghuhn's instructions can be found in Res.GG, April 5, 1846, no. 27. NL-HaNA, Koloniën, 1814–1849, 2.10.01, inv.no., 2681.

satisfactorily resolved, until the royal decree of February 10, 1839, finally settled matters.[45] This decree paved the way towards the publication in 29 parts between 1839 and 1847 of the *Verhandelingen*. In a letter to Joannes Andreas Susanna (1795–1859), the administrator of 's Rijks Museum, Forsten complains about these arrangements. He was not interested at all in contributing to the *Verhandelingen*, but was looking forward to publishing a separate volume on Celebes after his return to the Netherlands. Forsten confides to Susanna that he will not communicate anything scientific anymore, in order to prevent his findings from being published in the *Verhandelingen*.[46]

The most important motive here, of course, is rather egotistical: Forsten is not interested in seeing his findings disappear anonymously in a larger work on colonial natural history. Publishing his own book under his own name is clearly more interesting. It would give him a better chance of a good scientific career back in the Netherlands, and might even result in royal honours, as had happened to fellow Commissie members Müller and Korthals.[47] Forsten's personal interests were clearly not in line with the plan for the *Verhandelingen* that the government and 's Rijks Museum finally had come up with, and he sounds prepared to take drastic action. Forsten shows that he can be quite subversive when it suits his own purposes. He confirms his critical mood when writing this letter by complaining about the king's financial policies and remarking that funds from the Indies will soon dry out 'because Holland deals with the colony as a leech that sucks blood until she [the colony] will expire.'[48] Forsten has clearly decided to not look for an immediate confrontation, since he spends more than half a letter to Temminck on zoological details.[49] He probably realises that it would be rather unwise to bite the hands that feed him while still in the field. In the end, Forsten's untimely death prevented the matter from coming to a full confrontation.[50] He may not have been happy with

45 KB, February 10, 1839, no. 101. NL-HaNA, Staatssecretarie, 2.02.01, inv.no. 4450.

46 Letter Forsten to Susanna, January 26, 1841. NCAO, 1230_002.

47 Letter Forsten to Susanna, April 30, 1840. NCAO, 1229_002.

48 'Want met de kolonie handelt Holland als een bloedzuiger welke zoo lang zuigen zal tot dat zij den geest geeft.' Letter Forsten to Susanna, January 26, 1841. NCAO, 1230_002.

49 Letter Forsten to Temminck, May 20, 1841. NCAO, 1231. A letter to Hermann Schlegel (1804–1884) from the same date (Letter Forsten to Schlegel, May 20, 1841. NCAO, 1232) is also filled with zoological details. Forsten neatly divides his information: he communicates his ornithological findings to Temminck, and his herpetological findings to Schlegel, according to their personal interests.

50 In contradiction to Forsten, his fellow (provisional) Commissie member Franz Junghuhn (1809–1864) publicly flaunted the authority of the European centres by publishing many of his findings outside of Museum circles. See Goss, *Floracrats*.

the fact that, despite his misgivings, some of his specimens were published in the *Verhandelingen* by Müller without his knowledge.

4 The Long Arm of 's Rijks Museum van Natuurlijke Historie

During his years at university, Forsten had already been an avid collector of natural objects, like any good nineteenth-century naturalist.[51] In the *Nieuwe Amsterdamsche Courant* of August 7, 1838, we find an announcement for the sale of a 'beautiful and rich collection of natural objects.'[52] The objects were collected by Forsten himself and consisted of marine life: mainly shells, but also sea urchins, sea stars and coral. A catalogue of the collection was published and made available through bookstores in Utrecht (where the sale would take place) and Leiden, Forsten's home from 1829 to 1838.[53] The sale of his collection coincided with his departure to the Indies, so Forsten clearly could not find a reliable place to store it. Unfortunately, the sale did not bring in the expected profit. Forsten writes to Susanna: 'The sale of the shells has been an enormous miscalculation. I had expected at least twice as much, but everyone is always too taken with their own hobby horse.'[54]

This collection was undoubtedly inspired by the proximity of 's Rijks Museum van Natuurlijke Historie. As we have seen, Temminck was responsible for the founding of the Commissie, handpicked the members and their assistants, initially on his own, but later with his fellow triumvirate members. Temminck set out his intentions for the Commissie members when he wrote to the minister on March 21, 1822:

> Taking good notes, collecting everything and taking good care of the collected objects, sending shipments to the Motherland as often as possible, that is the primary instruction for a nature researcher in the tropics.[55]

---

51 See MacGregor, *Naturalists in the Field*.

52 'Eene schoone en rijke verzameling naturalien.' *Nieuwe Amsterdamsche Courant*, no. 2108 (August 7, 1838): 3, repeated three times in the *Opregte Haarlemsche Courant* on August 9, August 18 and September 1, 1838.

53 I have not been able to locate a copy of this catalogue.

54 'De schelpenverkoop is eene geweldige misrekening geweest. Ik had ten minsten eens zoo veel verwacht dog ieder is altoos te veel met zijn stok paardje ingenomen.' Letter Forsten to Susanna, January 26, 1841. NCAO, 1230.

55 'Goede aanteekeningen maken, alles verzamelen en voor het verzamelde de meeste zorg dragen, zoo dikwerf het mogelijk is bezendingeu naar het Moederland doen, ziedaar de

For the Commissie members, 's Rijks Museum should serve as their scientific point of reference. This was also the case for Forsten, especially because his personal collecting desires aligned with Temminck's interests. Temminck wanted exotic specimens to fill his Museum and Forsten needed them to make his name. 's Rijks Museum was the spiritual centre of the Commissie. The members were trained in taxonomy and specimen preparation at 's Rijks Museum. Museum staff would have been very well aware of the activities of the members in the field. They would receive shipments of specimens, publish some of the species, receive correspondence and reports from the field, and live in anticipation of what was perhaps their future as well.

All of the twelve letters of Forsten from the archive of the Natuurkundige Commissie are addressed to people at 's Rijks Museum – two each to Temminck and Hermann Schlegel (1804–1884), eight to Susanna. Schlegel started at the Museum in 1825 as preparator and replacement of recently departed Commissie member Boie. In 1828 he was appointed as curator of vertebrates and after Temminck's death in 1858 he would become director, a post he would hold until his death in 1884.[56] Susanna was the Museum administrator from 1825 to 1859. His central position and kind character made him the perfect confidante for the entire staff, including Forsten.[57] In all the correspondence of the members of the Natuurkundige Commissie, Temminck, Schlegel and Susanna are the most regular addressees. The centrality of Susanna in the life of the Commissie members is confirmed by the extant correspondence of Macklot. Of the fifty-one letters sent to Leiden by Macklot, thirty-two are addressed to Susanna, fourteen to Temminck and only one to Schlegel. In contrast, all but six of the twenty-six extant letters of Junghuhn are directed to Temminck, and none to Susanna or Schlegel. This is not surprising given the fact that Junghuhn had no previous relationship with Temminck or anyone else at 's Rijks Museum before his appointment. In fact, most of Junghuhn's letters in the archive date from after his return to the Netherlands in 1848, showing that his link to 's Rijks Museum while he was working in the field was a very tenuous one.

After Forsten's open application to the king, he was instructed to gain some experience and prove himself to be up to the required standard at 's Rijks Museum. Even though his stay at the Museum (1837–1838) seems to have been

hoofdinstructie voor een natuur onderzoeker in de tropische gewesten.' Veth, *Overzicht van hetgeen gedaan is*, 29.

56 Holthuis, *Rijksmuseum van Natuurlijke Historie*, 30–31, 43–51.

57 Holthuis, *Rijksmuseum van Natuurlijke Historie*, 26.

of a rather informal nature,[58] many of the other members spent some time in more official positions. Both Boie and Macklot worked there as curator, while Van Gelder and Overdijk served as preparators. The status of the work of some of the other members at 's Rijks Museum is somewhat ambiguous, like Forsten's. Müller and Van Oort spent some time working at the Museum and it is not inconceivable that Van Raalten, Keultjes and Schwaner went the same route. Only Korthals and Horner did not have such a preparatory stint at 's Rijks Museum. As a botanist, Korthals prepared with Reinwardt and Blume and did not have much to gain from a stay at 's Rijks Museum. Horner's absence from 's Rijks Museum can easily be explained by the fact that his appointment was a rather hurried affair and there simply was no time for some embedded preparation at 's Rijks Museum. The fact that only Boie, Macklot, Van Gelder and Overdijk are mentioned as employees by the biographer of the 's Rijks Museum and its staff, Lipke Holthuis, is a good indication that the association of the other members and assistants with 's Rijks Museum was quite informal.[59]

The Museum was also the recipient of what is perhaps Forsten's greatest legacy: his collections. We know of a total of three shipments, of which the first one (consisting of six barrels and six crates) and the third one (in four instalments and consisting of fourteen crates and eight barrels) surely reached the Leiden Museum.[60] The Naturalis BioPortal,[61] which provides an easy search experience for the biological layperson of data holdings of the Netherlands Biodiversity Data Services,[62] lists 2498 specimens attributed to Forsten.[63] This is not an accurate count of what survives of Forsten's original collections, since BioPortal only lists the digitised portion of the collection, currently about a quarter of the entire holdings of Naturalis. In addition, through 150 years of collecting, cataloguing, reorganising, and amalgamating, some specimens will have lost their association with Forsten, or perhaps even incorrectly gained an

58 In a letter to Schlegel, (Letter Forsten to Schlegel, August 2, 1838. NCAO, 1078_002), Forsten, rueing the fact that he missed the opportunity to personally say goodbye to Schlegel, reminisces their joint hunting trips that were more successful than his on-board cockroach hunting expeditions at night.

59 Holthuis, *Rijksmuseum van Natuurlijke Historie*.

60 Letter Temminck to the minister of internal affairs, January 20, 1848. NL-HaNA, Van Beresteyn, 2.21.018, inv.no., 1525.

61 https://bioportal.naturalis.nl.

62 https://docs.biodiversitydata.nl/en/latest.

63 Search performed on July 27, 2019 with Forsten as keyword in the Specimens/Collector field.

association with Forsten, while other specimens will have been lost (to decay) or exchanged with other institutions.[64]

Most of the first shipment was not in good shape when it arrived in Leiden and had to be discarded, but a packing list of the collection has survived.[65] In it, Forsten gives a detailed overview of the specimens that he had packaged up in each barrel and crate. The short 'table of contents' is as follows: Barrels I and II contained mammals; Barrel III mammals and creeping animals; Barrel IV creeping animals; Barrels V and VI fishes, molluscs and marine animals; Crate VII contained birds; Crate VIII birds and eight smaller boxes with insects; Crate IX skeletons, eleven boxes with insects and three with shells; Crate X and XI minerals and rocks; Crate XII dried plants and a box with crustaceans. Further packing lists are available in the archive, but this one already gives us a good impression of the size and scope of Forsten's collections.

And it was not only Forsten's collections that made their way to Leiden. Much of the collections of the other members also ended up in Leiden. After the death of Van Hasselt in September 1823 (Kuhl had already died in September 1821), Van Raalten was put in charge of the Kuhl/Van Hasselt papers and collections. It took Van Raalten two full years to process all the work that Kuhl and Van Hasselt had done. In the end, Van Raalten shipped four tin boxes of drawings and 90 crates of specimens in nine shipments. The fourth and final tin box was given to the Indies government for shipping to the Netherlands on September 22, 1825,[66] while the ninth and final shipment of specimens, consisting of two crates, was ready for transport to the Netherlands on November 25, 1825.[67] It seems that Van Raalten preparations and packaging skills were better than Forsten's, because 'no less than 90 crates with naturalia have arrived for 's Rijks Museum in such a good condition, that he [Van Raalten] received for this the most flattering tokens of gratefulness from the director of said Museum.'[68] This even resulted in a salary raise for Van Raalten.

Even more specimens were sent to the Netherlands in later years, with Müller and Korthals taking 76 crates, two suitcases and seven barrels of specimens and thirteen crates of living plants with them when they departed for the

64 For a recent discussion of some of the specimens collected by Forsten, see Van Grouw, Dekkers, and Rookmaaker, *Third Time Lucky*.

65 Bijlage f. Korten inhoud der kisten en vaten. NCAO, 1461_015.

66 Res.GG, September 22, 1825, no. 5. NL-HaNA, Koloniën, 1814–1849, 2.10.01, inv.no., 2487.

67 Res.GG, November 25, 1825, no. 13. NL-HaNA, Koloniën, 1814–1849, 2.10.01, inv.no., 2489.

68 'Niet minder dan 90 kisten met naturalia voor 's Rijks Museum zijn aangekomen in eenen zoodanigen goeden staat, dat hij daarvoor van de directeuren van dat Museum de vleijendste dankbetuiging heeft ontvangen.' Res.GG, April 10, 1827, no. 5. NL-HaNA, Koloniën, 1814–1849, 2.10.01, inv.no., 2800.

Netherlands in April 1837.[69] Another impression can be gained from the lists of the collections of the Commissie on New Guinea as published in two reports based on Macklot's correspondence.[70] Unfortunately, there is no complete overview of all the collections gathered by the members of the Natuurkundige Commissie, but I believe that Weber's estimation of 100,000 objects, may even be overly cautious.[71]

Forsten not only sent things back to the Museum, but also received things from them. He was heavily dependent on Susanna to keep his travel library up to date on the latest developments. In several letters to Susanna, Forsten communicated his wish lists and comments on books received. Because his departure from Leiden was rather rushed,[72] Forsten took the opportunity to write to Susanna on August 8, 1838, while waiting for favourable winds in Hellevoetsluis, asking for the purchase and delivery of a mathematical work and two Portuguese language related works.[73] On August 19, 1838, still waiting for the weather to turn, Forsten wrote to Susanna to confirm receipt of the requested texts.[74] In April 1840 Forsten complained to Susanna that he had received nothing at all from him, not even a letter, 'for 20 months.' In jest, Forsten suggested that since his departure maybe no new books have been published, since he had not received anything at all.[75] Only in December 1840 (more than two years after his departure from the Netherlands) did Forsten receive his first letter from Leiden,[76] while the first shipment of books finally arrived in December 1841, including works by Temminck and fellow Commissie member Müller.[77]

On July 19, 1842, Forsten sent another book order to Susanna, referencing a bookseller's catalogue that must have been included with the latest letter

---

69 Res.GG, April 19, 1837, no. 1. NL-HaNA, Koloniën, 1814–1849, 2.10.01, inv.no., 2569.

70 Macklot, "Uittreksels uit brieven"; Macklot, "Verslag van het Land."

71 Here lies a wonderful task for a future researcher. All shipments included a shipping list that was sent to the ministry. Many of them ended up in the archive of the Natuurkundige Commissie, but it is possible that some of them were lost in transition and are still available in the ministerial archives.

72 Forsten received his instructions in early July and his ship was set to depart in late July, so he only had a couple of weeks to arrange his affairs.

73 A version of Callet, *Tables portatives de logarithmes*, a Portuguese dictionary (either French-Portuguese, or Dutch-Portuguese), and a book in Portuguese (Forsten suggested *Os Lusíadas* by Luís Vaz de Camões). Letter Forsten to Susanna, August 8, 1838. NCAO, 1224_001.

74 Letter Forsten to Susanna, August 19, 1838. NCAO, 1225_001.

75 '20 maanden', letter Forsten to Susanna, April 30, 1840. NCAO, 1229_001.

76 Letter Forsten to Susanna, January 26, 1841. NCAO, 1230_001.

77 Letter Forsten to Susanna, January 27, 1842. NCAO, 1235_001.

from Susanna. He was looking for access to several zoological works (including works by Buffon and Cuvier), while asking Susanna to consult Schlegel for additional recommendations.[78] Clearly, in regard to newly published material, Forsten was heavily dependent on what 's Rijks Museum (mainly through Susanna) was willing and able to provide. This was what he needed to build his own career. From an earlier letter to Susanna,[79] it becomes clear that there was a Commissie library on Java, but it was out of reach to a traveller in the Outer Islands.

5 Conclusion

Despite the geographical distance, both the Dutch government and 's Rijks Museum heavily influenced the practice in the field of the members of the Natuurkundige Commissie. The government appointed them and issued them with the instructions for their practical work, while 's Rijks Museum served as the scientific centre for the Commissie. All members that were appointed in the Netherlands had some form of previous engagement with the Museum and most of them developed a warm bond with Susanna, who would be the happy recipient of informal reports from the field from a variety of members. In return, he supplied them with information on the latest developments in natural history. Because the members were all keen naturalists, it was natural for them to invest in their relationship with 's Rijks Museum. After all, upon their return they were hoping to develop a career in natural history and 's Rijks Museum was an important partner for this. Loyalty to their own career aims was translated to a strong sense of loyalty to Temminck and his goals. The Dutch government stood more at a distance from the members, especially because contact with the government mostly took place through the mediation of Temminck and the triumvirate. However, Temminck did not stand between the members and the Indies government. Contact between members and the Indies government was mostly unmediated, which created a reality in the field that was barely visible from the Netherlands.

78 Letter Forsten to Susanna, July 19, 1842. NCAO, 1236_002.

79 Forsten to Susanna, January 27, 1842. NCAO, 1235_001.

CHAPTER 3

Layered Loyalties: the Netherlands Indies

As we have seen in the previous chapter, the members of the Natuurkundige Commissie were driven by a strong personal motivation to collect specimens for 's Rijks Museum. Chapter 2 outlined how the interests of the Dutch government and 's Rijks Museum aligned and created a strong start for the Natuurkundige Commissie, which was unfortunately heavily weakened by the many deaths in the field. And it was in the field that the members encountered the governor-general under whose direction they were supposed to perform their mission. In this chapter, I look at how this direction functioned in practice, following the career of Forsten and drawing parallels and examples from the experience of other members.

1 In the Field under the Direction of the Governor-General

A significant phrase in article 1 of the instructions that Forsten received from the Dutch ministry of internal affairs is 'onder de order van den Gouverneur Generaal' (under the direction of the governor-general). This is a recurring phrase in the appointment decrees and instructions of the Commissie members. It also appears in the royal decrees of 1820, 1823, 1830, 1835. The Dutch government set out the path, while the governor-general became the one to make decisions in the field. It could take months to deliver a letter to the Indies and then further months before the reply was delivered. This severely hindered communication, so it was only logical and practical to give the governor-general freedom to act. An added difficulty was that governors-general usually did not stay in function for more than a couple of years. Forsten had to deal with three separate governors-general during his five years in the Indies. When he arrived, Dominique Jacques de Eerens (1781–1840, Figure 27) was governor-general. When De Eerens died on May 30, 1840, he was replaced by governor-general ad interim Carel Sirardus Willem van Hogendorp (1788–1856), who stepped aside for governor-general ad interim Pieter Merkus (1781–1844, Figure 28) on January 5, 1841, who would hold office until after Forsten had died.

Between 1816 and 1851, ten different people served as governor-general or commissioner-general (Table 2). Van der Capellen held office for almost ten years. The average time in office for the other nine was less than three years. The governor-general was supported by the Council of the Indies [Raad van

 | DOI:10.1163/9789004696877_005

FIGURE 27 Portrait of Dominique Jacques de Eerens, governor-general of the Netherlands Indies, copy after Cornelis Kruseman, oil on canvas, 1855–1858, 133.5 × 101.5 cm, SK-A-3800, Rijksmuseum, Amsterdam. Public domain HTTPS://ID.RIJKSMUSEUM.NL/20028659

Indië], who would advise on policy matters. Until 1836, the decrees or resolutions of the governors-general could be made either *in rade*, after consulting the Council, or *buiten rade*, without consulting the Council. After 1836, all resolutions were made after consulting the Council.

The period from 1826 to 1830 seems to have been an exceedingly problematic one. Lieutenant-governor-general De Kock presided over the Council, while commissioner-general Léonard Pierre Joseph Du Bus de Gisignies (1780–1849,

FIGURE 28 Portrait of Pieter Merkus, governor-general of the Netherlands Indies, copy after Jan Willem Pieneman, oil on canvas, 1844–1851, 141.5 × 109.5 cm, SK-A-3801, Rijksmuseum, Amsterdam. Public domain
HTTPS://ID.RIJKSMUSEUM.NL/20028660

Figure 29) kept himself at a distance.[1] This caused a great deal of administrative delay, with the added problem that Du Bus and De Kock did often not see eye to eye, especially when it came to the affairs of the Java War.[2] Du Bus was

1 De Prins, *Voor keizer en koning*, 189–203.

2 De Kock, as lieutenant-general, was in charge of the military campaign during the Java War (1825–1830). Du Bus often disagreed with De Kock's tactics. See Bossenbroek, *Wraak van Diponegoro*.

FIGURE 29 Portrait of Léonard Pierre Joseph Burgrave du Bus de Gisignies, commissioner-general of the Netherlands Indies, by François Joseph Navez, oil on canvas, 1836, 113.5 × 98 cm, SK-A-3797, Rijksmuseum, Amsterdam. Public domain
HTTPS://ID.RIJKSMUSEUM.NL/20028656

the highest authority in the Indies, but on matters that did not belong to his core tasks, he still needed to ask the advice of the De Kock and the Council. During these years, there is a triple set of resolutions: resolutions of the commissioner-general, resolutions of the lieutenant-governor-general *in rade*, and resolutions of the lieutenant-governor-general *buiten rade*.[3] Resolutions regarding the Natuurkundige Commissie exist in all three sets during these

3 When quoting the resolutions of the governors-general, I have not made a difference between *in rade* or *buiten rade*. The status of each resolution can be derived from the inventory numbers. For the resolutions made between 1820 and 1836, all resolutions in NL-HaNA, Koloniën,

years, so the back and forth between the staff of Du Bus and the staff of De Kock occasionally made for serious delays in the decision-making process.

The same situation existed in the period when Baud served as governor-general ad interim and Van den Bosch as commissioner-general (1832–1833), but their personal relationship was of a much better nature than the one between Du Bus and De Kock, while it was also much clearer how the responsibilities were divided. Sumatra was Van den Bosch' specific responsibility, so he left much of the work on Java in the capable hands of Baud.

In August 1838, Forsten left the Netherlands for the Indies, never to return to his homeland. When he arrived in Batavia, the heart of the Dutch colonial administration, on December 14, 1838, he immediately petitioned the governor-general to be allowed to start his expedition and leave for Celebes. However, Forsten's patience was severely tested. Forsten's arrival almost coincided with Horner's death on Sumatra on December 7, 1838. Governor-general De Eerens decreed on January 10, 1839, that Horner's assistant, Overdijk, needed to return to Java in order to accompany Forsten.[4] Overdijk did not arrive until March 7, 1839.[5]

Another cause for delay was the plan of directing member Diard to travel through other colonial empires in order to study improvement in sugar manufacturing in Europe and the British territories that could then be implemented in the Indies. This meant that Diard's status as directing member of the Natuurkundige Commissie would become vacant, and he already had his mind on other things. De Eerens asked Diard on February 25, 1839, for a proposal regarding future leadership of the Natuurkundige Commissie,[6] which was only received on August 2, 1839.[7] This proposal included a section on Forsten's future work in the Indies.

A third reason for the delay was the uncertain status of Junghuhn. Junghuhn should have left service on July 1, 1839, but Diard had kept him in service, going against De Eerens' resolution of September 29, 1839.[8] This required close attention from the side of the governor-general, drowning out the need for a quick resolution of Forsten's future.

A fourth reason for the delay, and perhaps the most significant at this stage, was the direction from the Netherlands that the incredibly large sum of f36,000

1814–1849, 2.10.01, inv.no. 2435–2725 are *buiten rade,* while all resolutions in NL-HaNA, Koloniën, 1814–1849, 2.10.01, inv.no. 2770–2863 are *in rade.*

4 Res.GG, January 10, 1839, no. 4. NL-HaNA, Koloniën, 1814–1849, 2.10.01, inv.no. 2590.

5 *Javasche Courant*, no. 20 (March 9, 1839): 6.

6 Res.GG, February 25, 1839, no. 3. NL-HaNA, Koloniën, 1814–1849, 2.10.01, inv.no. 2591.

7 See Res.GG, December 22, 1839, no. 3. NL-HaNA, Koloniën, 1814–1849, 2.10.01, inv.no. 2601.

8 Res.GG, September 29, 1839, no. 12. NL-HaNA, Koloniën, 1814–1849, 2.10.01, inv.no. 2598.

needed to be contributed to the publication of the *Verhandelingen* over a period of five years (two of which had already passed) from the annual budget of the Natuurkundige Commissie, which stood at f40,364. This significantly reduced De Eerens' financial liberty regarding the Natuurkundige Commissie, especially with recent pressure from the Netherlands regarding the budget. All in all, Forsten became the victim of an unfortunate pile-up of delays. It was not so much that there was no interest in Forsten's work, but more so that the continuing financial pressures and more urgent matters regarding the Commissie caused an impending expedition to drop on the list of priorities, also because there was no urgent imperial need that Forsten's expedition could be latched on to.[9]

Finally, on December 22, 1839, the situation was resolved for Forsten.[10] In resolution no. 3 of that day, De Eerens arranged Forsten's salary, and decreed that Forsten needed to travel to Manado on Celebes to execute what he was charged to do by the Dutch government, with the provision that he should be ready to travel back to the Netherlands on February 1, 1843, at the latest. Forsten would be accompanied by Overdijk. When Forsten objected against Overdijk, and Overdijk objected against his salary, this could have been another cause for delay, but the issue was quickly resolved by De Eerens. On January 19, 1840, two days after Forsten had been personally received by De Eerens,[11] Overdijk was honourably discharged. He had asked for a salary raise or a discharge if his salary raise was denied, so De Eerens had no choice, given the recent budgetary issues, to comply. Forsten had proposed Von Gaffron as new expedition companion and De Eerens appointed him at 75% of the salary that Overdijk would have received.[12]

This episode shows that, while the governor-general had a lot of freedom, he still was heavily dependent on what was decided in the Netherlands, especially when it came down to finances or the practical instructions for work in the field. In addition, unexpected situations in the field, like the death of Horner,

---

9 My previous statement that perhaps a lack of interest from De Eerens was the cause of these delays was clearly wrong. See Van Wingerden, "Science on the Edge of Empire," 810. Also, governor-general ad interim Van Hogendorp's term only started after Forsten had departed for Celebes and he had very limited involvement in Forsten's mission.

10 Res.GG, December 22, 1839, no. 4. NL-HaNA, Koloniën, 1814–1849, 2.10.01, inv.no. 2601. A copy of this resolution is available NCAO, 1079_001 to 1079_004.

11 Kort verslag van het in Indie verrigtte van 14 December 1838 tot April 1840. NCAO, 1076_006.

12 Res.GG, January 19, 1840, no. 1. NL-HaNA, Koloniën, 1814–1849, 2.10.01, inv.no. 2602.

or Diard's obstinacy in continuing to hire Junghuhn, could throw a spanner in the works.[13]

Forsten was not the only member who had his patience tested. When Boie, Macklot, Müller and Van Oort arrived in June 1826, they found that a recent change in leadership, with Van der Capellen stepping down for Du Bus, had changed the world substantially. Instead of an experienced and enlightened leader with a strong predilection for science, they had to deal with a new leader with no Indies experience who suddenly faced the newly erupted Java War, while he also had the task to reorganise the government finances in the Indies. As a result, they had to sit around penniless until Du Bus finally awarded them the salaries that they were due on February 27, 1827.[14] In contrast to Forsten, they were able to start their work in the immediate environment of Buitenzorg.

Likewise, Schwaner spent a year on Java, before departing for Borneo. Veth suggests that this was due to a lack of funds.[15] However, Schwaner arrived on August 6, 1842, and was awarded a salary on September 8, 1842, when he was also asked to write a proposal regarding how and where he intended to continue Horner's geological work.[16] It was only on May 30, 1843, that governor-general Merkus charged Schwaner with a mission to Borneo.[17] From this decree, it becomes apparent that Schwaner had not been sitting on his hands and had occupied himself with some coal surveying in Banten (Java). Because Schwaner's assignment from the Netherlands did not say anything about where he was supposed to continue Horner's work, the year that Schwaner spent on Java will most likely not have been experienced as an unnecessary delay. No delays were experienced by Kuhl, Van Hasselt, Van Raalten and Keultjes, and Korthals, Van Gelder and Overdijk, who were awarded their salaries and given their tasks within two months after arrival.

Horner's experience is somewhat ambiguous. His salary was awarded on August 31, 1835,[18] within two months of his arrival in July 1835. Only on October 1,

13 Diard obviously expected De Eerens to ratify Junghuhn's continued employment. It seems not to have been a rare occurrence in the Indies administration that an appointment was made by the governor-general to ratify an already existing arrangement made on lower administrative level.

14 Res.GG, February 27, 1827, no. 35. NL-HaNA, Koloniën, 1814–1849, 2.10.01, inv.no. 2501. See Veth, *Overzicht van hetgeen gedaan is*, 45–49.

15 Veth, *Overzicht van hetgeen gedaan is*, 108.

16 Res.GG, September 8, 1842, no. 1a. NL-HaNA, Koloniën, 1814–1849, 2.10.01, inv.no. 2637.

17 Res.GG, May 30, 1843, no. 10. NL-HaNA, Koloniën, 1814–1849, 2.10.01, inv.no. 2645. The resolution contains extensive written instructions for Schwaner.

18 Res.GG, August 31, 1835, no. 30. NL-HaNA, Koloniën, 1814–1849, 2.10.01, inv.no. 2856.

1835, was he ordered to join the rest of the Commissie on Sumatra.[19] When this fell through, due to both health issues experienced by Horner[20] and the order of November 14, 1835, to the Commissie to return from Sumatra,[21] a new plan was adopted on February 27, 1836, that would take Horner to Borneo.[22] In September 1835, Horner reports that he was already on his second trip on Java in the company of governor-general Baud,[23] so it seems that his time waiting for his instructions was not wasted.

2 Going Beyond Collecting 1: Mapmaking

In the end, the delay in Forsten's departure to Celebes was due to a very unfortunate combination of circumstances. Most of the time, matters regarding the Natuurkundige Commissie were dealt with promptly. The cases of Horner and Schwaner show that, even though official resolutions may have been slow to materialise, this did not leave them with nothing to do. Forsten also did not spend his year on Java idly, although he must have been tremendously relieved when he could finally commence his long-awaited expedition to Celebes, more than a year after his arrival in the Indies. Even though Forsten's instructions from the Netherlands were all about collecting objects of natural history, when he was finally given the formal assignment for his expedition, it turns out that it was not all about animals, plants, and rocks. De Eerens seems to have explicitly added a task that did not feature in the instructions from the Netherlands. This becomes clear from the introduction of Forsten's diary:

> On Wednesday, January 29, 1840, I boarded the brig Charles, Captain Prosser, to leave on the same from Batavia to Manado in order to find the treasures of the realm of nature and show them to the world according to the decree of his excellency the governor-general of December 22, 1839, No. 4. For assistance I was given the draughtsman H. Von Gaffron, not only to picture the new objects of zoology that I might find, but also to transfer to paper displays of nature that might astound our eyes, and to make topographical measurements to improve existing maps.[24]

19 Res.GG, October 1, 1835, no. 1. NL-HaNA, Koloniën, 1814–1849, 2.10.01, inv.no. 2857.
20 Horner, "Briefe und Tagebuchskizzen," 185.
21 Res.GG, November 14, 1835, no. 2. NL-HaNA, Koloniën, 1814–1849, 2.10.01, inv.no. 2857.
22 Res.GG, February 27, 1835, no. 12. NL-HaNA, Koloniën, 1814–1849, 2.10.01, inv.no. 2858.
23 Anonymus, "Untitled," 4.
24 'Woensdag 29 january 1840 begaf ik mij aanboord van de bark Charles, Kapt. Prosser, om met dezelve van Batavia naar Menado te vertrekken, ten einde aldaar volgens besluit van

When we compare this mission statement with Forsten's actual instructions, there is one thing that immediately draws our attention: the improvement of maps. The only assignment that the governor-general gives Forsten is 'to complete the task that has been given to him in the Netherlands.'[25] Most of the mission statement falls into this category, but nowhere is any mention made in Forsten's instruction of improving maps.[26] This task is specifically connected to Heinrich Von Gaffron (1813–1880, Figure 30), a former Prussian officer who took service in the Dutch colonial army in Harderwijk on July 25, 1838.[27] He was sent to the Indies, where he met Forsten in late 1839, who, upon discovering Von Gaffron's hunting and (modest) drawing skills, suggested him as his assistant-artist. According to his instructions from the Netherlands, Forsten was supposed to have been accompanied by Van Gelder. Since Van Gelder had left government service before Forsten arrived, this was not possible anymore. Overdijk was left without a member to accompany after Horner's death, so De Eerens recalled him from Sumatra in order to accompany Forsten. However, the relationship between Overdijk and Forsten was strained. Forsten writes that Overdijk 'has a wilful, proud, and lazy character, is unable to prepare better than the indigenous hunters and is entirely without scientific knowledge.'[28] Of course, these were not the ingredients for a successful working relationship on an expedition. When De Eerens finally appointed Forsten to go on his expedition, Overdijk was still on the roll. However, within a month the tables had been turned, Overdijk had been fired and Von Gaffron had been appointed. It is likely that the private audience that Forsten enjoyed with De Eerens on

Zijne Excellentie den Gouverneur Generaal, dato 22 december 1839 No. 4 de schatten van het Rijk der natuur optesporen en wereldkundig te maken. Tot behulp was mij als teekenaar medegegeven H. von Gaffron; om niet alleen de nieuwe voorwerpen der zoologie welke ik mogt vinden aftebeelden, maar ook natuurtaferelen die ons oog mogten verbazen zoo veel mogelijk op dat papier overtebrengen, en topographische opnemingen te doen, tot het verbeteren der bestaande kaarten.' Reisverhaal van Dr. Forsten naar Menado. NCAO, 1075_003.

25 'Ten einde aldaar de hem in Nederland opgedragen taak te volbrengen.' Res.GG, December 22, 1839, no. 4. NL-HaNA, Koloniën, 1814–1849, 2.10.01, inv.no. 2601.

26 Forsten himself also does not say anything about which maps he is supposed to improve. Fortunately, David Henley, (Henley, "Minahasa Mapped") gives us a good idea of the maps that were available in Forsten's days.

27 Bastin and Brommer, *Nineteenth Century Prints*, 317; Maronier, *Pictures of the Tropics*, 24–32. For his enlisting date, see military registry card Heinrich von Gaffron. NL-HaNA, Koloniën / Stamboeken Militairen KNIL Oost- en West-Indië, 2.10.50, inv.no. 321, f. 223.

28 'Daar hij een eigenzinnig, trots en lui karakter bezit, niet beter kan prepareeren als de inlandsche jagers en volstrekt zonder eenige wetenschappelijke kennis is.' Kort verslag van het in Indie verrigtte van 14 December 1838 tot April 1840. NCAO, 1076_003.

FIGURE 30 Portrait of a certain v.Gaffron, most likely Heinrich von Gaffron, photograph from Woodbury & Page (Batavia), September 9, 1871, 8.5 × 5.4 cm, TM-60021251, Collectie Wereldmuseum (Leiden). Public domain
HTTPS://HDL.HANDLE.NET/20.500.11840/28007

January 17, 1840,[29] helped to sway things in Forsten's favour, because two days later everything was arranged according to Forsten's wishes. While both Forsten and Overdijk did most likely not have the skills to draw maps, the addition of the military-trained Von Gaffron opened up new possibilities.

29 Kort verslag van het in Indie verrigtte van 14 December 1838 tot April 1840. NCAO, 1076_006.

The fact that mapmaking was not part of Forsten's original instructions, does not mean that it was a very odd phenomenon in the world of natural history. Since natural history often took place in an expeditionary context, it is by no means rare to see both maps and collections of specimens emerge from one and the same expedition. As Felix Driver, Dane Kennedy and Martin Thomas have sufficiently shown, expeditionary science was multi-faceted in nature.[30] Even though many expeditions had geographical exploration as its main impetus, it is only to be expected that expeditions with natural history as their main impetus, such as those of the Natuurkundige Commissie, adopted a component of geographical exploration as well, especially when they went to the less well-known areas on the outskirts of empire. It remains open to interpretation whether the mapmaking skills that Von Gaffron possessed (and Overdijk did not) played a significant role in the decision to replace Overdijk with Von Gaffron. Given the prominence of the task of mapmaking in Forsten's personal mission statement in his diary, it is not inconceivable that Forsten used Von Gaffron's skills in this area to convince De Eerens to make the appointment.

So why did De Eerens ask Forsten to contribute towards mapmaking? 'To govern territories, one must know them.'[31] During the days of the VOC, governing had never been a main goal. The VOC maintained a network of trade posts and forts throughout the archipelago and never intended to govern what would become the Netherlands Indies as part of a Dutch empire. However, basic knowledge of the archipelago was required to be able to maintain a trading network. Many maps were produced of all the trading posts, forts, and other important locations in the trade network.[32] When the Dutch took over from the British in 1816, the available cartographic material was fragmented. The first overview map was only published in 1842, and the first atlas was published between 1853 and 1862.[33] While 'the connection between cartography and the exercise of imperial power' had always been there,[34] the government in the

30 Driver, *Geography Militant*; Kennedy, *Last Blank Spaces*; Kennedy, *Reinterpreting Exploration*; Thomas, *Expedition into Empire*.

31 Edney, *Mapping an Empire*, 1.

32 See Van Diessen, *Grote atlas van de Verenigde Oost-Indische Compagnie*, for the plethora of maps that was produced during these days. See Sutton, *Capitalism and Cartography*, for how early printed maps of the West Indies legitimised Dutch expansion in the seventeenth century.

33 Von Derfelden van Hinderstein, *Algemeene kaart van Nederlandsch Oostindië*; Melvill van Carnbee and Versteeg, "Algemeene atlas van Nederlandsch Indie." See Van Diessen, *Grote atlas van Nederlands Oost-Indië*, 6.

34 Akerman, "Introduction," 1. Many studies are available that show the intimate connection between mapmaking and empire: Edney, *Mapping an Empire*, and Akerman, *Imperial Map*, are just two examples. Further material is available in the series Publications of

Netherlands Indies only took this fully on board in 1853, when they started with a cartographic survey on a trigonometric basis, something that the British had already been doing for decades in India.[35] Until that time, maps (and especially topographical maps as opposed to hydrographical maps) were produced on an occasional basis without central direction from the government.[36] It is no surprise that much of the archipelago was still unmapped. After all, in the early nineteenth century, 'effective Dutch colonial authority outside Java was restricted to the Moluccas and a few ports and their immediate hinterlands.'[37] Forsten's destination, the city of Manado in Northeast Celebes was one of these ports and, as such, experienced a limited form of control by the Batavia government.[38] It was all too easy for De Eerens to make use of Forsten's expedition to this outpost and gain a better understanding of what was actually going on there by having Forsten generate a map of the areas he would visit. So instead of only obtaining natural history objects for a faraway museum, De Eerens was able to squeeze a useful map out of the costly project that was the Natuurkundige Commissie.[39]

---

the International Cartographic Association (ICA) (e.g. Storms et al., *Mapping Asia*) and The Kenneth Nebenzahl, Jr., Lectures in the History of Cartography (e.g. Akerman, *Cartographies of Travel and Navigation*; Akerman, *Imperial Map*; Akerman, *Decolonizing the Map*).

35 Diessen, *Grote atlas van Nederlands Oost-Indië*, 7; Ormeling, "De Topografische Dienst." For the British Great Trigonometrical Survey in India, see Edney, *Mapping an Empire*. I know of only one mention of trigonometric techniques used during the work of the Natuurkundige Commissie. See Jacobs, "Heinrich Christian Macklot," 30.

36 There was a naval school in Semarang (Java) where between 1782 and 1826 hydrographic maps were produced for which the measurements were taken by the colonial navy. This school did not survive De Bus' budget cuts and did not systematically produce topographical maps. See Diessen, *Grote atlas van Nederlands Oost-Indië*, 10.

37 Lindblad, "Outer Islands," 93. This also becomes apparent in the circumstances of the Sarasin cousins' expeditions in the 1890s and 1900s. See Schär, "Earth Scientists as Time Travelers."

38 The town of Manado (usually called Menado by the Dutch), which gives it name to the official Dutch residency of Menado, is the main town of the area that was also known as the Minahasa (Minahassa in Dutch, and usually preceded by an article). In *c.*1860 the Minahasa area consisted of the administrative units of Manado, Kema, Tondano (Forsten's main base of operations), Amurang and Belang (Henley, *Fertility, Food and Fever*, 40). For more on this area as part of the Netherlands Indies see Henley, *Fertility, Food and Fever*; Schefold, *Minahasa Past and Present*; Van Spreeuwenberg, "Blik op de Minahassa." Van Spreeuwenberg was a fellow passenger of Forsten's on the bark *Charles* that took them from Batavia to Manado.

39 Raj, *Relocating Modern Science*, draws out the importance of local agency in mapmaking in chapter 2. Unfortunately, Forsten's diaries do not include any evidence for this, even though the locals must have played an important role in shaping the map.

Forsten adopted the mapmaking project as an integral part of his mission, even though he left the practical work almost entirely to Von Gaffron.[40] He confirms this when he writes to Susanna:

> My draughtsman has made a very handsome map of the residency of Manado, also of Gorontalo and Ternate. Over here I have been allowed the opportunity to let him copy a handsome map of the island of Amboina,[41] as far as drawing animals allows him the time for this. I hope to turn all of this in Holland into a handsome whole.[42]

Forsten clearly intended to include the map(s) into his final published work (Map 2 and Map 3). Even though he left from the Netherlands before the publication of the *Verhandelingen*, he would have been aware of the extent of the work of Müller and Korthals and he must have known that they were planning to publish maps as part of their travel reports.

Convincing De Eerens to appoint Von Gaffron also very much served Forsten's own purposes and his plans for his own Celebes publications. While De Eerens may have used this expedition of the Natuurkundige Commissie for imperial purposes, Forsten, in equal measure, used De Eerens imperial needs to get rid of Overdijk and enhance his future publication with skilfully drawn maps. Forsten occasionally adapted his travel schedule to allow for topographical measurements and observations. Pages of his diary are devoted to exact descriptions of roads between villages, precise numbers of houses and inhabitants of countless villages, and barometer measurements used to calculate the heights of mountains and volcanoes. During some of these surveying trips, there was hardly time to hunt and collect, so some of these surveying passages are devoid of natural history observations.

---

40 At least this was something that Von Gaffron could do well. According to Bastin & Brommer (Bastin and Brommer, *Nineteenth Century Prints*, 317), 'he was a better artist of scenery than of people' and his drawings, 'compared with similar drawings by other artists connected with the Natural Sciences Commission […], can only be described as mediocre.'

41 Amboina was the name the Dutch used for Ambon (both the island and the administrative unit).

42 'Mijnen teekenaar heeft een zeer fraaije kaart van de residentie Manado gemaakt, ook van Gorontalo en Ternate. Alhier ben ik in de gelegenheid gesteld hem eene fraaije van het eiland Amboina te laten copieren, voor zoo verre het dieren teekenen hem de tijd daartoe gunt. Ik hoop van dit alles in Holland een fraai geheel te kunnen maken.' Letter Forsten to Susanna, July 19, 1842. NCAO, 1236_003.

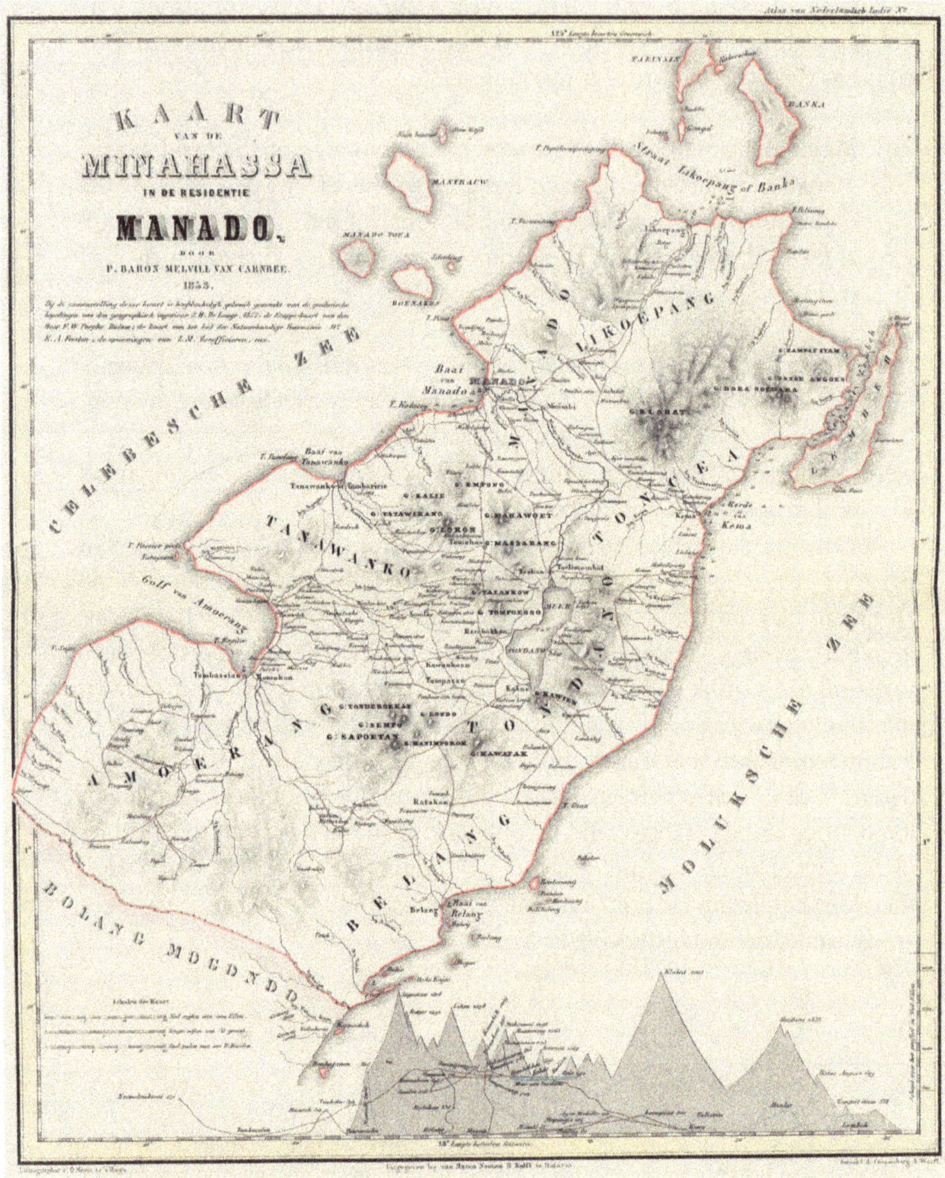

MAP 2 Map of the Minahasa in Manado. Map 22 from Melvill van Carnbee and Versteeg. *Algemeene atlas van Nederlandsch Indie*. Public domain. In compiling this map, Van Carnbee made use of Forsten's map.
HTTP://NLA.GOV.AU/NLA.OBJ-230938308

MAP 3 Map of Ambon and several neighbouring islands. Map 24 from Melvill van Carnbee and Versteeg. *Algemeene atlas van Nederlandsch Indie*. Public domain. In compiling this map, Van Carnbee made use of Von Gaffron's map of Ambon. HTTP://NLA.GOV.AU/NLA.OBJ-230938528

At several points, Von Gaffron was allowed to undertake separate journeys to map certain areas that Forsten would not be able to travel to because of his ill health. The mapmaking was Von Gaffron's project. His skill in this art is visible in a later map of South Borneo, which Von Gaffron most likely produced during his travels with Commissie member Carl Anton Ludwig Maria Schwaner (1817–1851).[43] Forsten and Von Gaffron's map of the Manado area was known

43 Heinrich von Gaffron, (1847). Kaart van de districten Mattam, Mendawie, Sampit, Pemboean, Kotta-Waringin en Pontianaksche Kapoeas van Sintang naar zee en het Z.W. gedeelte van Borneo. NL-HaNA, Koloniën / Kaarten en Tekeningen, 4.MIKO, inv.no. 642A-I. In 1860 a map by Von Gaffron accompanied Von Gaffron and Pijnappel, "Beschrijving van het westelijke gedeelte." A geological map of the same area (*Geologische Karte des südlichen Borneo. Gezeichnet von H. Von Gaffron 1843–1848*) comes as a foldout accompaniment to Martin, "Neue Fundpunkte."

by military physician and naturalist Pieter Bleeker (1819–1878) (who does not seem to have personally seen it).[44] Bleeker's knowledge of Forsten's map is based on the work of naval officer and cartographer Pieter baron Melvill van Carnbee (1816–1856), who quotes Forsten's map as one of the sources for his published map of the Minahasa (Map 2).[45] Unfortunately, the current whereabouts of Von Gaffron's map of the Minahasa (or any of his other maps from this expedition) are unknown.[46]

3 The Maps of Salomon Müller

Forsten and Von Gaffron were not the only participants of the Natuurkundige Commissie to produce maps. While Von Gaffron seems to have been the only expedition member or assistant of the Commissie who had received military training and, as such, had the required skills to draw maps, several other members developed mapmaking skills themselves during their work in the Indies. It seems that especially Müller had a strong affinity for mapmaking and actively worked on improving his skills. The geographical and ethnographical volume of the *Verhandelingen* contains several of Müller's maps.

This volume of the *Verhandelingen* consists of five main parts, accompanied by a total of 86 plates and three maps. The first part covers New Guinea (where the Commissie stayed for just four months from May to August 1828), the second part describes some of the islands visited on the way to and from New Guinea, the third part covers Timor (where the Commissie stayed for more than a year from late 1828 to late 1829), the fourth part covers Borneo (where Müller, Korthals and Horner travelled from July to September 1836). The first four parts are extensive travel narratives that include zoological and botanical, but mostly geographical and ethnographical notes. The fifth and final part does not contain a travel narrative and only briefly covers some highlights on Java and Sumatra. Müller writes that, since the first four parts are much more extensive than originally envisioned, for this part, he is forced to limit himself to a short commentary on the plates. He expresses the hope and

44 Bleeker, *Reis door de Minahassa*, 66, 146–147. Reference from Abendanon, *Geologische en geographische doorkruisingen*, vol. 4, 1877–1880, who suggests that the Forsten-Von Gaffron map was used by later mapmakers.

45 Melvill van Carnbee, "Kaart van de Minahassa."

46 Even Abendanon, writing in 1917, wonders: 'The question is, where is the manuscript map of Forsten?' ('De vraag is waar de m.s.k. van Forsten is gebleven.'), Abendanon, *Geologische en geographische doorkruisingen*, vol. 4, 1880. I believe it is very well possible that these manuscript maps are still present somewhere in the Arsip Nasional Republik Indonesia in Jakarta.

expectation that the government may be inclined to decide to publish a sequel to the *Verhandelingen*, in which he can more extensively fulfil the goal of the Natuurkundige Commissie (namely, to expand the knowledge on the Netherlands Indies).[47] Even though Müller was kept on in government service until 1850, he never published full travel narratives of his work on Java and Sumatra.[48]

The extent of Müller's mapmaking skills can be traced by looking at his four published maps in some more detail. Three maps (of New Guinea, Timor, and Borneo) were included in the *Verhandelingen* and an additional map of Sumatra was published by Müller in 1837 in his *Berigten over Sumatra*.[49]

3.1 *Map of a part of the South and Southwest Coast of New Guinea, according to the survey, done by His Majesty's ships Dourga, Triton, Sireen and Postillon, between the years 1826 and 1835; compiled by Salomon Müller, as illustration of his Contributions to the knowledge of New Guinea. 1840.*[50]

The title of this map (Map 4A–B) states that this map has been 'compiled' by Müller. In the accompanying text, Müller sketches a history of the cartography of New Guinea and gives more details regarding the four survey journeys that the government organised between 1826 and 1835.[51] The *Dourga* expedition, led by Dirk Hendrik Kolff (1800–1843), took place in 1826. The *Triton* and *Iris* expedition, under command of Jan Jacob Steenboom (1786–1828), took place in 1828, and included Müller himself and four other Commissie participants (Macklot, Van Raalten, Van Oort, and Zipelius).[52] The ship *Sirene*,[53] led by A. de Boer, visited Merkusoord (the Dutch settlement founded in 1828) in

47 Susanna, *Verhandelingen*, vol. 3, 449–450.

48 Unfortunately, Müller's personal diary and travel reports are not present in the archives of the Natuurkundige Commissie. Perhaps he took them with him when he moved back to Germany.

49 Müller, *Berigten over Sumatra*. This 35-page extract from Müller's diary was edited for publication by Susanna when Müller was still in the Indies.

50 'Kaart van een gedeelte der zuid- en zuidwest-kust van Nieuw-Guinea, volgens de opnemingen, gedaan door Zr. Ms. schepen Dourga, Triton, Sireen en Postillon, tusschen de jaren 1826 en 1835; zamengesteld door Salomon Müller, tot toelichting zijner Bijdragen tot de kennis van Nieuw-Guinea. 1840'.

51 See Mörzer Bruyns, *Met de Triton en Iris*, 73–87, and Mörzer Bruyns, "De Nederlandse hydrografische opnemingen," for a review of the history of Dutch cartography of New Guinea in this period.

52 See Mörzer Bruyns, *Met de Triton en Iris*, for two diary accounts of this journey by lieutenant Modera and government official Van Delden. Other publications on the New Guinea expedition are Mees, "Vogelkundig onderzoek"; Mörzer Bruyns, "Taking Possession"; Overweel, "Keep Them Out!".

53 Müller calls the ship the *Sireen*. I follow Mörzer Bruyns in using the spelling *Sirene*.

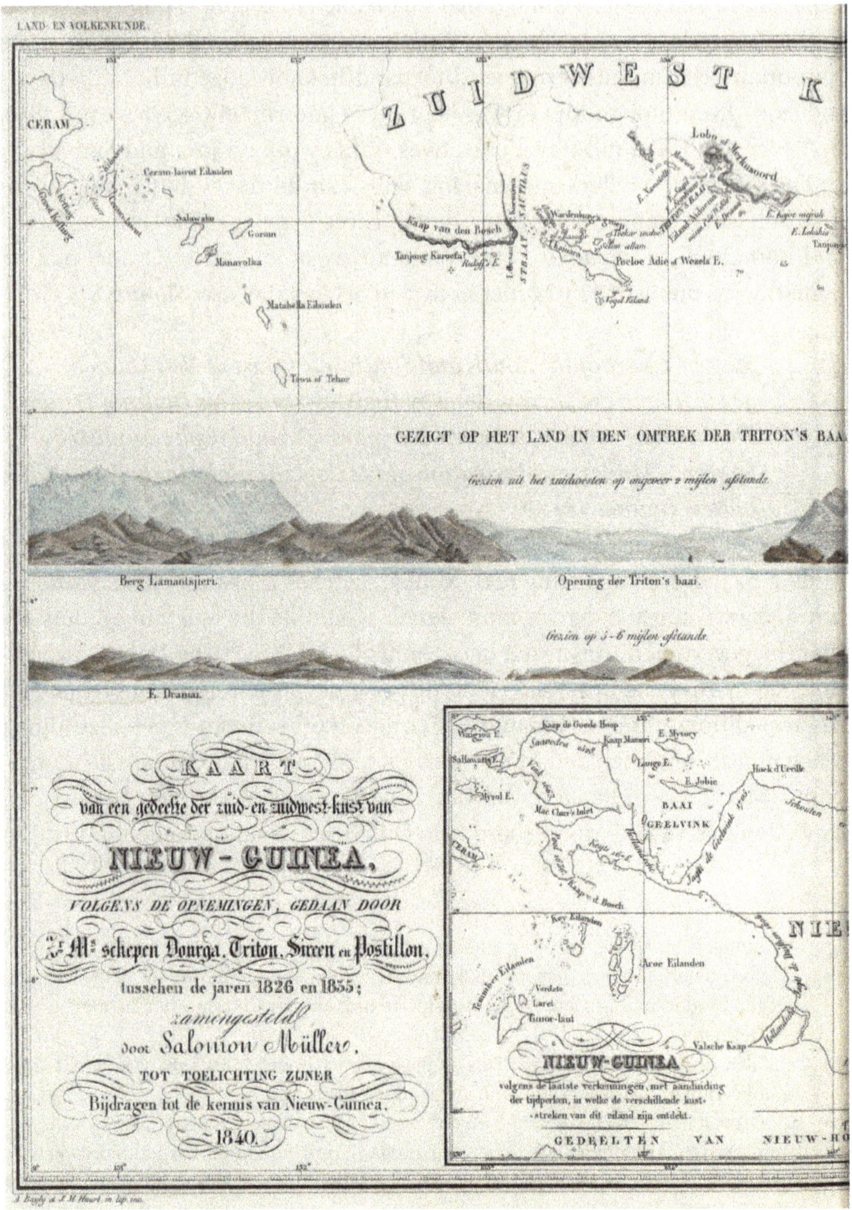

MAP 4A Map of a part of the South and Southwest-coast of New-Guinea (left panel) by Salomon Müller. From the volume *Land- en volkenkunde* of the *Verhandelingen*. NCAO, 871_087. Public domain
HTTPS://DH.BRILL.COM/NCO/VIEW/NCO_NNM001000871_087/MAKINGSENSE

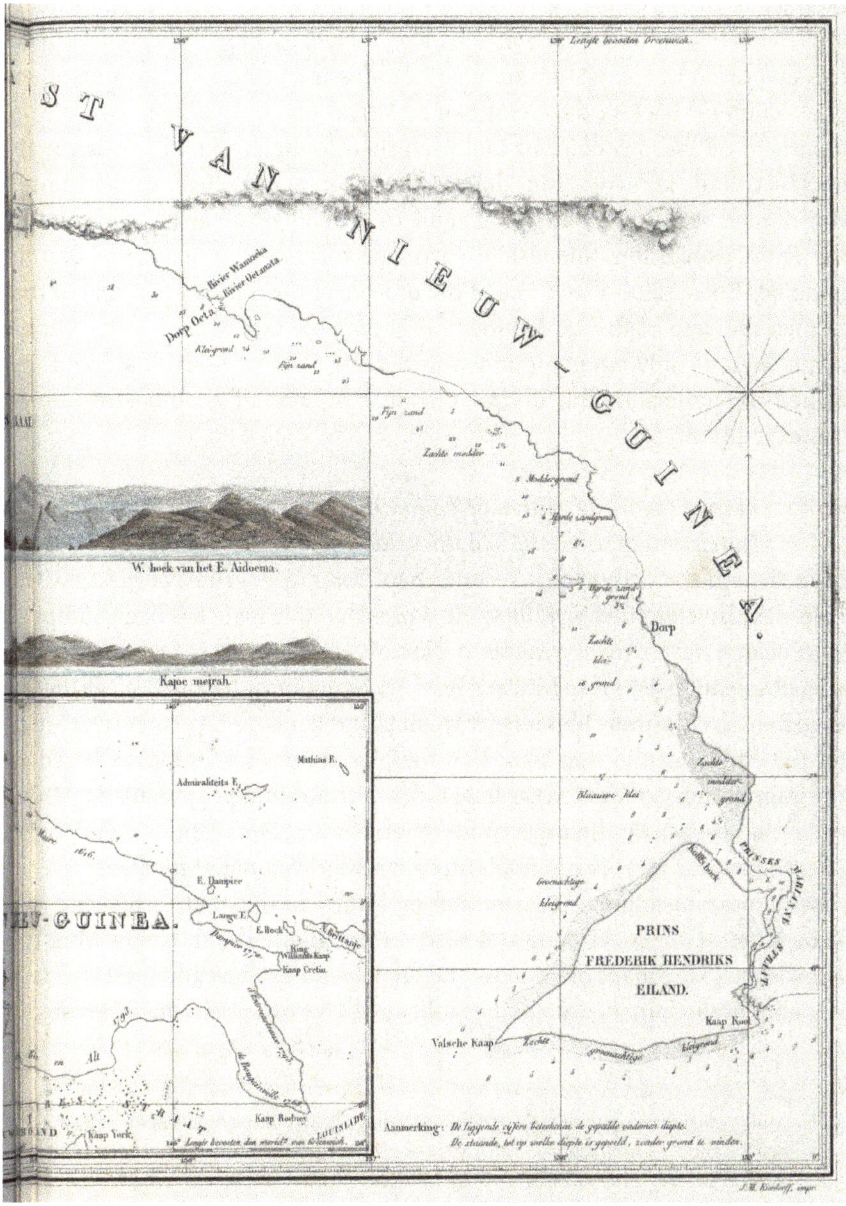

MAP 4B Map of a part of the South and Southwest-coast of New-Guinea (right panel) by Salomon Müller. From the volume *Land- en volkenkunde* of the *Verhandelingen*. NCAO, 871_088. Public domain
HTTPS://DH.BRILL.COM/NCO/VIEW/NCO_NNM001000871_088/MAKINGSENSE

1832, while the *Postiljon*[54] and *Sirene* expedition, led by Michiel Langenberg Kool (1796–1845), took place in 1835 and included three indigenous collectors employed by the Natuurkundige Commissie but no European members or assistants.[55] Back in the Netherlands, Müller had been able to consult the reports of all four expeditions and used their findings and all existing maps of New Guinea known to him to compile this map. It is clear Müller himself contributed very little, if anything at all, to the actual surveying work. During the *Triton* expedition, this task of surveying was given to lieutenant Carolus Johannes Boers (1801–1829), one of the officers of the military contingent that formed the core of the expedition.[56] It is possible that Müller, at the time a young man of only twenty-four, witnessed the work of Boers first hand and gained an understanding of the practical aspects of mapmaking during this expedition.

3.2 *Map of the western area of Timor including the possessions of the Dutch and of their allies in that place; by Sal. Müller. 1841.*[57]

Even though the title of this second map (Map 5) does not suggest so, this map was also compiled by Müller. He writes that this map 'has been compiled with great care, partly according to the best sources, partly according to our own observations'.[58] From Müller's further comments it becomes clear that he used maps by Matthew Flinders (1774–1814) and Louis de Freycinet (1779–1841) for the outline of the coast, but that the placenames on Timor are based on the information gathered in the field by the Natuurkundige Commissie. Müller notes that especially the orography (relief of mountains) and potamography (the course of rivers) leave much room for improvement.[59] It seems that the most important original contribution by Müller to this map is his extensive knowledge of the local names of places throughout Timor. Connecting this knowledge to the maps of Flinders and Freycinet must have required a systematic and careful form of notetaking in the field. Unfortunately, the whereabouts

54 Müller calls the ship the *Postillon*. I follow Mörzer Bruyns in using the spelling *Postiljon*.

55 See Susanna, *Verhandelingen*, vol. 3, 9–12, for Müller's overview of these expeditions.

56 Willem Mörzer Bruyns (Mörzer Bruyns, *Met de Triton en Iris*, 33.) assumes that Boers was appointed as member of the Natuurkundige Commissie, but I have found no proof for this. Boers was part of the military contingent of the expedition and only happened to travel together with the members of the Natuurkundige Commissie from Batavia to Makassar to join the *Triton*.

57 'Kaart van het westelijke gedeelte van Timor bevattende de bezittingen der Nederlanders en van hunne bondgenooten aldaar; door Sal. Müller. 1841.'

58 'Dezelve is met groote zorg, deels volgens de beste bronnen, deels volgens onze eigene waarnemingen zamengesteld.' Susanna, *Verhandelingen*, vol. 3, 137–138.

59 Susanna, *Verhandelingen*, vol. 3, 138.

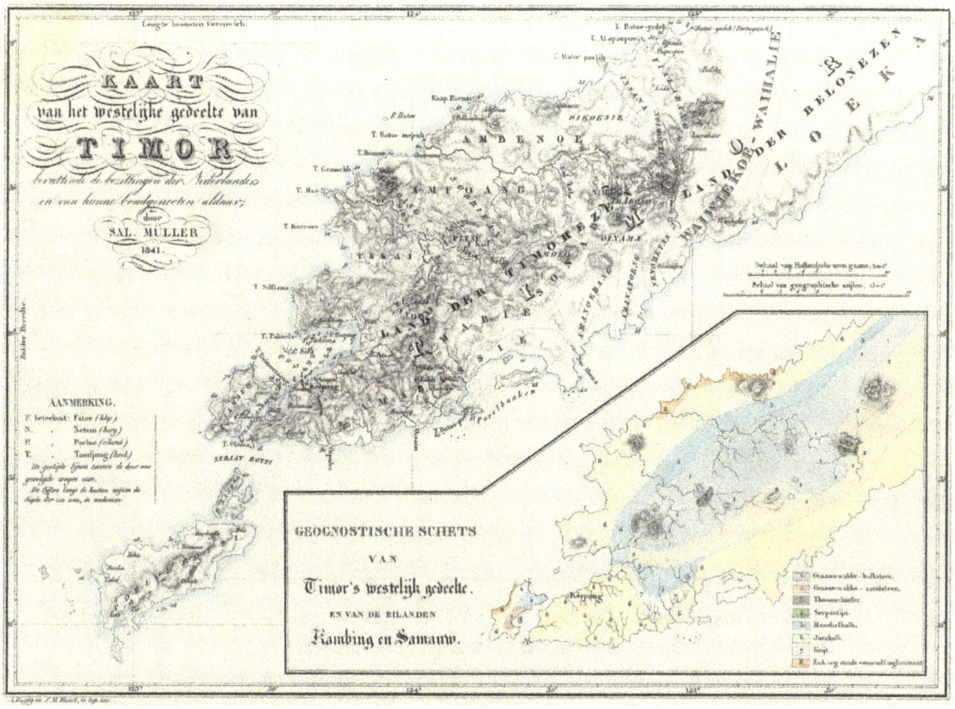

MAP 5 Map of the western area of Timor by Salomon Müller. From the volume *Land- en volkenkunde* of the *Verhandelingen*. NCAO, 871_089. Public domain
HTTPS://DH.BRILL.COM/NCO/VIEW/NCO_NNM001000871_089/MAKINGSENSE

of Müller's original notes of the Timor expedition are unknown, so we are denied a glimpse into the process that led from field notes to the incorporation of placenames on this map.

3.3 *Figurative map of an area of Sumatra mostly denoting the road and rivers that flow from the inland of Padang to the east coast, surveyed by S. Müller and H. Hellbach, 1834.*[60]

This map (Map 6) accompanied the small 35-page publication *Berigten over Sumatra*, published in Amsterdam in 1837, before Müller actually returned from the Indies. The publication contains an extract from Müller's diary (most likely selected and edited for publication by Susanna) in which he describes the circumstances that led to the creation of this map. Müller had been asked by the resident of Sumatra's Westkust to travel to the uplands in order to map

60 'Figuratieve kaart van een gedeelte van Sumatra voornamelijk aantoonende de wegen en rivieren, welke uit de Padangsche binnenlanden naar de oostkust aflopen, opgenomen door S. Müller en H. Hellbach. 1834.'

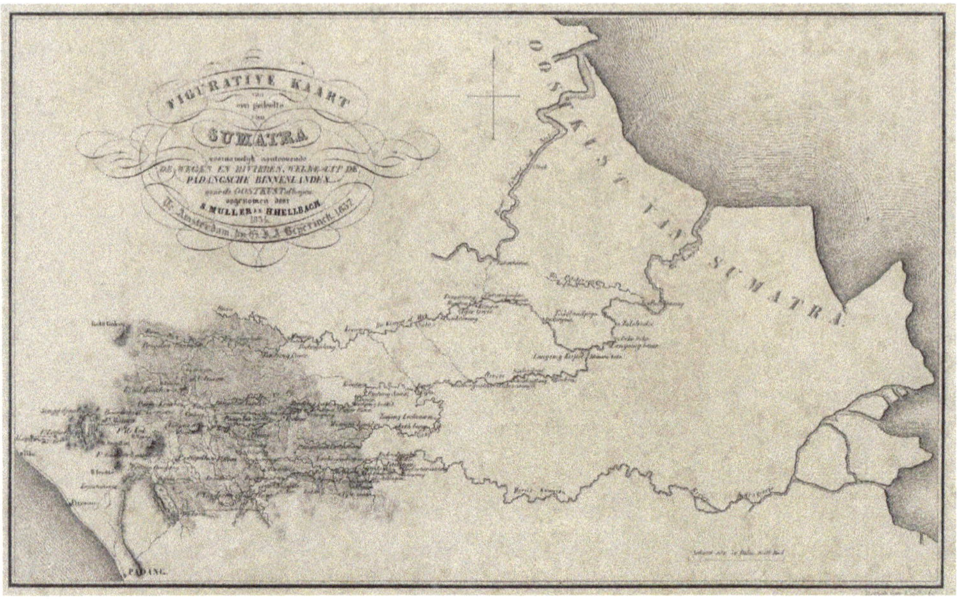

MAP 6 Figurative map of an area of Sumatra by Salomon Müller and Antonie Hermanus Helbach. From Salomon Müller, *Berigten over Sumatra*. Public domain
HTTP://OBJECTS.LIBRARY.UU.NL/READER/INDEX.PHP?OBJ=1874-366710

the local trading posts there, which would help to gain control over trade routes in the area. Because Korthals was ill, Müller set out on this expedition on his own. A chance meeting with his co-author, led to the latter joining him.[61] His co-author is designated on the map as H. Hellbach, while his name is spelled Helbach in the accompanying text. This H. Hel(l)bach in reality was Antonie Hermanus Helbach (1808–1866), who was a lieutenant in the colonial army at the time.[62] Together, Müller and Helbach surveyed the area, much helped by Helbach's intimate knowledge of the terrain and his knowledge of the local Malay dialect. The skills of these two men led to the map as it is published in Müller's *Berigten over Sumatra*. Müller acknowledges the cooperative nature of this enterprise to such an extent that it does not become clear how much of the practical work in creating the map is actually his. However, his close association and association with Helbach must have given him a direct insight into the inner workings of mapmaking.

61 Müller, *Berigten over Sumatra*, 17–18.

62 For Helbach's career, see military registry card Antonie Hermanus Helbach. NL-HaNA, Oorlog / Stamboeken Officieren KL, 2.13.04, inv.no. 383, f. 21.

3.4 *Map of the coast and inland of Banjarmasin, belonging to the Journey in the southern area of Borneo, by Salomon Müller, 1845. & Banjarmasin, Martapura and a part of the Lawut-lands, 1845.*[63]

After the deaths of Boie, Van Raalten, Macklot and Van Oort, Müller had become the longest serving participant of the Natuurkundige Commissie in the field. He had seen the far reaches of the archipelago from New Guinea to Sumatra and gained an incredible amount of experience from his tireless work in the field. The map of Sumatra and the land survey undertaken by Korthals and Bürger in order to plan a road on Sumatra had put the Natuurkundige Commissie on the map with the governor-general as a vehicle that could be used for mapmaking as well.[64]

When doubts began to be voiced about the quality of the results of a longstanding expedition to Borneo by army officer Henri Albert Henrici (1782–1838), the Natuurkundige Commissie was called in to evaluate the work that had been done there. Henrici was born in Rome in 1782, had had a decorated military career during the Napoleonic wars in Europe and ended up in Dutch military service in 1817. In 1820, he was sent to the Indies where he was promoted to major in 1825, and lieutenant colonel in 1832.[65] In 1830, he was tasked with an expedition to West Borneo in order to find out more about the terrain, trade routes, population, and defence. The goal was to create a map of the area in order to evaluate the possible outcome of a military expedition to the area.[66] Henrici returned from his initial expedition in 1832 but lost all his reports at sea. On December 27, 1832, he was allowed to undertake a new expedition to Borneo to make up for the loss of the outcomes of the first expedition.[67] After a couple of years, doubts began to creep in regarding the

63 'Kaart van de Kust- en Binnenlanden van Banjermasing, behoorende tot de Reize in het zuidelijke gedeelte van Borneo, door Salomon Müller. 1845. & Banjermasing, Martapoera en een gedeelte der Lawut-landen, 1845.'

64 Weber and Van Zanen, *Kunstenaar op Sumatra*, 23. Colombijn, "Hoe de Nederlanders op Sumatra," does not mention the involvement of the Natuurkundige Commissie in the development of road building on Sumatra.

65 Details of Henrici's career and the spelling of his first names are taken from his military registry card. NL-HaNA, Oorlog / Stamboeken Officieren KL, 2.13.04, inv.no. 383, f. 6. No recent substantial work on Henrici's career seems to be available. Short reports, most with some inaccuracies, can be found in Posewitz, *Borneo*, 17; Van Steenis-Kruseman, "Malaysian Plant Collectors," 226; Gebhardt, *Ornithologen Mitteleuropas*, 149–150, and Glas, "Een Oostenrijker op Borneo." Henrici's archives (including maps, field notes, and correspondence) are available at NL-HaNA, Koloniën / Kaarten en Tekeningen, 4.MIKO, inv.no. III.4.

66 A summary of Henrici's original assignment can be found in Res.GG, February 6, 1836, no. 8. NL-HaNA, Koloniën, 1814–1849, 2.10.01, inv.no. 2858.

67 Res.GG, December 27, 1832, no. 4. NL-HaNA, Koloniën, 1814–1849, 2.10.01, inv.no. 2845.

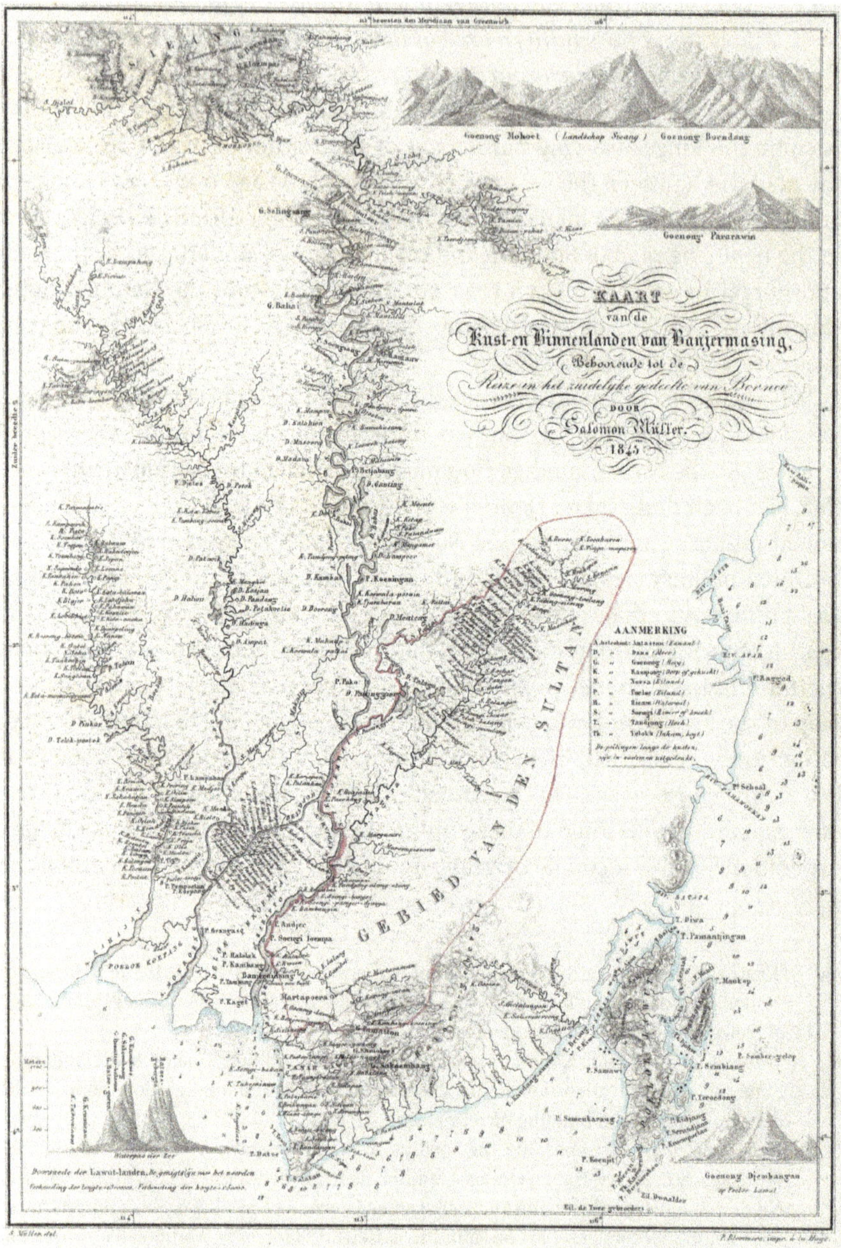

MAP 7 Map of the coast and inland of Banjarmasin by Salomon Müller. From the volume *Land- en volkenkunde* of the *Verhandelingen*. NCAO, 871_090. Public domain HTTPS://DH.BRILL.COM/NCO/VIEW/NCO_NNM001000871_090/MAKINGSENSE

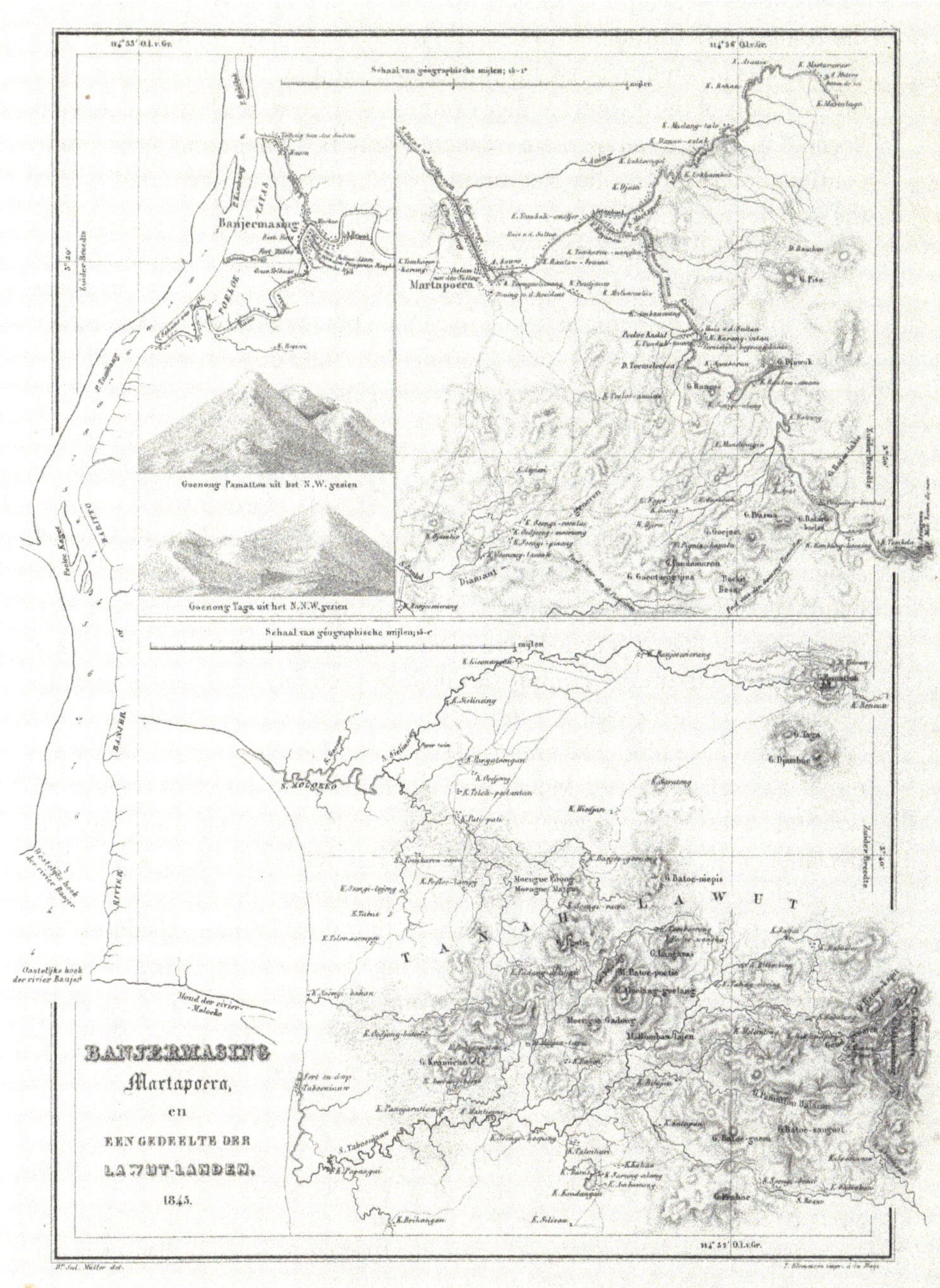

MAP 8 Banjarmasin, Martapura and a part of the Lawut-lands by Salomon Müller. From the volume *Land- en volkenkunde* of the *Verhandelingen*. NCAO, 871_091. Public domain
HTTPS://DH.BRILL.COM/NCO/VIEW/NCO_NNM001000871_091/MAKINGSENSE

quality of Henrici's reports. Massive amounts of money had been invested but the results were considered to be quite meagre.[68] Finally, on March 12, 1836, Henrici was ordered to cease his expedition, hand over all his documents to the members of the Natuurkundige Commissie (who were already set to go to Borneo) in order to allow them evaluate Henrici's findings and advise on its continuation.[69] Clearly, the Natuurkundige Commissie was expected to have the experience to deal with such an assignment.[70]

In the original assignment for the Borneo expedition, no mention is made of Henrici. Müller, Korthals and Horner are to create collections of specimens, with zoology being assigned to Müller, botany to Korthals and mineralogy to Horner. They are ordered to only take the necessary notes in order to maximise the time spent on collecting. This assignment, given to them on February 27, 1836, was issued by governor-general ad interim Baud, who would later prove to be a warm advocate for the natural history work of the Natuurkundige Commissie.

It is interesting to note that the following decree of March 12, 1836, which assigned the Commissie the task of verifying Henrici's work, was issued by Baud's successor De Eerens, who was also responsible for adding the mapmaking mission to Forsten's expedition. Clearly, De Eerens had a different idea of what the Natuurkundige Commissie could be used for. As a military man, De Eerens had a much closer interest in surveys like the ones that Henrici was supposed to have produced and used the Natuurkundige Commissie for this purpose as well. In the end, Müller, Korthals and Horner undertook this expedition and traversed the designated area based on Henrici's map from August to December 1836. Based on their evaluation of Henrici's work, De Eerens advised the minister in the Netherlands to discontinue Henrici's Borneo expeditions.[71]

In the *Verhandelingen*, Müller indicates that their Borneo expedition had as its primary goal to complete and unify the cartographic work of Henrici.[72] Diplomatically, Müller refrains from saying anything about the part where they had to report on the accuracy of the results of Henrici's work. Based on Henrici's map, that he had been forced to hand over to the Commissie, they

68 See Res.GG, February 6, 1836, no. 8. NL-HaNA, Koloniën, 1814–1849, 2.10.01, inv.no. 2858 for a concise overview of the reasons for the doubts that had arisen.

69 Res.GG, March 12, 1836, no. 1. NL-HaNA, Koloniën, 1814–1849, 2.10.01, inv.no. 2859. The Natuurkundige Commissie had already received their orders in Res.GG, February 27, 1836, no. 12. NL-HaNA, Koloniën, 1814–1849, 2.10.01, inv.no. 2858.

70 For a report of the Borneo expedition and its place in Dutch imperialistic policies on Borneo, see Van Wingerden, "Op reis vanuit een drijvende stad."

71 Res.GG, November 3, 1837, no. 14. NL-HaNA, Koloniën, 1814–1849, 2.10.01, inv.no. 2576.

72 Susanna, *Verhandelingen*, vol. 3, 323.

decided to split the Borneo expedition into two smaller trips, one to the inland area north of Banjarmasin, and another to the mountainous area to the south and east of Banjarmasin. This resulted in the two-folio map (Map 7 and Map 8) that is present in the *Verhandelingen*, with one folio each dedicated to the area of each trip. Müller writes that he has added very little to Henrici's map and that his own contribution mostly consisted of redrawing the larger maps of Henrici on a smaller scale. Müller only made some minor changes, mainly of the nature of spelling changes to indigenous names, just as with the Timor map. In addition, some of the edges of the map contain information from a map published by the expedition of Jules Dumont d'Urville (1790–1842). Both leaves of the Borneo map carry the note '(Dr.) S. Müller del.' [Dr. Salomon Müller delineavit] at the bottom, signifying indeed that Müller drew these maps himself.

All in all, these four maps and their backstory, and the inclusion of mapmaking in Forsten's expedition, show that an intimate relationship existed between the expeditions of the Natuurkundige Commissie and the surveying of the areas visited for purposes other than purely natural history. In fact, the maps were used in controlling trade routes and predicting the results of possible military expeditions. Von Gaffron's later mineralogical map of Borneo, made during his association with Schwaner, also assisted the government in understanding where certain mineral deposits could be found, and which ones were located suitably for exploitation. Likewise, Schwaner's own map of Borneo showed the connection between two important river systems and revealed the fact that further up inland they were actually not connected, which made it difficult to find an inland trading route.

4 Going Beyond Collecting 2: Mineral Deposits and Mining

It was not only mapmaking that was an imperially useful by-product of the expeditions of the Natuurkundige Commissie. A second aspect of Forsten's mission that shows the hand of the governor-general is the inspection of the gold mines of Northeast Celebes. These mines had been under Dutch influence since VOC times and operated under a system of annual compulsory deliveries. As soon as the Dutch regained control of the Indies in the aftermath of the Napoleonic wars, this system was revitalised.[73] Forsten was not the first member of the Natuurkundige Commissie to occupy himself with gold mines.

73 Henley, *Fertility, Food and Fever*, 42.

Müller published a book on Sumatra in 1846 based on his travels there from 1833 to 1836.[74] Most of the book is devoted to the origins of the name Sumatra and various customs of the indigenous peoples, but the final chapter of his book deals with the gold mines at Salida.[75] Müller states his reasons for doing this, when he writes in the preface:

> Of the last three pieces there is no doubt that the one about the gold mines at Salida is the most important. The therein communicated notes, based on authentic documentation, on the profits of those mines, which, besides those of the Parang Mountains, in the Priangan area on Java, were the most important that were exploited under the rule of the former East India Company in all of the Indies Archipelago, can serve as a measure of the advantages and disadvantages that can be expected from similar exploits for the government.[76]

Müller clearly indicates that the direct use of their inspection of the gold mines in Salida was not of a natural historical nature. It was a full inspection of the management and profitability of the mine that would allow the government to understand what the yields of similar mineral exploitations would be for the government.

It is not as clear as this in Forsten's case. His instructions from the Netherlands show that he was also charged to take every opportunity to learn more about the geology and mineralogy of the areas travelled. From that point of view, it is not a surprise that Forsten was interested in gold samples from the various areas he visited, as he confirms when he writes to Susanna: 'I visited the gold mines in the same area and I bought samples of gold for the Museum,'[77] and later to Temminck: 'Gorontalo is only interesting where mineralogy and

74 Müller, *Bijdragen tot de kennis van Sumatra*.

75 Müller, *Bijdragen tot de kennis van Sumatra*, 164–184.

76 'Van de drie laatste stukken is ongetwijfeld dat over de goudmijnen bij Salida van het meeste belang. De daarin medegedeelde, op authentieke bescheiden gegronde bijzonderheden, aangaande de opbrengsten dier mijnen, welke, benevens die van den berg Parang, in de Preanger-Regentschappen op Java, de voornaamste waren, onder het bestuur der voormalige Oost-Indische Compagnie, in geheel den Indischen Archipel bewerkt, kunnen tot maatstaf dienen der voor- en nadeelen, welke uit soortgelijke ondernemingen, voor rekening van het Gouvernement, te wachten zijn.' Müller, *Bijdragen tot de kennis van Sumatra*, x–xi.

77 'De goudmijnen in dezelve gelegen heb ik gezocht en staaltjes goud voor het museum opgekocht.' Letter Forsten to Susanna, January 26, 1841. NCAO, 1230_003.

ethnography are concerned. I have collected twenty different types of gold in small quantities.'[78]

But Forsten did more than just collect gold samples, and followed in Müller's footsteps. He undertook two inspection trips of the local gold mines (October and November 1840 to Belang in the Minahasa area, and November 1841 to Pagojama and Pagowat in the Gorontalo area, Figure 31). On his visit to the Belang mines, Forsten suggests that he was charged by his superior to make this inspection visit:

> After we had sat down and rested for a while, I acquainted him [the local supervisor of the gold mines] with my person saying that I had come to inspect the gold mines that are present here in all their parts in order to send a report to my *raja*, for which purpose I requested from him every help and assistance.[79]

This help is duly given and Forsten travels around for a week or two, visiting all the gold mines in the area and noting down details on their working operations.[80] When summarising his findings, Forsten writes:

> The government has now signed a treaty with this *raja*, who needs to employ every year in all these [seven] mines together 750 people and produce 400 Spanish mats of pure gold for fourteen guilders per mat. However, the pox has had its disadvantageous influence here as well, so that last year, when the term within which the *raja* was obliged to deliver the gold was long past, the *raja* was lured to Belang and held prisoner there until all the gold was delivered. How they will do it this year I do not know, because in the six mines that I visited there were perhaps in total only ten people working.[81]

78 'Alleen voor het mineralogische en ethnographische gedeelte is Gorontalo interessant. Ik heb eene 20 verschillende goudsoorten verzameld dog in kleine hoeveelheden.' Letter Forsten to Temminck, January 22, 1841. NCAO, 1233_003.

79 'Na wat gezeten en uitgeblazen te hebben maakte ik hem met mijn persoon bekend zeggende, gekomen te zijn om de goudmijnen welke alhier zijn te bezigtigen in alle zijne deelen, ten einde er rapport van aan mijnen Radja te maken, waartoe ik hem alle hulp en bijstand verzocht.' Reisverhaal van Dr. Forsten naar Menado. NCAO, 1075_047 (p. 88).

80 For a narrative of Forsten's gold mine inspection, see Van Wingerden, "Goudkoorts in de gordel van smaragd."

81 'Het Nederlandsch Gouvernement heeft nu een verdrag met dezen Radja gesloten, die aan deze mijnen tezamen jaarlijks 750 menschen moet laten werken en 400 Spaansche matten zuiver goud tegen f 14 de Spaansche mat moet afleveren, doch de kinderpokken hebben ook alhier hunnen nadeeligen invloed gehad, zoo dat verleden jaar de termijn

FIGURE 31 'Goudmijn te Gorontalo' ('Gold mine in Gorontalo'). Lithograph after a drawing by Jannes Theodorus Bik, 1821–1822. Plate 13 of De Vriese and Pijnappel, *Reis naar het oostelijk gedeelte van den Indischen Archipel*. Public domain. Universiteitsbibliotheek, Rijksuniversiteit Groningen, 54B 8722

Undoubtedly, this would have been extremely valuable information for the government in Batavia, just like Müller's findings on the profitability of the Salida mines on Sumatra. Forsten does not further identify the *raja* who charged him with this inspection mission, but he is obviously using a locally familiar term to explain his relationship to governor-general De Eerens and the status of his mission.

As we saw earlier, De Eerens was the one who had used the expedition to Borneo to verify the findings of Henrici's expeditions and had attached a mapmaking element to Forsten's expedition. What was intended as a natural historical collection expedition, gained an imperial dimension that would result in strategic commercial and military intelligence for the government in Batavia. Likewise, including inspections of commercial mines also provided the government with essential information regarding the economic possibilities of mining existing mineral deposits. At an earlier stage, Müller had obviously already performed a similar task, which was not an isolated incident. Van Oort mentions several occasions when the Commissie members inspected gold mines on Sumatra.[82]

Just as mapmaking, as a logical component of expeditionary science, was not a stretch for the Natuurkundige Commissie, mining inspections were very close to home. The Commissie even had three members that were designated as geologists and mineralogists and as such, inspecting and looking for mineralogical deposits was part of their task. From there, it was only a small step to reporting on the exploitability of such deposits. Macklot, Horner and Schwaner were the mineralogists/geologists of the Commissie, and all of them were involved in some form in commercial geological exploration in their capacity as members of the Natuurkundige Commissie.

Already in 1827, Macklot had been asked by the government to inspect a coal deposit in Lebak in Banten (West Java). The inspection took place in August 1827.[83] Macklot concluded that there was enough coal present in the deposit to make commercial exploitation viable. The downside was that there were no geographical maps available of sufficient quality and river transport was going to be an issue. The river could be made navigable, but there was not enough bamboo in the area to use in constructing transportation rafts. Since

reeds lang voorbij zijnde waarop den Radja het goud moest afleveren hem naar Belang had weten te lokken en hem daar zoo lang gevangen hield, tot dat het goud was binnen gekomen, hoe of men dit jaar doen zal weet ik niet daar er in de zes door mij gereizen mijnen misschien te zamen slechts 10 man aan het werk waren.' Reisverhaal van Dr. Forsten naar Menado, NCAO, 1075_053.

82 Weber and Van Zanen, *Kunstenaar op Sumatra*, 107, 109–112.

83 Weber and Van Zanen, *Kunstenaar op Java*, 70–71.

there were no military mapmakers available at the time, and there was no great need for coal exploitation, the governor-general decided not to start with the exploitation of this coal deposit but keep it in mind for a better moment in the future.[84] Samples of this coal deposit were sent to the Netherlands and the coal turned out to be of excellent quality. Almost ten years later, on February 28, 1836, Horner was ordered by governor-general ad interim Baud to return to Banten and inspect the coal deposit there. The first government-owned steamboat Willem I had recently arrived in the Netherland Indies[85] and two more were expected to arrive soon, which necessitated the mining of coal in the Indies in order to become self-sufficient.[86]

Macklot's inspection of the coal deposit in Lebak may not have resulted in any immediate tangible result, but it put the Commissie on the map as a vehicle that could be used for mining inspections. On December 24, 1827, lieutenant-governor-general De Kock suggested to commissioner-general Du Bus, spurred on by a report from the faraway island of Timor, that perhaps Diard could visit Timor to investigate the natural world.[87] Diard was not yet a member of the Commissie, but since he had already successfully completed similar missions to Bangka and Borneo, he was a believable candidate for the job.

When Du Bus refused to send Diard to Timor, De Kock suggested that maybe the Natuurkundige Commissie could take on this mission on the way back from the expedition to New Guinea.[88] Du Bus responded positively to this proposal on January 31, 1827, and De Kock ordered the Natuurkundige Commissie on February 5, 1828, to visit Timor on the way back from New Guinea in order to 'set up a natural historical and specifically geological investigation [...] as directed by the resident.'[89] The Commissie had departed to Makassar to join up with the expedition just days earlier, so the official confirmation of their mission to Timor had to follow them. From the *Verhandelingen* it becomes apparent that the geological mission of the Commissie consisted of an inspection of the possibilities for gold and copper mining, according to Müller 'the actual goal' of their mission.[90] Unfortunately, the results of their research were rather disappointing. Müller writes:

---

84 Res.GG, December 20, 1827, no. 2. NL-HaNA, Koloniën, 1814–1849, 2.10.01, inv.no. 2511.

85 The *Willem I* had arrived on February 3, 1836. See https://www.marhisdata.nl/schip?id=8384.

86 Res.GG, February 28, 1836, no. 34. NL-HaNA, Koloniën, 1814–1849, 2.10.01, inv.no. 2561.

87 Res.GG, December 24, 1827, no. 29. NL-HaNA, Koloniën, 1814–1849, 2.10.01, inv.no. 2808.

88 Res.GG, January 29, 1828, no. 5. NL-HaNA, Koloniën, 1814–1849, 2.10.01, inv.no. 2809.

89 'Een natuurkundig, en meer bijzonder een geologisch onderzoek te doen [...] naar de aanwijzing van de Resident.' Res.GG, February 5, 1828, no. 1. NL-HaNA, Koloniën, 1814–1849, 2.10.01, inv.no. 2810.

90 'Het eigenlijke doel.' Susanna, *Verhandelingen*, vol. 3, 132.

> However, actual mines are nowhere to be found; these metals [gold and copper] are largely procured through panning it from the sand of the rivers and in other places from sand-like or clay-like soil.[91]

Likewise, Horner was sent to Sumatra in order to inspect deposits of gold, platinum, iron and especially coal and evaluate the commercial exploitation options of these deposits.[92] From June 1837 to his death in September 1838, Horner dedicated himself to this task. The presence of Overdijk and four indigenous hunters safeguarded the creation of collections, but the itinerary was now dictated by where these deposits were located.

Shortly after the Commissie had been dissolved, the king set up a new arrangement for mining operations in the Indies and specialised mining engineers arrived from the Netherlands.[93] These engineers were organised in an official government body (Mijnwezen) that was set up to deal with mining in the Indies.[94]

Some of the pioneering work for the coal mining industry lay in the hands of the Natuurkundige Commissie. As we have seen, Macklot and Horner had occupied themselves with inspecting coal deposits on Java, Borneo, and Sumatra. When Horner died in 1838, he was replaced by Commissie member Schwaner who arrived in the Indies in August 1842. Already on September 8, 1842, he was asked by governor-general ad interim Pieter Merkus to suggest a plan for where and how he intended to continue Horner's geological research.[95] When Schwaner received his instructions from Merkus on May 30, 1843, the first article reads:

> What Mr. Schwaner shall consider as the main goal of his investigations is the research regarding the existence and the nature of such coal deposits in the Netherlands Indies, the exploitation of which can be useful to the service of steam navigation or industry and trade in general.[96]

---

91 'Eigenlijke mijnen, echter, vindt men er nergens; die metalen worden grootendeels uit het zand der rivieren en op andere plaatsen, uit eenen zandigen of kleiachtigen grond, door middel van uitwassching verkregen.' Susanna, *Verhandelingen*, vol. 3, 306.

92 Res.GG, March 31, 1837, no. 7. NL-HaNA, Koloniën, 1814–1849, 2.10.01, inv.no. 2568; Res.GG, May 3, 1837, no. 1. NL-HaNA, Koloniën, 1814–1849, 2.10.01, inv.no. 2570.

93 Paulus, *Encyclopaedie van Nederlandsch-Indië*, vol. 4, 838.

94 See De Ruiter, "Het Mijnwezen in Nederlands-Oost-Indië." De Ruiter's evaluation of the geological work of the Natuurkundige Commissie is cursory and leaves much to be desired. A more extensive, but still cursory treatment of the geological work of the Commissie can be found in volume 1 (Scientific Explorers) of Van Gorsel, *Pioneers and Milestones*.

95 Res.GG, September 8, 1842, no. 1a. NL-HaNA, Koloniën, 1814–1849, 2.10.01, inv.no. 2637.

96 'De Heer Schwaner zal zich als het hoofddoel van zijne navorschingen voorstellen het onderzoek omtrent het bestaan en de geaardheid van zoodanige steenkolengroeven

The second article tells him to start on Borneo, while article 3 outlines all the details that are required to judge whether a deposit can be exploited. Article 4 adds gold and platinum to the list of mineral deposits of interest, while article 5 invites Schwaner to conduct an ethnographical and statistical survey and make note of local customs and commerce. It is only in article 6 and 7 that Schwaner is allowed to collect specimens, but then only those that are not yet known, or of sufficient rarity to warrant shipping them to Europe. Instructions regarding his notetaking and reports can be found in articles 8 and 9. Article 10 notes that all Schwaner's collections and notes are government property, while article 11 tells Schwaner to evaluate his work with the governor-general after three years to see if and where his research should be continued.

Schwaner was accompanied on his expedition to Borneo by Forsten's former assistant Von Gaffron and Belgian mining engineer Lambert Pierre Comblen. Schwaner and Von Gaffron were seen to have the theoretical and geological knowledge that was required to prospect for new coal deposits, while Comblen was expected to have all the practical knowledge required to set up mining experiments.[97] As such, Comblen was the first participant in the Natuurkundige Commissie who did not have any natural history background or knowledge. Schwaner's proposal to open the first coal mine on Borneo was supported by military engineer general-major Carel van der Wijck (1797–1852) and on October 26, 1845, the opening of the new mine was approved by governor-general Jan Jacob Rochussen.[98] In 1846, both Von Gaffron and Comblen left the Natuurkundige Commissie. Comblen was appointed as supervisor of all coal mining activities on Borneo,[99] while Von Gaffron continued exploring Borneo, but now in the service of the governor of Borneo.[100] Schwaner continued to be involved in coal exploration, but also was able to undertake a major and famous expedition through the interior of Borneo from November 1847 to February 1848 that made his name as the 'scientific discoverer of Borneo.'[101]

All in all, nearly all expeditions of the Natuurkundige Commissie, especially since the arrival of the first mineralogist of the Commissie in the person of Macklot in 1826, contained some aspect of geological research into mineral

---

in Nederlandsch Indië, wier ontginning voor de dienst der stoomvaart of de nijverheid en den handel in het algemeen nuttig kan zijn.' Res.GG, May 30, 1843, no. 10. NL-HaNA, Koloniën, 1814–1849, 2.10.01, inv.no. 2645.

97 Res.GG, August 2, 1843, no. 1. NL-HaNA, Koloniën, 1814–1849, 2.10.01, inv.no. 2648.

98 Res.GG, October 26, 1845, no. 3. NL-HaNA, Koloniën, 1814–1849, 2.10.01, inv.no. 2674.

99 Res.GG, February 22, 1846, no. 2. NL-HaNA, Koloniën, 1814–1849, 2.10.01, inv.no. 2679.

100 Res.GG, May 19, 1846, no. 6. NL-HaNA, Koloniën, 1814–1849, 2.10.01, inv.no. 2682.

101 Posewitz, *Borneo*, 18.

deposits. For a long time, this was just one of the aspects of the natural historical mission of the Natuurkundige Commissie. It was understood that geology and mineralogy were part of natural history, and such mineralogical research could serve both the purposes of 's Rijks Museum and those of the Indies government. Forsten collected samples of the different ways in which gold was found for 's Rijks Museum, but with equal pleasure inspected the general state of the mining operations for the government in Batavia. Many of the governors-general saw the Natuurkundige Commissie as a vehicle that could be used to assist in gathering information on the still relatively unknown islands of Sumatra, Borneo, and the Outer Islands, either in the form of maps, ethnographical and statistical information, or inspections of mineralogical deposits.

As long as the main instructions for the members continued to come from the Netherlands, collecting objects of natural history was the first and foremost task. The first instructions issued from the Indies by a governor-general (those to Schwaner by Merkus) immediately placed mineralogical exploration for commercial purposes and ethnographical and statistical research above collecting specimens. The long arm of 's Rijks Museum had lost its grip on the Natuurkundige Commissie and the governors-general were able to get a firm handle on the Natuurkundige Commissie without having to fear from ministerial interference. Along with the publication of the *Verhandelingen* (which accomplished most of what the Dutch government expected from the Commissie), this shifting balance to the Indies would prove to be another nail in the coffin of the Commissie.

5 Layered Loyalties in the Field

All members in the field experienced some form of loyalty to each of the three stakeholders (the Dutch government, 's Rijks Museum, and the Indies government). As I have shown at the beginning of chapter 2, all members had a strong affinity for natural history. For many, a collecting expedition had been a dream of adventure and was supposed to have functioned as a steppingstone for a future career in natural history. After their return, both Müller and Korthals continued to write based on their collecting career for years, with Korthals retiring in 1842 and Müller being retired in 1850 after the dissolution of the Commissie.[102] Müller's career with the Commissie lasted from 1823 to 1850, and he continued to publish after 1850, while Korthals' career with the Commissie

102 KB, February 11, 1843, no. 65. NL-HaNA, KdK 1841–1897, 2.02.04, inv.no. 148; KB, July 19, 1850, no. 10. NL-HaNA, KdK 1841–1897, 2.02.04, inv.no. 624.

lasted from 1830 to 1842. Both rose from amateur preparator (Müller) and apothecary assistant (Korthals) to university doctor, a career trajectory that was not a given in the first half of the nineteenth century. The fact that a person like Carl Ludwig Blume, who started out as assistant to Reinwardt, managed to have such a distinguished academic career,[103] created the expectations that at least some of the members, most of whom started with better academic credentials than either Müller, Korthals or Blume, could have become eminent naturalists in the Netherlands.

With a continuing eye on their own career prospects, all the members in the field had to negotiate between the three stakeholders, and they were all too well aware of the diplomacy and manoeuvring that this required, as is made obvious by the following example from Forsten's correspondence. In his letters, on occasion, he discussed his finances. During his travels, Forsten received a salary of f500 per month, plus a travel allowance of f150. In addition to this, he received a monthly allowance of f350 for collecting costs,[104] a total of f1,000 per month. Forsten confides in Susanna that he does not need all of it, 'even though I believe it to be politically correct to complain like the others. I continually keep complaining that it takes effort to make ends meet without getting into debt.'[105] Four months later, Forsten writes to Temminck: 'Concerning my financial circumstances, I have to candidly confess to your Honour that I have difficulties making ends meet because all required supplies are enormously expensive.' Forsten lists the prices he has to pay for arak, butter, wine, gin, transport and adds: 'and on top of this I need to provide food for myself and my people.'[106] Schlegel also receives the complaints as Forsten exclaims to him full of pathos: 'Add to this the salaries of my hunters of 97 guilders and you will understand how rich I will be when I return to Leiden.'[107] Forsten repeats his complaints to the governor-general in his report from June 1, 1842. He writes:

103 Van Steenis, "Dedication"; Den Ouden, "C. L. Blume." The absence of a complete biography of Blume's life and work is a glaring omission and shows how understudied the history of science in the Netherlands Indies in the early nineteenth century actually is.

104 These numbers are communicated by Forsten to Susanna (Letter Forsten to Susanna, January 26, 1841. NCAO, 1230_004) and repeated to Schlegel (Letter Forsten to Schlegel, May 20, 1841. NCAO, 1232_002). Forsten used part of the allowance to pay for the salaries of his local helpers.

105 'Hoewel ik het voor politiek houde om even als de anderen altijd te klagen, zoo klaag ik steeds dat ik werk heb om rond te komen en geen schulden te maken.' Letter Forsten to Susanna, January 26, 1841. NCAO, 1230_004.

106 'En dan moet ik buitendien nog voor het voedsel van mij en mijn volk zorgen.' Letter Forsten to Temminck, May 20, 1841. NCAO, 1231_003.

107 'Daarbij het traktement der jagers f 97 dan kunt u begrijpen hoe rijk ik in Leiden terug zal komen.' Letter Forsten to Schlegel, May 20, 1841. NCAO, 1232_002.

> Please allow me to present to your Excellency in the interest of my mission and of its regular completion several financial statements, from which it will become clear to your Excellency, that, through the circumstances I have been placed in, outside of my doing and with the most modest observance of our national financial interests, the sum of 350 guilders per month, allocated to me according to the existing provisions for reimbursement of the costs of my mission, has not in the least been adequate for that purpose.[108]

In the remainder of the report Forsten carefully explains why these costs were necessary and could not be avoided and in the final paragraph asks for reimbursement. While Forsten did not experience any financial difficulties in the field, his diplomatic stance was to always complain. Only his friend Susanna got to hear the real story, while Temminck, Schlegel and the governor-general were given the picture that Forsten believed he should project in order to best serve his own interests.

6 The Shifting Balance from the Netherland to the Indies

All in all, the members were not fully autonomous in the field. They were beholden to their collecting mission from 's Rijks Museum, but this usually was not a problem since it was closely aligned with their personal interests and was expected to enhance their career prospects. In the first two decades, the government in the Netherlands wrote the basic instructions for the members, allowing the governors-general little leeway in the big picture. Temminck was able, for example, to direct from the Netherlands through the Dutch government that Forsten had to go to Celebes. However, the Netherlands were far away, so once the members were in the Indies, the governors-general were supposed to direct their work in the Indies. The year of Forsten's delay shows

108 'Het zij mij vergund uwen Excellentie in het belang mijner zending en van het regelmatig volbrengen van dezelve, bezijde dezes eenige staten aan te beiden, waar uit het uwen Excellentie zal blijken, dat door de omstandigheden waarin in geplaatst ben geworden, geheel zonder mijn toedoen en met de schroomvalligste behartiging van 's lands geldelijke belangens, de som van f 350 's maands die mij volgens de bestaande bepalingen tot goedmaking der kosten van mijne kommissie is toegerekend, op verre na, daartoe niet toereikende is geweest.' Letter Forsten to the governor-general, June 1, 1842. NL-HaNA, Van Beresteyn, 2.21.018, inv.no. 1523. Forsten was personally acquainted with the then interim governor-general Pieter Merkus (see letter Forsten to Susanna, May 20, 1841. NCAO, 1231_001).

that strife between the Dutch government and, in his case, governor-general De Eerens could incapacitate the members. The animosity between De Eerens and the Dutch government centred on finances, and on who had the authority to award salary raises and appoint new people.

In the final decade, the balance of power visibly shifted from the Netherlands to the Indies. Temminck lost his grip on the Commissie. Already in 1830, Korthals had been given tasks that Temminck did not consider appropriate in the context of a Natuurkundige Commissie that was on a collecting mission for 's Rijks Museum. And in 1838, Forsten was the last member whose instructions were wholly commensurate with Temminck's idea of the Natuurkundige Commissie. Forsten was also the last member to receive his instructions directly from the Netherlands. It seems that De Eerens' struggle for control had paid off and created a higher degree of authority for the governor-general, proven by the fact that the instructions for Schwaner and Junghuhn were born and written in the Indies by the governors-general (Merkus for Schwaner, and Rochussen for Junghuhn).

Another illustration of the fact that Temminck had lost control of the Commissie is the story of the Sumatran elephant. In Europe there were discussions on whether the Sumatran elephant was a separate species or not. Temminck was involved in the discussion and wanted to settle the matter. However, there was no elephant specimen from Sumatra available at 's Rijks Museum, so he needed to source one. In previous years, Temminck could have written to his collectors of the Natuurkundige Commissie and ask them to procure a specimen. After all, many members had spent time on Sumatra. However, this matter only arose in 1843/1844, when no zoologists were available in the field anymore (Forsten had died in January 1843). The Natuurkundige Commissie was not going to be able to honour a request for an elephant from Sumatra from Temminck. There was no one on Sumatra anymore and the king had expressly decreed in 1842 that Schwaner was going to be the last appointment, so there was no hope of a new member visiting Sumatra.

Temminck wrote a letter to the minister of colonial affairs, Baud, on February 9, 1844, explaining about the debate and asking whether a specimen could be procured, with specific instructions. Baud immediately sent the letter to governor-general Merkus (on February 16, 1844), who forwarded the request to all four residents on Sumatra on June 30, 1844, asking them to comply with Temminck's request for a complete skeleton of an adult elephant from Sumatra, or a skull with the tusks.[109] Two of the four residents acted upon the request

109 Res.GG, June 30, 1844, no. 11. NL-HaNA, Koloniën, 1814–1849, 2.10.01, inv.no. 2658.

FIGURE 32 'Wachters in de vestibule' ('Guards in the hall'). The entry of the Rijksmuseum van Natuurlijke Historie in the Raamsteeg (1913–1990) was guarded by four elephant skeletons, two on each side. Among these were specimens of *Elephas maximus sumatranus*, Temminck, 1847. Public domain. On October 12, 2024, Topaz Gigapixel ai was used to increase the resolution of this image. HTTPS://WWW.ERFGOEDLEIDEN.NL/COLLECTIES/BEELDMATERIAAL/ZOEKEN-IN-BEELDMATERIAAL/DETAIL/B32854F8-26BC-11E3-BA65-3CD92BEFE4F8/MEDIA/54D6B819-BC3A-7132-5595-632B25B7DD17

and two elephants were caught in Lampong and one in Palembang. The one in Palembang had been caught by hunters who had been specially commissioned after having received the request from Temminck. All three specimens were shipped to the Netherlands and the other residents were informed that an elephant was no longer required (Figure 32).[110]

A year later, Temminck complained that none of the three elephant skeletons he had received was complete and requested a complete skin and a complete skeleton again. This time, the request was forwarded to the officers of the military health service on Sumatra.[111] Whether a specimen was procured this time, needs to be ascertained, but Temminck successfully published the Sumatran elephant (*Elephas maximus sumatranus*) in 1847 as a subspecies

110 Res.GG, January 22, 1845, no. 1. NL-HaNA, Koloniën, 1814–1849, 2.10.01, inv.no. 2665.

111 Res.GG, February 25, 1846, no. 12. NL-HaNA, Koloniën, 1814–1849, 2.10.01, inv.no. 2679.

of the Asian elephant (*Elephas maximus*).[112] This episode shows that the Natuurkundige Commissie was not able to function according to its original mission anymore and Temminck had to find other ways to fulfil the needs of 's Rijks Museum. The fact that this actually worked undoubtedly contributed to the lack of resistance from Temminck's side when the Commissie was finally dissolved.

The balance had decisively shifted to the governor-general and the loyalty of the members of the Natuurkundige Commissie to the original goals of 's Rijks Museum had shifted. A significant moment can also be discerned in the appointment of Junghuhn. In 1839, during the struggle between De Eerens and the ministry, the governor-general had lost out in the case of Junghuhn, whose provisional appointment had been discontinued. In the years after losing his position with the Commissie, Junghuhn had been charged by Merkus to undertake a mission to the Batak lands (Sumatra), much of the same nature as Schwaner's mission to Borneo. The results of this mission and Junghuhn's work on Java after this established his reputation as an able and useful explorer, which led to his eventual appointment as a member in 1845.

When Junghuhn, after his appointment, continued with his work as before, and asked the government for travel compensation, he was reminded by governor-general Rochussen that travel of the members of the Natuurkundige Commissie would only be paid for by the government when it concerned an officially sanctioned expedition. Since he had not been sent on an expedition, he did not receive the requested travel compensation. Junghuhn was invited to send in a proposal for an expedition.[113] When his proposal was translated into instructions, the first two articles of Junghuhn's instructions read as the final triumph of the governor-general over 's Rijks Museum.

> Article 1: Mr. Junghuhn will consider as the main goal of his activities a natural historical geographical investigation of the Dutch possessions in the Indies Archipelago, but he will especially devote himself to the knowledge of the geography, topography, geognosy, mineralogy and meteorology of these realms.
>
> Article 2: In addition, he will pay attention to the fact that, whatever fruits can be gained from his investigations for science in general, his work must in the first place be of service to the government.[114]

---

112 Temminck, *Coup-d'oeil général*, vol. 2, 91–92.

113 Res.GG, July 18, 1845, no. 25. NL-HaNA, Koloniën, 1814–1849, 2.10.01, inv.no. 2671.

114 'Artikel 1: De Heer Junghuhn zal als hoofdoel zijner werkzaamheden beschouwen een natuurkundig geographisch onderzoek der Nederlandsch bezittingen in den Indischen

As opposed to what we saw in Forsten's instructions, collecting specimens was not even part of the main goal of Junghuhn's expedition anymore. In article 4, he is still allowed to collect, but it is again made subservient to a utilitarian goal. While in his earlier appointment in 1838 he was reminded that all his collections were government property, this is not explicitly repeated in his instructions here, laying a seed for the conflict that would later erupt in the Netherlands between Blume and Junghuhn regarding the ownership of his collections. The last two members of the Natuurkundige Commissie, Schwaner and Junghuhn, were directed towards a purpose that would serve the needs of the Indies government. 's Rijks Museum had become an afterthought. And because the Natuurkundige Commissie never had been the only way in which expeditions had taken place,[115] the survival of the Commissie was not required for the further exploration of the archipelago.

7 Conclusion

All the members of the Commissie had to deal with the reality of a geographical dichotomy. Even though they had been appointed to serve in the Indies under the orders of the governor-general, the main mission of the Natuurkundige Commissie long remained the same, not in the least due to the fact that the members had a strong loyalty to this mission through their own intrinsic motivation and their personal loyalty to Temminck and 's Rijks Museum. The government in the Netherlands long supported Temminck's vision, but first introduced a utilitarian element in 1830 with Korthals' instructions to evaluate the work of the Hoofdkommissie van Landbouw. The power struggle between De Eerens and the ministry in 1839 showed that the long arm of 's Rijks Museum was gradually losing its grip and that the governors-general started to assert themselves and intended to make more use of the Natuurkundige Commissie for purposes of their own.

De Eerens introduced mapmaking and mining inspections into Forsten's expedition plans for Celebes in 1840, much along the lines of what

Archipel, maar bepaaldelijk zal hij zich toeleggen op de kennis der geographie, topographie, geognosie, mineralogie en meteorologie dezer gewesten.

Artikel 2: Hij zal daarbij vooral letten, dat, welke vruchten de wetenschap in het algemeen ook van zijne onderzoekingen moge trekken, zijn arbeid in de eerste plaats dienstig moet zijn voor het Gouvernement.' Res.GG, April 5, 1846, no. 27. NL-HaNA, Koloniën, 1814–1849, 2.10.01, inv.no. 2681.

115 We only have to remind ourselves of the Borneo expeditions by Diard and Henrici, and the Sumatra expedition by Junghuhn, to see the truth of this.

previous members had already been doing from 1827 onwards. While this was a respected part of expeditionary science, the balance gradually shifted from a purely natural historical point of view (as advocated by Temminck) to a much more utilitarian point of view (as advocated by the governors-general). Forsten was the last man standing from the point of view of 's Rijks Museum. Subsequent members Schwaner and Junghuhn were almost completely out of touch with Temminck (Junghuhn much more so than Schwaner) and the governors-general had gained complete control over the practical direction of the Natuurkundige Commissie. They had become the dominating object of loyalty for Schwaner and Junghuhn.

In the meantime, the publication of the *Verhandelingen* had taken the wind out of the sails of the Commissie in the Netherlands. Now that the long-awaited publication that would glorify the Dutch empire had been published (1839–1847), and the royal patronage of king Willem I had ended when he abdicated in 1840, there was no pressure from the king anymore to make the Natuurkundige Commissie a success and there was no political purpose in the Netherlands left for the Commissie to serve. The dwindling of royal interest could not be kept at bay by Baud, even though the combined persuasion power of Baud and the triumvirate procured one final appointment in 1845 (Junghuhn). When it turned out that Junghuhn had no loyalty whatsoever to 's Rijks Museum, there was nothing to stop the downward spiral for Temminck.

Because the Dutch government had lost interest and Temminck had lost his grip, the balance shifted to the Netherlands Indies where governors-general Merkus and Rochussen used the Commissie for practical goals, as previous governors-general had already been doing. And because these goals could be met (and had been met previously) through other means than the Natuurkundige Commissie, there was no pressure from the Netherlands Indies to the Netherlands to continue to operate the Natuurkundige Commissie.

The fact that the expeditions of the Natuurkundige Commissie were so successful in areas other than natural history collecting was also its downfall. The production of maps, the inspections of mineral deposits and mining, the provision of information on trade routes and the customs of the local population inspired the governors-general to strengthen this aspect of the expeditions of the Natuurkundige Commissie to such an extent that the collecting mission was forced to the background. Everything together resulted in the Commissie becoming surplus to requirements and its dissolution in 1850 was uncontested by all main stakeholders.

PART 2

Pierre-Médard Diard

CHAPTER 4

A Transimperial Naturalist: Pierre-Médard Diard between Empires

There are three Pierre-Médard Diards. The first one is the intrepid young traveller, 'courant après toutes les bêtes de la creation,'[1] who braved the tropics, and despite being thwarted by the double-crossing British managed to send valuable collections and notes to the Muséum national d'Histoire naturelle [National Museum of Natural History] in Paris, including the first scientific description of the Malayan tapir. The second one is the unreliable Frenchman in British service who did not keep his word when he had promised Raffles of Singapore that he would collect for him, and, among other 'crimes,' used his description of the living tapir from Raffles' menagerie to gain scientific fame at the expense of its original British discoverer. The third one is the invisible famous naturalist in service of the Dutch, whose 'deeds as president of the Natuurkundige Commissie, etc., seem to have been forgotten to such an extent that we have not been able to collect any material in order to provide an account of his life and deeds.'[2] According to the various accounts of his life and adventures, these three Diards barely resemble each other. French and British Diard are diametrically opposed, while Dutch Diard is almost an afterthought and entirely disconnected from French and British Diard. However, they are indeed one and the same person.[3]

In the previous chapters we briefly encountered Diard as one of only two members of the Natuurkundige Commissie being appointed in the Netherlands Indies rather than in the Netherlands. As we have seen, the Commissie existed of Dutch nationals and a substantial contingent of 'empireless' adventurous German-speaking Europeans. In addition to this, Diard was not just any ordinary member of the Commissie, but served as its directing member from 1830 to 1841. In total, Diard served more than thirteen years (1828–1841) on the

1 'Running after all the beasts of creation.' Diard, "Pierre-Médard Diard," 200.

2 'Hetgeen hij vroeger als president der natuurkundige commissie enz. heeft verrigt, schijnt zoodanig in 't vergeetboek geraakt te zijn, dat wij tot heden niet geslaagd zijn, eenig materiaal te verzamelen tot het opstellen eener schets van zijn leven en verrigtingen.' *Nieuwe Rotterdamsche Courant*, no. 102 (14 April 1863): 1.

3 There actually is a Pierre Diard that cannot be identified with Pierre-Médard. This Pierre Diard (1784–1849) was a French botanist. For a biography, see Gentil, "Notice sur Pierre Diard." For his sole posthumously published monograph, see Diard, *Catalogue raisonné des plantes*.

 | DOI:10.1163/9789004696877_006

Commissie in the Indies, longer than any other European member.[4] While he was relatively obscure in the Netherlands, he played an important role in the development of the Commissie during his time as directing member.

Since Diard had not been appointed by either of the Dutch stakeholders, he naturally occupied a different position in the landscape of layered loyalties. He shared a personal affinity for and motivation towards natural history with the other members, but did not have a strong connection to 's Rijks Museum. In fact, he did not even personally know Temminck before he became a member. He was appointed by the Indies government and his appointment as a member was only retrospectively ratified in the Netherlands, which means that he also had not been dependent on the Dutch government for his appointment.

In addition to this, Diard was not a young naturalist without any experience who depended on the Commissie as a career-propelling vehicle. When he joined the Natuurkundige Commissie, he was already a seasoned traveller and has served several masters as a *naturaliste-voyageur* in two other colonial empires. His promotion to directing membership of the Natuurkundige Commissie was not a planned career move, but his experience and his position made him the perfect candidate for the job.

1 Transimperial History in the Netherlands Indies

Given the fact that Dutch imperial studies are still a 'prospective field,'[5] it is no wonder that 'in the Netherlands, scholarship on the extraordinary presence of foreign Europeans in the Dutch Empire has remained scarce.'[6] We have to look elsewhere, for example in Switzerland. There, Bernhard Schär is one of the few scholars to have done serious work in this regard. Schär has taken the careers of Swiss naturalist cousins Paul Sarasin (1856–1929) and Fritz Sarasin (1859–1942) as a departure point and has shown how their research on Ceylon in the British empire was very much part of a much larger transimperial scientific landscape and how data gathered in the British empire crossed the boundary of empire and was incorporated in scientific theories in Switzerland and Germany.[7] As 'empireless' naturalists, the Sarasin cousins were able to cross imperial

4 It must be noted that Müller served on the Commissie for 27 years (1823–1850), but spent only 11 of those in the Indies (1826–1837).

5 Raben, "New Dutch Imperial History?"

6 Schär, "Switzerland, Borneo and the Dutch Indies." For an overview of Swiss colonial endeavours and actors, see Zangger, *Koloniale Schweiz*.

7 Schär, "From Batticaloa via Basel to Berlin." See Von Brescius, *German Science*, for the parallel case of the Schlagintweit brothers in the British empire.

boundaries in Asia and not only went on expeditions on Ceylon in the British empire, but also in the Netherlands Indies where they spent several years on Celebes.[8]

Just as the Sarasin cousins, Diard moved between different empires, while being a foreigner in two of them. In the span of only six years, he repeatedly crossed the geographical boundaries between the French, British, and Dutch empires in Southeast Asia. As a person originating from one of the classic imperial nations, his presence in the other empires was not always easily accepted and he was not always free to move when and where he liked. In contrast with Paul and Fritz Sarasin, Diard was not the one to take back his collections to Europe himself and gradually introduce them to the networks of naturalists that existed there. The ownership of the collections that he gathered were almost always a source of contention between the different imperial realms.

Diard was a mobile actor who moved from empire to empire and met with resistance and created controversy wherever he went. As such a mobile subject, Diard can inform us on the cooperation, competition and connectivity that existed between the British, French, and Dutch empires in the realm of natural history. Daniel Hedinger and Nadin Heé advocate a 'transimperial approach [that] seeks to discuss imperial competition, cooperation and connectivity not as separate phenomena but as entangled processes.'[9] In their view,

> transimperial history is about the movements of people, knowledge and goods across empires and about the formations of imperial alliances as well as anti-imperial networks and exchanges. Thus, transimperial history can be applied to both the macro and the micro level, to state or individual actors, from top down and bottom up.[10]

Diard is one of those individual actors who on the micro level moves over imperial frontiers (hence the *trans*imperial), and is one of those 'actors who feel at home in multiple imperial settings.'[11] Combining these tools with the concept of layered loyalties, will help us gain a better understanding of the transimperial aspects of natural history in this specific context. But before this can be done, a look at current research on Diard and a narrative of the transimperial adventures of the three Diards is in order.

8 Schär, *Tropenliebe*, and Schär, "Earth Scientists as Time Travelers."
9 Hedinger and Hée, "Transimperial History," 430.
10 Hedinger and Hée, "Transimperial History," 439.
11 Hedinger and Hée, "Transimperial History," 432.

2 The State of Research on Diard

After decades of silence, Diard has recently received some academic attention again. Danièle Weiler has done some work on Diard and his travelling partner Alfred Duvaucel (1793–1824) in the context of international relations between France and Singapore. Her most recent effort was published at the 200th anniversary of the founding of Singapore.[12] Logically, her work concentrates on Diard's presence in and around Singapore (an area he visited twice in 1819) and only provides in-depth coverage of the first couple of years of his colonial career. During the vOilah! France Singapore Festival in 2019, Duvaucel and Diard were prominently featured in a public media presentation by contemporary artist Julien Nonnon.[13] The National Museum of Singapore included a drawing by Duvaucel and a letter by Diard, both on loan from the Muséum national d'Histoire naturelle in Paris, in the 2019–2020 temporary exhibition *An Old New World: From the East Indies to the Founding of Singapore, 1600s–1819*.[14] In 2021 the Embassy of France in Singapore and the National Library Board of Singapore jointly published *Diard & Duvaucel: French Natural History Drawings of Singapore and Southeast Asia, 1818–1820*. This bilingual publication (English and French) features reproductions of the 120 natural historical drawings that were sent back to the Muséum in Paris by Diard and Duvaucel. This set includes what may be the earliest known drawing of an animal from Singapore (a spiny turtle, *Heosemys spinosa*, Gray 1831), making it a landmark collection for the history of natural history in Singapore.[15]

However, these publications focus on a period in Diard's life of not even two years. It has been a long time since the whole Diard was put in the limelight of science. During his lifetime, a report on Diard's activities was published by Noel Champoiseau, a member of the Société d'Agriculture, Sciences, Arts et

12 Pilon and Weiler, *French in Singapore*; Weiler, "Pierre Médard Diard and Alfred Duvaucel."

13 Leslie Colin, 'Hommage aux naturalistes Diard en Duvaucel', https://lepetitjournal.com/singapour/hommage-aux-naturalistes-diard-et-duvaucel-270082. See Oury, "Diard, Duvaucel and the Muséum National d'Histoire Naturelle," 33.

14 https://www.nhb.gov.sg/nationalmuseum/-/media/nms2017/documents/media-releases/media-release-national-museum-of-singapore-presents-an-old-new-world-from-the-east-indies-to-the-founding-of-singapore-1600s-1819.pdf. The virtual version of this exhibition can be found here: https://www.nhb.gov.sg/nationalmuseum/virtual-gallery/virtual-exhibitions/an-old-new-world.

15 The original drawings are kept at the Muséum national d'histoire naturelle: https://bibliotheques.mnhn.fr/medias/doc/EXPLOITATION/IFD/MNHN_MS604_2/figures-peintes-d-oiseaux-et-de-reptiles-envoyees-de-l-inde-par-duvaucel-et-diard, and are available online in open access at the website of the National Library Board (Singapore): https://www.nlb.gov.sg/main/book-detail?cmsuuid=c908d4fd-4e4b-441d-bdab-48c9eb69d0c1.

Belles-Lettres of the Department of Indre-et-Loire, Diard's home region.[16] Fifty years later, Charles-François-Hippolyte Diard (1796–1877), published a biographical account of his older brother for the same Société.[17] Both these publications were the main sources for the most extensive treatment of Diard's career that even included a short section on Diard's work in the Netherlands Indies. This was published in 1935 by Jean-Henri-Eugène Peyssonnaux (1888–1937), a French civil servant in colonial Vietnam.[18] Peyssonnaux's account relies heavily on the work of Champoiseau and Hippolyte Diard. In addition, he had access to archival material held by the family, but the whereabouts of this material is, unfortunately, not known to me.

In contrast, the British have only ever felt the need to treat Diard in context with Raffles. Dutch treatment has been even scantier and only a few biographical entries have been written.[19] In fact, a clear timeline of his work as a civil servant in the Indies is completely absent.[20] The most extensive research in this regard was done by Marie-Odette Scalliet, but she hid it well in a footnote of the text edition of the diary of artist Antoine Payen (1792–1853).[21]

3 Diard in Three Empires

It is not my intention to write an exhaustive biographical account of Diard's life and work. Nevertheless, it is necessary to recount Diard's early life in order to understand the development of his transimperial adventures. Pierre-Médard Diard was born on March 19, 1794, in Château de la Brosse (or de Labrosse) in Chenusson, canton Châteaurenault in the Indre-et-Loire Department in France. Already in his youth, like many members of the Natuurkundige

16 Champoiseau, "Voyages aux Indes Orientales."

17 Diard, "Pierre-Médard Diard." An obituary for Hippolyte Diard was published in 1878: Chevalier, "Compte-rendu annuel du secrétaire perpétuel."

18 Peyssonnaux, "Vie, voyages et travaux." Other French publications on Diard include Cuvier, "Notice sur les voyages"; Brébion, "Diard, naturaliste français"; Claeys, "Deux Français," and Oury, "Diard, Duvaucel and the Muséum National d'Histoire Naturelle."

19 The two most extensive are Van Steenis-Kruseman, "Malaysian Plant Collectors," 136, and Fransen, Holthuis, and Adema, "Type-catalogue of the Decapod Crustacea," 227–229.

20 Veth, *Overzicht van hetgeen gedaan is*, devotes ample time to Diard's involvement with the Natuurkundige Commissie, but disregards most of Diard's agronomic career outside of the Commissie. In the other two early standard treatments of the Natuurkundige Commissie, Sirks, *Indisch natuuronderzoek*, and Gijzen, *'s Rijks Museum van Natuurlijke Historie*, Diard features even less prominently and only in his capacity as member of the Natuurkundige Commissie.

21 Scalliet, *Antoine Payen*, 668-670n409.

FIGURE 33 Portrait of Pierre-Médard Diard in the uniform of the honour guard, where he served in 1813 and 1814. Plate 2 (after page 18) from Peyssonnaux, "Vie, voyages et travaux." The original plate is reproduced from a lithograph that was in the possession of Paul Viot from Amboise. Current whereabouts unknown. Public domain

Commissie, Diard showed a strong affinity for natural history and exploration and took up the study of medicine in order to enhance his chances of being able to travel.[22] His studies were interrupted when he enlisted in the army in 1813 in the honour guard, where he served in 1813 and 1814 (Figure 33).[23] After the defeat of Napoleon, Diard resumed his medical studies in Paris.

22 Diard, "Pierre-Médard Diard," 198.

23 Champoiseau, "Voyages aux Indes Orientales." Champoiseau suggests Diard jumped at the opportunity to defend his country (Champoiseau, "Voyages aux Indes Orientales,"

Through a local doctor from Châteaurenault, a certain Dutrochet, Diard was introduced to Georges Cuvier (1769–1832), who took the promising and ambitious young student under his wings. In 1815, Diard assisted in the then famous, now notorious dissection of Sarah Baartman, also known as the Hottentot Venus.[24] A year later, in 1816, Diard was set to accompany Louis de Freycinet (1779–1842) on the *Uranie*, but unnamed circumstances prevented him from joining the expedition.[25] His plans to travel the world did not diminish with this lost opportunity. In 1817, Diard was planning to depart for Bengal on the ship *Alouette*. He had already obtained the title of correspondent of the Muséum national d'Histoire naturelle in Paris and carried several letters of recommendation, including one by Cuvier. But on the eve of his departure, fortuitously, an elephant in the menagerie of the Muséum died. Diard stayed behind for the dissection and did not board the *Alouette*, which would perish in sight of the Cape with the entire crew. Finally, on August 20, 1817, Diard departed from Bordeaux with destination Calcutta.[26]

Diard spent most of his adult life in Southeast Asia (Map 9), from his arrival in Calcutta on January 5, 1818, to his death in Batavia on February 16, 1863, working, living and travelling in French India, British Malaysia and Sumatra, and what was to become French Indochina. He settled in the Netherlands Indies in 1824, where he would stay until 1841. On August 28, 1841, he left for Bordeaux and spent almost seven years away from the Indies, probably mostly in France. He returned to the Indies in early 1848 and stayed there until his death in 1863, except for a brief period from November 10–12, 1857,[27] to April 21–26, 1859,[28] in which he undertook an expedition to Mauritius, Bourbon (La Réunion) and Ceylon.

---

223), while Peyssonnaux puts Diard's enlistment in the context of the forced levy of 180,000 soldiers to compensate for the losses of the Russian campaign (Peyssonnaux, "Vie, voyages et travaux," 9–10).

24 Champoiseau suggests that Diard was the person who performed the actual dissection (Champoiseau, "Voyages aux Indes Orientales," 224). Both Qureshi, "Displaying Sara Baartman," and Crais and Scully, *Sara Baartman and the Hottentot Venus*, make it clear that Cuvier was in charge of the dissection and do not mention Diard's involvement. I believe it is possible that Diard, as one of a team of naturalist-assistants to Cuvier, did indeed take part in performing the actual dissection.

25 'Des obstacles imprévus et indépendants de sa volonté.' Champoiseau, "Voyages aux Indes Orientales," 224.

26 Champoiseau, "Voyages aux Indes Orientales," 224. For Diard's departure date, see Peyssonnaux, "Vie, voyages et travaux," 11.

27 *Java-Bode*, no. 91 (November 14, 1857): 8. The precise date of Diard's departure is not given.

28 *Bataviaasch Handelsblad*, no. 61 (April 27, 1859): 8. The precise date of Diard's arrival is not given.

MAP 9 Map of Farther India. Map 35 from Bartholemew, *The Century Atlas*. Public domain
HTTPS://CATALOG.HATHITRUST.ORG/RECORD/001286097

Near the end of his life, Diard developed a method of preparing specimens that perfectly preserved suppleness and lustre. Unfortunately, the method involved the lavish use of arsenic and Diard died as a result of arsenic poisoning on February 16, 1863. During his colonial career, Diard was a dedicated naturalist, collecting for the Muséum national d'Histoire naturelle in Paris, for the British East India Company, and for 's Rijks Museum van Natuurlijke Historie in Leiden. His itinerary from India, to Sumatra, Java, Vietnam and back to Java may not seem extraordinary, but it becomes much more interesting when one realises that Diard crisscrossed three different empires and one independent nation in the space of only six years. And Diard was not just a private traveller, but worked in service of the empires and nations he traversed. The complex political situation in South and Southeast Asia heavily influenced Diard's itinerary and serves as a backdrop for the narratives of French, British and Dutch Diard. A brief introduction to the political situation in the colonial world in South and Southeast Asia will certainly be of use to the reader.

After the Napoleonic wars, that formally ended in 1815, the colonial world was in flux. In Europe, international order was restored through the Congress of Vienna (1814–1815). However, the colonial world required a different solution. Even though Napoleon was far more ambitious in Europe than in the colonies, the Indian Ocean was still an arena of war and conflict. The conquest of Portugal and the establishment of the Batavian Republic in the Netherlands as an ally to France threatened to give France power of the colonial areas of the European imperial centres. The British took the opportunity to consolidate their power in India, and from there pre-emptively occupy the possessions of other colonial powers that had fallen into France's sphere of influence. In practice, this meant that places like Macau, Ceylon, Mauritius, and Java exchanged colonial masters during the Napoleonic wars. By 1812, the British controlled the entire Indian Ocean and France had no power there to speak of.[29]

After the defeat of Napoleon, some territories were restored to their original colonial powers, even though the British held on to strategic locations like Mauritius and Ceylon. The Dutch East Indies were returned to the Netherlands according to the Anglo-Dutch Treaty of 1814, and in 1815 the Treaty of Paris restored the five *comptoirs* (including Pondicherry and Chandernagor) in East India to the French. When Diard arrived in Asia in 1818, the French did not have an empire to speak of and in Asia actually only controlled the five *comptoirs* in India.[30] It was very much business as usual for the British who had had to

29 See Mikaberidze, *Napoleonic Wars*, 469–500, for a detailed history of South and Southeast Asia in the context of the Napoleonic wars.

30 Aldrich, *Greater France*, 23.

relinquish some of the conquered possessions, but were still without doubt the quintessential imperial power of the area.

The Dutch had only just started to re-stake their claims in the East Indies with the first post-Napoleonic governor-general Van der Capellen arriving in Batavia on Java in 1816.[31] The British and Dutch spheres of influence had not yet crystallised, and borders and demarcations between them were still somewhat fluid. Only in 1824 did a second Anglo-Dutch Treaty put an end to the uncertainty and create the border between the British and the Dutch that still serves as the border between Malaysia and Indonesia today.[32] It was in those uncertain years between 1818 and 1824 that Diard moved between the three imperial powers that were getting to grips with change and new beginnings.

4 French Diard 1 (1818–1820): Cheated by Raffles

Diard arrived on January 5, 1818, in Calcutta, the headquarters of the British East India Company. Already in February, he was welcomed at a meeting of the Asiatick Society, the first and foremost scientific society in the area. With the recommendations of his patron Cuvier in hand, he was also given free access to the local menagerie and botanical garden for research purposes. In May 1818, Diard was joined by his friend, compatriot and former study companion Alfred Duvaucel, stepson of Georges Cuvier.[33] Duvaucel had left France a couple of months after Diard. The young men decided to settle in the French *comptoir* of Chandernagor, fifty kilometres upriver from Calcutta. They rented a small house there that was turned into part museum, part menagerie, just because of the sheer amount of material in their collections, and they hired local personnel to assist in their collecting project.

On one of their rare visits to Calcutta (in November 1818) they made the acquaintance of Thomas Stamford Raffles, the newly appointed governor of Bengkulu, an area on the west coast of Sumatra.[34] Raffles had an affinity for natural history and was impressed by the work of the two young Frenchmen.[35]

31 See Jeurgens, "Op zoek naar betrouwbare informatie."

32 Locher-Scholten, *Sumatran Sultanate and Colonial State*, 51–52. See Tarling, *Anglo-Dutch Rivalry*, 133–177, for an extensive report on the Anglo-Dutch Treaties.

33 For an overview of Duvaucel's career, see Duvaucel, Jacquemont, and Petit, *Impressions de voyages en Inde*.

34 For more on Raffles, see Glendinning, *Raffles and the Golden Opportunity*, and Bastin, *Sir Stamford Raffles and Some of His Friends*.

35 For Raffles' patronage of the arts and sciences, see Bastin, "Sir Stamford Raffles and the Study of Natural History."

Subsequently, he invited them to work for the East India Company and accompany him on his journey to Bengkulu. On the way there they would receive ample opportunity to explore hitherto unexplored areas. Diard and Duvaucel did not have to think long. They abandoned their planned expedition in the Bengal area, and accepted Raffles' offer of employment. Peyssonnaux, ominously precedes his account of the Raffles' expedition with:

> Unfortunately, the two Frenchmen did not deal directly with Lord Raffles and left it to a member of the Supreme Council of the East India Company[36] to arrange their participation in the mission in the best of their interests. They would later regret having had faith in simple verbal promises.[37]

The expedition left Calcutta on December 7, 1818. They made a stop on the west coast of the Malaysian Peninsula, all the while collecting and exploring. What Diard and Duvaucel may not have been aware of is that it was not just a leisurely journey to Bengkulu with the occasional break for collecting trips. Raffles had received permission from Lord Hastings, the governor-general of the Calcutta Presidency, to find a suitable location in the area for a trading post, as long as he would not antagonise the Dutch. On January 28, 1819 Raffles' party landed in Singapore, where negotiations with the Sultan of Johore led to a treaty that was signed on February 6, 1819, which signified the founding of modern Singapore. Shortly after, the party departed again and took landfall in Aceh (Sumatra) where 'heroic' Diard during a collecting trip managed to escape with his life and three assistants, but without the fruits of his labour, from the midst of a 200-man strong throng of enraged locals. The expedition further visited Malacca and Singapore again to finally reach their destination of Bengkulu on July 31, 1819.

But trouble would rear its ugly head in the form of a letter from Raffles to Diard and Duvaucel in which he announced that he was going to send off their collections to London and if they would be so kind as to write a full catalogue

36 John Bastin (in Bastin, *Sir Stamford Raffles and Some of His Friends*, 276) identifies this member of the Supreme Council as Charles Milner Ricketts (1776–1867). For more on Ricketts, who retired from India due to health reasons in January 1819, see Fisher, "Ricketts, Charles Milner."

37 'Malheureusement les deux Français ne traitent pas directement avec Lord Raffles et laissent au membre du Conseil Suprême de la Compagnie des Indes, ami de Diard, le soin de régler, au mieux de leurs intérêts, leur participation à la mission. Ils regretteront par la suite d'avoir eu foi en de simples promesses verbales.' Peyssonnaux, "Vie, voyages et travaux," 16.

of their collections in order to publish their findings under Raffles' supervision (and presumably under his name).[38] The young naturalists protested and reminded Raffles that they trusted his verbal promises that they could keep the duplicates for themselves and they would be allowed to personally publish their discoveries under their own names. But it was to no avail. Their collections and notes were impounded and they were left with the occasional second duplicates from their impressive collections. Discouraged, the two young men left Bengkulu on April 1, 1820, each on a different ship, never to see each other again. Duvaucel went to India and died there four years later, weakened by a rhinoceros attack.[39] Diard set sail for Batavia and still had a long career ahead of him.

5 British Diard (1818–1820): Cheating Raffles

The British account of the conflict with Raffles has given rise to British Diard. His career lasted not even two years and had been intricately tied up with the plans and fortunes of Sir Thomas Stamford Raffles. He hired Diard and Duvaucel in Calcutta in November 1818, and, as we can see from the following paragraph from a letter to the Duchess of Somerset, Raffles cared very much about natural history and did everything in his power to advance the growth of knowledge of the natural world:

> I take down from hence a medical man of the name of Jack, who will be entrusted with the botanical part of my researches;[40] and I have two Frenchmen, M. Diard and M. Duvausel [sic], the former the pupil, and the latter the step-son of Cuvier; so that in comparative anatomy I shall be strong. These three *savans*, with a missionary clergyman, who takes charge of a printing-press, form my equipment from Calcutta, so that I hope we may do something.[41]

This suggests that his entourage was not just meant as pleasant company *en route* to Bengkulu (Sumatra), but that he seriously wanted to 'do something,'

38 Both Champoiseau and Peyssonnaux do not precisely date this letter and as a result, the French chronology of the conflict that ensued becomes somewhat muddled.

39 The jury is still out on which species of rhino it was that hastened Duvaucel's demise: Rookmaaker, "Mauled by a Rhinoceros." A photograph of the mounted alleged culprit is available in Rookmaaker, "Rhinoceros of South Asia," 357.

40 Jack is Scottish botanist William Jack (1795–1822).

41 Raffles, *Memoir of the Life*, 372.

namely to catalogue the natural riches of Bengkulu and start rolling off announcements of 'his' findings to the world as soon as possible. The itinerary of this trip is worked out in more detail in John Bastin's article on Raffles and natural history[42] and the time devoted to natural historical research confirms the dedication and seriousness with which Raffles approached the topic.

The two main accounts of the conflict from a British point of view have come to us through Sophia Raffles and William Jack. Together, these two accounts paint a clear picture of British Diard. Sophia published her memoirs of the life and work of her husband in 1830 as an apology for his (also in Britain) criticised independent behaviour. She also seems to have been much vexed by the Diard/Duvaucel case, since she encloses a whole set of correspondence between Raffles and (mostly) Diard and Duvaucel in an appendix.[43]

Sophia includes a letter from Raffles to Diard and Duvaucel from March 7, 1819, in which the conditions of their arrangement are clearly laid out. In this letter, Raffles indicates his plan to request permission of the Governor General to make the following offer to the young naturalists:

> 1st. Your researches to be confined to Sumatra and the smaller Islands in its immediate vicinity.
>
> 2dly. The draftsmen, &c. engaged by you to be entertained at the charge of Government, who will also defray all incidental and necessary expenses to which you may be subjected in the prosecution of your researches, *'on condition that such researches are made for and on account of the Honourable the East India Company, and that your collections, &c. are considered as their property.'*
>
> 3dly. An estimate to be framed of your monthly expenses for such establishment, &c, in which a fixed sum will be paid to you to cover all charges of every description, *'and the arrangement and transmission to Europe, or to Bengal, of such information and materials as you may collect, to be subject to the orders of Government.'*[44]

Sophia helpfully highlights the bones of contention in italics to prove that Raffles' position was legitimate and put down in writing to Diard and Duvaucel already a year before the conflict erupted.

---

42 Bastin, "Sir Stamford Raffles and the Study of Natural History." See also Dorai and Low, *Diard & Duvaucel*, 35.

43 Raffles, *Memoir of the Life*, 702–723.

44 Raffles, *Memoir of the Life*, 372–373. The full letter without italics is reproduced on page 703 of the same publication.

William Jack was a Scottish army surgeon in service of Raffles as botanist on this trip and his account comes to us in the form of a letter to Nathaniel Wallich (1786–1854), the Danish superintendent of the Royal Botanic Garden, Calcutta.[45] Jack considered the agreements between Raffles and the two Frenchmen as laid out in the March 1819 letter to be extremely generous, and proclaims to Nathaniel Wallich (1786–1854) on March 15, 1820, that Diard and Duvaucel 'have been patronised by that princely man, as never men were patronised before.'[46] After their arrival in Bengkulu on July 31, 1819, all was well, and Diard and Duvaucel worked hard. Raffles had to go back to Calcutta in October 1819, taking Jack with him. When they returned in March 1820, they also brought the bad news that the East India Company had not approved the continuation of the arrangement from the letter of March 1819. However, 'on our arrival here we found that they had been very industrious,' and Raffles offered to continue the arrangement paid for from his own pocket.

Jack confides to Wallich that only the reading of the full correspondence can communicate 'the measure of ingratitude' of the two Frenchmen,[47] and even accuses them of planning to abscond during Raffles' absence with their collections. The correspondence Jack refers to that has partly (and perhaps selectively) come to us through Sophia runs from March 7 to March 29, 1820. In the short space of only three weeks the conflict completely spun out of control and led to a definitive breach of trust between Raffles and Diard and Duvaucel, leading Raffles to write: 'I request to dispense with a personal visit from either yourself or Mr. Duvaucel before your departure.'[48]

On March 7, 1820, Raffles informed Diard and Duvaucel that their arrangement had not been approved of by the East India Company and would end on April 1, 1820. Based on their agreement from March 7, 1819, Raffles asked Diard and Duvaucel to hand over their collections and documents. When Diard and Duvaucel objected to this, Raffles appointed a committee of three (Captain Thomas C. Watson, William T. Lewis, Esq., Lieutenant N. S. Trueman) to arrange the transfer of the collections and documents. Under pressure, Diard and Duvaucel signed an agreement and handed over their collections and documents, except for a box of Duvaucel's papers that had been stolen the night before.[49] The documents would be restored to them after a year, when

45 Gage and Burkill, "William Jack's Letters."

46 Gage and Burkill, "William Jack's Letters," 201.

47 Gage and Burkill, "William Jack's Letters," 202. Gage and Burkill made the editorial decision to transcribe Jack's letters 'with the ommission [...] of certain criticisms passed on Diard and Duvaucel.' (Gage and Burkill, "William Jack's Letters," 150).

48 Raffles, *Memoir of the Life*, 720.

49 Raffles, *Memoir of the Life*, 718.

they would be free to publish them. They were allowed to select some duplicates from their collections that could be sent to France after the main collection had been sent to Britain.[50] Shortly after this, a sealed box with Duvaucel's papers was found and they were forced to hand over all documents relating to natural history. Jack asserts that Diard and Duvaucel still managed to keep behind some documents from this box, 'so that they have given their word of honor to what the whole court know to be untrue.'[51]

The British version of the story differs markedly from the French version. What stands out most is that it is much more detailed. The French version is chronologically somewhat muddled, obscuring the fact that a full year of excellent cooperation followed the letter of March 7, 1819. Even the fact that Raffles asked for a full catalogue of the collections, identified by Peyssonnaux as the cause of the conflict, was not the reason for the dispute, since this was asked on September 18, 1819, in the context of Raffles reporting to the East India Company on the progress of the work of Diard and Duvaucel.[52] It was only in March 1820, when Raffles returned with the bad news that the arrangement would be discontinued, that a conflict arose.

In the British version, Jack and Sophia Raffles clearly come to Raffles' defence and leave no stone unturned in trying to make it clear that Raffles is entirely blameless. The official reports of the committee carry much weight, also because it outlines clearly that Diard and Duvaucel had been given the chance to voice their concerns. In the end, the agreement from March 7, 1819, swayed the verdict in Raffles' favour and Diard and Duvaucel were only left with the documents they wilfully withheld and a limited number of duplicates.

It seems that early on Diard and Duvaucel had been given to understand by Raffles that they would be allowed to keep the duplicates and to publish their findings in French a year after they had been published in English.[53] Because this oral agreement had not been put in writing, the initial request to hand over all their collections and documents must have felt deeply unfair, inspiring them to resist Raffles' request. The final arrangement was not very different from the oral agreement, even though forced separation from their notes, when they perhaps would have expected to only have to hand over copies, and a very limited collection of duplicate specimens may not have been what they had envisioned. In the end, mistakes were made on both sides, allowing both

50 Raffles, *Memoir of the Life*, 719.

51 Gage and Burkill, "William Jack's Letters," 206.

52 The letter is part of the appendix of Raffles, *Memoir of the Life*, 703–705.

53 Gage and Burkill, "William Jack's Letters," 204.

the French and the British to tell a rival version of the same story, creating a cheated French Diard and a cheating British Diard.

6 Slander and Rehabilitation in the British Press

What may have further rankled the French is the fact that the publication of the Singapore/Sumatra collections in the *Transactions of the Linnean Society* in two parts[54] 'established Raffles' reputation as a zoologist.'[55] In the introduction to the first part Raffles explains that the collection was assembled by 'two French gentlemen,' who 'advanced pretensions diametrically opposed to the spirit and letter of their engagement, and altogether inconsistent with what I had a right to expect from them, or they from me.'[56] He explains that their notes were useless to him and that he had to undertake the writing of the descriptions himself. It did not sit well with the French that an important French collection was published by an Englishman without even acknowledging the collecting naturalists by name.

And this is not all. Diard would soon become one of the subjects (without his knowledge or participation) of a naturalist's brawl in the public zoological press in Britain in the early 1830s. He was in the firing line of a vicious debate with Nicholas Aylward Vigors (1785–1840) and William Swainson (1789–1855) as its main protagonists. Vigors and Swainson were not on the best of terms, to put it mildly. Vigors used his privileges as secretary of the Zoological Society of London (founded by Raffles in 1826) to limit Swainson's access to their natural history collection.[57] Even though Vigors and Swainson were the forerunners of quinarianism, a short-lived movement that advocated a rigid fivefold hierarchical system in taxonomy, they made no secret of their mutual loathing.

The public part of the broil started with the publication by Vigors of an extensive critique of the work of Anselme Gaëtan Desmarest (1784–1838) in *The Zoological Journal*,[58] followed by a second publication with Thomas Horsfield (1773–1859),[59] critiquing the work of René Primevère Lesson (1794–1844) and

54 Raffles, "Descriptive Catalogue."

55 Bastin, "Sir Stamford Raffles and the Study of Natural History," 16.

56 Raffles, "Descriptive Catalogue," 239–240.

57 See Desmond, "Making of Institutional Zoology," for a full narrative of the troublesome relationship between Vigors and Swainson.

58 Vigors, "Reply to some Observations."

59 Horsfield was a protégé of Raffles in the Indies and under his patronage travelled in the Malay Archipelago, resulting in Horsfield, *Zoological Researches in Java*. An extensive

Diard and Duvaucel, albeit only in a footnote.[60] These two works rankled Swainson, who had received ample support from the French (and full access to the collections of the Muséum) on a previous visit to Paris, while Vigors kept him away from the collections of the Zoological Society.

The bone of contention in the case of Diard was that the majority of the French naturalists (including Desmarest,[61] Frédéric Cuvier[62] and even the great Georges Cuvier)[63] attributed the discovery of the Malayan tapir (*Tapirus indicus*, Desmarest, 1819) to Diard and Duvaucel rather than to William Farquhar (1774–1839), who first published a description in the *Asiatick Researches* of 1820 (Figures 34 and 35).[64] This article contained an original letter by Farquhar from January 19, 1816, but it was only published in 1820, in an Asian journal with limited distribution in Europe. When Farquhar's letter was published, it contained an additional note by Diard with some anatomical remarks. Diard had based his observations on a tapir at the menagerie in Barrackpore (India), which was close to their base in Chandernagor.[65] Diard also sent his own description back to France, which led to Desmarest being the first to publish the species in 1819 based on Diard's report, which led the French to identify Diard as its discoverer instead of Farquhar. Vigors did not know Farquhar's publication, but knew that the description by Desmarest was based on Diard's description and that Diard was in the service of Raffles. He heavily opposed this attribution to Diard and wrote:

> These naturalists [Diard and Duvaucel] were the hired collectors of Sir Stamford Raffles. They acted immediately under his orders, and were amply remunerated for their subordinate services by his liberal pay. As well might we attribute the discoveries of Capt. Cook to the sailors who worked his vessel, as the scientifick discoveries of Sir Stamford to his hired dependants.[66]

---

account of Horsfield's life and works by John Bastin is included in a recent reissue: Horsfield and Bastin, *Zoological Researches in Java.*

60 Vigors and Horsfield, "Observations on Some of the Mammalia," 106.

61 'La connaissance de cet animal en France est due à M. Diard.' Desmarest, *Mammalogie*, vol. 2, 411.

62 Geoffroy Saint Hilaire and Cuvier, *Histoire naturelle des mammifères*. The entry on the Malayan tapir, 'Le maiba,' can be found in volume 2.

63 Cuvier, *Discours sur les révolutions*, 68; Cuvier, *Histoire des progrès*, vol. 3, 388.

64 Farquhar et al., "Account of a New Species of Tapir." For more on Farquhar's work as a naturalist, see Dozier, *Natural History Drawings*.

65 See Dorai and Low, *Diard & Duvaucel*, 162, for a species of turtle that was described based upon a drawing of a specimen in the menagerie at Barrackpore.

66 Vigors and Horsfield, "Observations on Some of the Mammalia," 106.

FIGURE 34 Tapir of Malacca, unknown (Chinese) artist, watercolour on paper, 1819–1823, 37.6 × 54.1 cm, SHM1995.2885, William Farquhar Collection of Natural History Drawings (gift of Mr. G. K. Goh), National Museum of Singapore. Public domain. On October 12, 2024, Topaz Gigapixel ai was used to increase the resolution of this image. HTTPS://WWW.ROOTS.GOV.SG/COLLECTION-LANDING/LISTING/1083462

FIGURE 35 'Tapir de l'Inde ou Maïba.' Lithograph by Langlumé, after a drawing by Werner. Unnumbered plate from Étienne Geoffroy Saint Hilaire and Frédéric Cuvier, *Histoire naturelle des mammifères*, vol. 2. Public domain. Universiteitsbibliotheek, Rijksuniversiteit Groningen, uklu M1 F---- 1

A short while later, in a description of the American tapir that lived in the menagerie of the Zoological Society, Vigors wrote about the French attributions of the discovery of the Malayan tapir to Diard in *The Gardens and Menagerie of the Zoological Society Delineated*:

> The history of this transaction affords too striking an illustration of the injustice of certain among the French zoologists to the merits of our countrymen to be passed over without observation.[67]

Swainson responded to all of this with a lengthy article in *The Magazine of Natural History*, in 'Defence of "Certain French Naturalists."'[68] Swainson states that British zoology was still young and that the British should not 'arrogantly imagine we have grasped the whole; [...] and that there exists a conspiracy among the naturalists of France against those of England, merely, forsooth, because they choose to study nature in their own way.' Swainson exonerated Lesson and Desmarest and then moved on to defend Diard. It all hinged on the fact that the *Asiatick Researches* in which Farquhar published the description and image of the tapir was very difficult to access (which indeed it was), and probably unknown to the French naturalists in question. They would not have knowingly ignored Farquhar's publication, so they must not have known about it. Swainson concludes by arguing against a 'national jealousy' of 'certain French naturalists' and stresses his good experiences in the Muséum in Paris.

The matter regarding Diard was quickly settled when a further response by Edward T. Bennett (who turned out to be the actual author of *The Gardens and Menagerie* and not Vigors) clarified that Diard was not to blame,[69] since Farquhar's publication was accompanied by some observations by Diard which Diard concludes by writing:

> To Major Farquhar alone belongs the honor of having first given, with the trifling exception I have noticed, a correct description of the interesting animal which forms the subject of these observations. I have only to add, that the young male tapir which is in the menagerie at Barackpore, is in all respects the same, with that described by Major Farquhar.[70]

---

67 Bennett, *Gardens and Menagerie*, 194–195.
68 Swainson, "Defence of "Certain French Naturalists"."
69 Bennett, "Evidences in Proof of Certain Statements."
70 Farquhar et al., "Account of a New Species of Tapir Found," 423–424.

The vituperative exchanges between Vigors and Swainson continued for a while, but Diard had managed to successfully escape the barrages and was completely exonerated, and all of this without his input or knowledge.

7 French Diard II (1820–1824): a Career Nipped in the Bud

Before this whole episode erupted, Diard had already moved on. After his departure from Bengkulu, he arrived in Batavia on April 7, 1820.[71] He spent some time rebuilding his collections there. Champoiseau quotes a letter from Diard from October 1820 in which we read a report of his first agricultural experiments in the Netherlands Indies. Experiments with indigo cultivation were already taking place on Java. When Diard visited such a plantation, he noticed the planters were unable to make it profitable. After some experimentation, Diard managed to improve the yields by one third.[72] This brought Diard to the attention of governor-general Van der Capellen, who invited Diard to submit a proposal for an experimental *indigoterie*. Diard was allowed to choose a plot and would be given direction of this plantation.

But a letter from France would change all this. Diard finally received word that Duvaucel and himself had received an indemnity of 6,000 Francs, to be turned into an annual allowance of 3,000 Francs, from the French government and that both were appointed as *naturaliste du roi*. Diard dropped the indigo project and started looking for possibilities to serve French interests, which his allowance of course obliged him to do.

His search came to an end when he met the French consul to Cochinchina, Jean Baptiste Chaigneau (1769–1832, Figure 36).[73] Chaigneau had been appointed in October 1820 as *agent de France*, *consul*, and *commissaire du roi*[74] and was travelling via Batavia to Vietnam when he met Diard there. Chaigneau was also charged to bring with him 'some Frenchmen with a proven ability in the navy, military engineering or other aspects of science.'[75] He quickly

71 *Bataviasche Courant*, no. 16 (April 20, 1820): 1.

72 Diard's letter, as quoted by Champoiseau, "Voyages aux Indes Orientales," 232, unfortunately does not indicate the nature of these experiments.

73 For more on Chaigneau, see Salles, *Un mandarin breton*. This book contains a reissue of two articles from the *Bulletin des Amis du vieux Huế* with a new introduction by Bernard le Nail. See Cordier, *Le Consulat de France à Hué*, for a series of archival documents on the French consulate in Huế.

74 Cordier, *Le Consulat de France à Hué*, 38.

75 'Quelques Français d'une capacité éprouvée dans la marine, le génie ou d'autres parties scientifiques.' Cordier, *Le Consulat de France à Hué*, 44.

FIGURE 36
Portrait of Jean-Baptiste Chaigneau as Grand Mandarin. Photograph by G. Garitan of a copy by Paul Sarrut of an anonymous painting from 1805, 1900–1950, 61 × 48 cm. 75.7070, Musée du Quai Branly – Jacques Chirac (Paris). Public domain
HTTPS://COMMONS.WIKIMEDIA.ORG/WIKI/FILE:CHAIGNEAU_JEAN_BAPTISTE_04993.JPG
ORIGINAL PAINTING: HTTPS://WWW.QUAIBRANLY.FR/EN/EXPLORE-COLLECTIONS/BASE/WORK/ACTION/SHOW/NOTICE/373560-PORTRAIT-DE-JEAN-BAPTISTE-CHAIGNEAU-OU-LE-QUOTGRAND-MANDARINQUOT-FRANCAIS-NGUYEN-VAN-THANG

identified Diard as one of them and recruited him. Diard set sail again for his second stay in the French sphere of influence on April 29, 1821,[76]and arrived in Vietnam as part of a small diplomatic and commercial mission on May 17, 1821.[77] An 1824 report on the work of Duvaucel mentions Diard as 'his partner having long since gone to Cochinchina from where one only occasionally receives news.'[78] Fortunately, we have the benefit of hindsight and the publication of much of Chaigneau's reports, which allows us to get some insight into Diard's activities in Cochinchina.

Upon arrival in Huế, then the capital of the reigning dynasty, Chaigneau presented himself at court and received permission to reside in Huế together with his nephew and Diard, but they were not allowed to leave the capital. This was not to Diard's liking, but, in Chaigneau's words:

> It is very difficult in this country for Europeans to travel to the interior of the kingdom and outside of the established routes.[79]

76 *Bataviasche Courant*, no. 18 (May 5, 1821): 1.

77 Peyssonnaux, "Vie, voyages et travaux," 31, based on Chaigneau's report as published in Cordier, *Le Consulat de France à Hué*, 55.

78 'Son compagnon s'étant depuis long-tems rendu en Cochinchine, d'où l'on ne reçoit que rarement de ses Nouvelles.' Anonymus, "Notice sur les voyages," 201.

79 'Il est fort difficile dans le pays, d'obtenir pour les Européens de voyager dans l'intérieur du royaume et hors des routes ordinaires.' Cordier, *Le Consulat de France à Hué*, 70.

FIGURE 37
'King of Cochinchina, in His Dress of Ceremony,' 'drawn by a Chinese.' Unnumbered plate from John Crawfurd, *Journal of an Embassy*, opposite page 314. Public domain. This is a portrait of the Vietnamese emperor Minh Mạng.
HTTPS://ARCHIVE.ORG/DETAILS/B29350815_0002/PAGE/N345/MODE/2UP

And it was not without reason that it proved to be difficult to get permission to travel. The political situation in Vietnam, and with it the attitude towards the French, had recently changed. In 1802, the territory of what is now modern Vietnam had become united for the first time under Nguyễn Phúc Ánh, who would be the founder of the Nguyễn dynasty and assumed the throne as emperor Gia Long (1762–1820, r. 1804–1820). Gia Long had a very positive attitude towards the French, so much so that his intended successor was sent to France to be educated there. Activities of French missionaries were tolerated. Chaigneau had played a role in Gia Long's rise to power and was rewarded with a high position in court as a mandarin.

In 1819, Chaigneau visited France, which resulted in his appointment as the king's consul. However, when he returned to Vietnam, with Diard as part of his entourage, it was to find a new emperor in charge. Minh Mạng (1791–1841, r. 1820–1841, Figure 37) was the fourth son of Gia Long.[80] His accession led to an immediate change in policy towards the French. Minh Mạng was not as favourably disposed towards the French as his father was and he started to limit the freedom for foreign nationals in his empire. This is what made it so difficult for Diard to receive permission to travel. What may have helped considerably in

80 See Choi, *Southern Vietnam*, for an extensive discussion of Minh Mạng's reign.

obtaining some freedom of movement is that Minh Mạng 'respected the scientific materials and discoveries of the Western world.'[81]

Only in October 1821 was Diard permitted to undertake his first two-week expedition outside the capital. Finally, in early 1822 Diard received permission to freely roam the empire, partly due, it seems, to the fact that the emperor saw no harm in this strange person who occupied himself with picking insects from elephant dung for display in palaces in Europe.[82] Diard's travels took him south and he travelled all the way to Saigon, where he arrived in May 1822. He then travelled to Cambodia, at the time a vassal state of Vietnam,[83] where he visited that famous temple complex of Angkor Wat, one of the earliest recorded visits to that city of a western traveller.[84]

Upon his return to Saigon in August 1822, he ran into a British diplomatic mission under leadership of John Crawfurd. One of the members of this expedition was fellow naturalist George Finlayson. In his report of the mission (posthumously published by Raffles), we read the following description of Diard:

> M. Diard is evidently a man of great enterprise and acuteness, and admirably qualified for the arduous pursuit in which he is engaged. He is fond of adventure, and ingenious in overcoming obstacles.[85]

Finlayson is evidently somewhat amused by the 'lively and well-educated Frenchman'[86] and takes his demeanour to also reflect some flaws in his character:

81 Choi, *Southern Vietnam*, 62.

82 The anecdote, as related by the 19-year-old son of Chaigneau, is available in full in Peyssonnaux, "Vie, voyages et travaux," 41–42.

83 This is a somewhat obscure period in Cambodian history. Some facts can be gleaned from Choi, *Southern Vietnam*, while most histories of Cambodia in the nineteenth century start in 1859, the year when Cambodia become part of the French colonial empire. See Osborne, *French Presence*. The most extensive account of Cambodia during Diard's visit is Chandler, "Cambodia before the French."

84 Peyssonnaux, "Vie, voyages et travaux," 46, assumes Diard was the first Westerner to visit Angkor Wat and extensively discusses the veracity of this. Even in his time, there were no primary sources extant that confirmed this visit, but a relative of Diard testified to 'the scientific rigor' ('la rigueur scientifique') of Hippolyte Diard, 'an extremely precise man' ('un homme extrêmement précis'), who is our source for Diard's visit in Diard, "Pierre-Médard Diard," 203.

85 Finlayson, *Mission to Siam*, 308.

86 Finlayson, *Mission to Siam*, 308.

> If there be any thing amiss in the character of Diard, it is, (and it is with hesitation and doubt that I make the remark,) perhaps, a disposition to over-rate the number, extent, and value of his discoveries; and perhaps too, an ardour of zeal, which may be apt to lead one beyond the precise limits of accurate observation.[87]

The quote above is the final paragraph of a page-long positive review of Diard's achievement. Given the fact that Finlayson's travel report was edited for publication by Raffles, it may be possible that Raffles' opinion of Diard comes filtering through a little, but this cannot be ascertained without Finlayson's original manuscript.

The success of Chaigneau's mission to Huế did not live up to the expectations, especially because Minh Mạng was decidedly less pro-French than Gia Long who was still in power when Chaigneau's mission parameters were set. Being part of the diplomatic mission, Diard was dependent on its developments. In June 1823, an embassy from the king of Pegu (a Burmese kingdom) had reached Saigon. Diard got so well acquainted with the members of the mission that he was invited to come back with them to Pegu. But in December 1823 Diard was recalled to Huế by the emperor to take part in a political meeting at court that would decide the relationship with Pegu. The result was that Diard was appointed as a member of the embassy to be dispatched to Pegu to make friends with the king there.

Here the *naturaliste du roi* turns into a diplomat of the Vietnamese emperor and the contours of a Vietnamese Diard emerge. Choi Byung Wook indicates that there are extensive Vietnamese sources available on the reign of Minh Mạng.[88] Given the growing political dimensions of Diard's sojourn in Vietnam, it is not unlikely that these sources could contribute to completing our image of a Vietnamese Diard.[89]

After his appointment to the embassy, Diard moved back to Saigon, where he arranged to meet up with the Burmese embassy at a later moment. When Diard was on his way to join them, he was stopped by the British, who had just invaded Burma, and he was forced to alter his plans.[90] He went to Malacca

87 Finlayson, *Mission to Siam*, 308–309.

88 Choi, *Southern Vietnam*, 12–13.

89 A lack of the required language skills prevents me from using these sources and from finding out whether a colleague has perhaps already written on Vietnamese Diard in Vietnamese. If not, I hope that this chapter will inspire a colleague to take up the challenge and do this.

90 See Kitzan, "Lord Amherst and the Declaration of War," for the circumstances leading to the start of the Anglo-Burmese war in 1824.

instead, determined to procure a live Asian tapir for the menagerie of the Muséum in Paris. His 'hunt for tapirs,' which lasted for six months, resulted in failure.[91] Of the four live tapirs that he managed to acquire, three did not survive their captivity, while the fourth broke from his cage and disappeared into the marshland while it was about to be shipped to Paris.

Diard's stay in Vietnam, when considered from a natural history point of view, was constantly battered by disappointment and failure. He suffered repeated setbacks in his exploits and was restricted in his movements by nature of him being French. A move back to British-influenced territory was blocked by the British themselves. All his plans came to nothing, and the one enduring element of his stay in Cochinchina would only see the light of day forty years later. During his extensive travels through Cochinchina and Cambodia, Diard had taken the time to write a report on the commercial and agricultural opportunities the land offered. He described the climate, analysed the landscape and came up with recommendations for how to best grow coffee, pepper, nutmeg, clove, cinnamon, cotton, indigo, sugar and silk, all colonial cash crops.

Diard's original report from the early 1820s either never reached the French ministry or was filed away into obscurity. Forty years later, in March 1863, a fourteen-page summary of his findings was published, but only when his brother Hippolyte had brought the original report to the attention of the ministry.[92] Incidentally, in the introduction of the published report Diard is still hailed by the French as the discoverer of the tapir: 'It is to him in particular that we owe this [discovery] of the Asian tapir.'[93]

8 Dutch Diard (1824–1863): a 'Versatile' Career

As we have seen, Diard had already been to the Netherlands East Indies, so when he arrived in Batavia on December 7, 1824, it was to reunite with old acquaintances. But it remains a matter of controversy how friendly that reunion actually was. Dutch Diard emerges somewhat murkily from the sources. We have already seen that Diard first arrived in the Netherlands Indies on April 7, 1820, after the Raffles debacle. According to H. J. Veth, Diard was seized

91 'Chasses des tapirs.' Peyssonnaux, "Vie, voyages et travaux," 56.

92 'C'est à lui notamment qu'on doit celle [découverte] du tapir asiatique.' Diard, "Chine et Indo-Chine."

93 Diard, "Chine et Indo-Chine," 3.

by the Dutch authorities on Sumatra and transported to Batavia to be incarcerated there as a 'suspicious person' but was saved from this fate by Reinwardt.[94]

However, according to the shipping news in the *Bataviasche Courant*, Diard arrived from Padang to Batavia on the British ship the *Indiana*. In my opinion, it is very unlikely that the Dutch, who were in control of Padang at this time,[95] would have used a British ship for a prisoner transport. Fransen, Holthuis and Adema follow Veth's account,[96] while Van Steenis-Kruseman does not mention any arrest at all.[97] Peyssonnaux is also not aware of an arrest, but explains at length that only the Dutch king could grant residency in the Netherlands Indies to foreign nationals, and that these could be held if permission was lacking. Claeys mentions that, when Diard arrived in Batavia for the second time on December 7, 1824,[98] he was regarded with suspicion by some because he had worked for the British.[99] In the end, I have to conclude that Diard was never arrested by the Dutch authorities, neither in 1820, nor in 1824. On the contrary, his arrival was welcomed by Van der Capellen and Reinwardt.[100]

Diard would stay in the Netherlands Indies from 1824 to 1841 and again from 1848 to 1863. In these forty years of 'versatile' service,[101] Diard occupied himself with a wide variety of activities in the realms of natural history and agricultural experimentation and improvement. It is impossible to overstate the wide-ranging nature of Diard's work as a man of science and experiment in the Netherlands Indies. He was duly recognised for his work in the Indies when governor-general De Eerens wrote in his recommendation to the crown in 1839 to honour Diard with membership of the Order of the Dutch Lion:

> Diard [...] has not only made himself worthy of the attention of the government by an assiduous fulfillment of all the scientific commissions assigned to him, but has increased his merits by his disinterest in executing these.[102]

94 'Verdacht persoon.' Veth, *Overzicht van hetgeen gedaan is*, 57.

95 Locher-Scholten, *Sumatran Sultanate and Colonial State*, 51.

96 Fransen, Holthuis, and Adema, "Type-catalogue of the Decapod Crustacea," 228.

97 Steenis-Kruseman, "Malaysian Plant Collectors," 136.

98 *Bataviasche Courant*, no. 50 (December 11, 1824): 1.

99 Claeys, "Deux Français," 1006.

100 Van Heiningen, *Correspondence of Caspar Georg Carl Reinwardt*, 485.

101 I borrow this qualification from governor-general Baud, who actually seems to have used it negatively when he wrote about 'the well-known versatile character of Mr. Diard' in a discussion with governor-general De Eerens on Diard's possible involvement in an experimental sugar mill, as quoted by Leidelmeijer, *Van suikermolen tot grootbedrijf*, 160.

102 'Heeft niet alleen door eene ijverige vervulling van alle hem opgedragen wetenschappelijke kommissien zich der aandacht van het bestuur waardig gemaakt, maar heeft zijne

The king followed De Eerens' recommendation and Diard was appointed as member of the Order of the Dutch Lion. These words of praise were written fifteen years after Diard settled in the Indies, and eleven years after he had been appointed as member of the Natuurkundige Commissie in 1828. It is to the window between these two events that we now turn, because these years saw the final conflicts that followed in the wake of Diard's transimperial movements.

9 Diard and the Natuurkundige Commissie

In Dutch historiography, Diard is mostly known for his membership of the Natuurkundige Commissie. During his first stay in the Indies, from April 7, 1820 to April 29, 1821, he had got acquainted with Reinwardt and had learned about the existence of the Natuurkundige Commissie. The first members would arrive during Diard's first stay and it is more than likely that they met each other regularly given the close association with Reinwardt for both Kuhl and Van Hasselt (who worked directly under Reinwardt) and Diard.[103]

When Diard returned on December 7, 1824, it was only to discover that Reinwardt had returned to the Netherlands, Kuhl and Van Hasselt had died and only one of their assistants, Van Raalten, was still alive and well. Even though replacements had already been appointed in the Netherlands, they had not yet arrived and the Natuurkundige Commissie seemed dead in the field, especially because Van Raalten's only task was to prepare the Kuhl/Van Hasselt papers and collections for shipment and no serious efforts were undertaken to create new collections. Due to the sorry state the Commissie was in, and the fact that appointments were the privilege of the king back in the Netherlands, it was unlikely that Diard would become a member. During the next four years, the situation would change and opportunities opened up which allowed Diard to become a member of the Commissie in 1828.

The road to membership, however, was not smooth, and was, as had almost became customary in Diard's life, littered with conflict. Within two months of his arrival, Diard was appointed as corresponding member of the Bataviaasch Genootschap van Kunsten en Wetenschappen. At the general meeting on February 2, 1825, within two months of his arrival, he held a lecture on the

verdiensten nog vermeerderd door zijne belangloosheid in het volvoeren derzelve.' KB, June 9, 1839, no. 61, NL-HaNA, Staatssecretarie, 2.02.01, inv.no. 4488.

103 Kuhl and Van Hasselt mention meeting Diard in February/March 1821 in Buitenzorg. See Kuhl and Van Hasselt, "Uittreksels uit brieven van de heeren Kuhl en Van Hasselt," *Algemeene konst- en letter-bode*, no. 6 (February 8, 1822): 85.

'deer of the Indies.'[104] On April 5, 1825, he was appointed to join Jan Isaac van Sevenhoven (1782–1841) on the island of Bangka.[105] Van Sevenhoven had been appointed to restore some order to Palembang (Sumatra) and Bangka and had proposed an inspection of the tin mines of Bangka, since the productivity of the mines was falling. Governor-general Van der Capellen appointed Diard as the expert to accompany Van Sevenhoven and suggest improvements. A later published profile of Bangka indicates that Diard most likely did not collect specimens on Bangka, like Horsfield had done before him, but limited himself to 'statistical subjects.'[106] The same report indicates that Diard determined the tin content of the various mines, information that will have undoubtedly been used to increase profitability.[107]

After his return, Diard was appointed as 'cultivation inspector' on December 31, 1825, by Van der Capellen,[108] but this position was abolished again on June 6, 1827, mostly because it was unclear what the position exactly entailed. On July 21, 1826, when he was still 'cultivation inspector,' Diard was appointed to undertake an expedition to Borneo,[109] assisted by F. J. Perret.[110] He left for Borneo from Surabaya (Java) on August 21, 1826, in the company of resident Hartmann,[111] and returned to Semarang (Java) on January 7, 1827.[112]

This mission to Borneo turned out to be a source of irritation for the members of the Natuurkundige Commissie. Commissie members Boie and Macklot, together with assistants Müller and Van Oort, had finally arrived on June 6, 1826. Despite the petitioning of directing member Boie, none of the members were allowed to join Diard's Borneo expedition. The Commissie request was declined with the reassurance that Diard would only temporarily accompany resident Hartmann and would return soon. What turned this reassurance void was the fact that Diard hired local hunters and preparators to accompany him during the Borneo expedition, 'in order,' as Boie would have it, 'to better serve

104 'Indische herten.' *Bataviasche Courant*, no. 6 (February 5, 1825): 4.

105 Res.GG, April 5, 1825, no. 9. NL-HaNA, Koloniën, 1814–1849, 2.10.01, inv.no. 2788.

106 Anonymus, "Korte schets," 125–126. Based on the ornithological collections in Leiden and Paris, G. F. Mees comes to the same conclusion. See Mees, "List of the Birds," 7–9.

107 Anonymus, "Korte schets," 146.

108 'Inspecteur over de cultures.' Res.GG, December 31, 1825, no. 21. NL-HaNA, Koloniën, 1814–1849, 2.10.01, inv.no. 2791.

109 Res.GG, July 21, 1826, no. 11. NL-HaNA, Koloniën, 1814–1849, 2.10.01, inv.no. 2494. Specific instructions can be found in Res.GG, August 1, 1826, no. 35. NL-HaNA, Koloniën, 1814–1849, 2.10.01, inv.no. 2495.

110 Res.GG, August 4, 1826, no. 23. NL-HaNA, Koloniën, 1814–1849, 2.10.01, inv.no. 2495.

111 *Bataviasche Courant*, no. 34 (August 23, 1826): 12.

112 *Bataviasche Courant*, no. 3 (January 17, 1827): 7.

France by first examining this land here that is new to naturalists.'[113] It did not sit well with Boie that the Commissie was not allowed to travel to Borneo to bring glory to the Dutch empire, while this French outsider could freely go and send his collections to France. Even the fact that Diard's official mission was 'to find out, through a precise local investigation, which cuts in the fairly high administration costs of the Dutch possessions over there should be implemented in order to bring the normal expenditure back into proportion with the income'[114] did nothing to alleviate this suspicion.

Meanwhile, back in Europe, the Dutch mastermind behind the Commissie, Coenraad Jacob Temminck, suffered from similar sentiments regarding Diard. Temminck was already somewhat annoyed by a comment by Georges Cuvier in his 1824 lecture 'Sur les progrès des sciences naturelles depuis le retour de la paix maritime' [On the progress of the natural sciences since the return of maritime peace] where he maintained that thanks to the work of Diard and Duvaucel the riches of the Dutch colonial empire were already present in the Paris Muséum before the Dutch had even considered doing some scientific research there themselves.[115] Temminck states that his intercession with Cuvier prevented this passage from becoming part of the printed lecture.[116]

Reports from Paris made Temminck believe that amongst the collections that were received at the Muséum in 1826 were also specimens collected on Diard's Bangka expedition.[117] This is why Temminck asked the ministry 'if it might not be of importance to try and coax this very capable, but also very ambitious man to our interests.'[118] But it was not until disaster struck that

113 'Afin de mieux servir la France en examinant le premier cette terre nouvelle pour les naturalistes.' Quoted in Veth, *Overzicht van hetgeen gedaan is*, 62.

114 'Om door een naauwkeurig plaatselijk onderzoek nategaan, welke verminderingen in de aanzienlijke administratie-kosten van de nederlandsche bezittingen aldaar zullen moeten worden ingevoerd, om de gewone uitgaven aldaar in eene meer geëvenredigde verhouding tot de ontvangsten te brengen.' *Rotterdamsche Courant*, no. 8 (January 18, 1827): 2.

115 Veth, *Overzicht van hetgeen gedaan is*, 58. See Deleuze, *Histoire et description*, for the prominent presence of Diard and Duvaucel's collections in the Muséum in the early nineteenth century. Even in the current galerie de Paléontologie et d'Anatomie comparée of the Muséum national d'Histoire naturelle, many objects collected by Diard and Duvaucel are still on display.

116 A later version of this paper (without Cuvier's contested remark and with a slightly altered title) appeared as 'Discours sur l'état de l'historie naturelle et sur ses accroissements, depuis le retour de la paix maritime' in: Cuvier, *Histoire des progrès*, vol. 4, 345–372.

117 Temminck to the minister, September 19, 1826, quoted in Veth, *Overzicht van hetgeen gedaan is*, 58.

118 'Of het niet van belang kon zijn te beproeven dezen zeer kundigen maar tevens zeer eergierigen man in onze belangen over te halen.' Temminck to the minister, March 27, 1827, quoted in Veth, *Overzicht van hetgeen gedaan is*, 61.

Diard was called to serve in the Commissie. Directing member Boie fell gravely ill in August 1827 and died soon after on September 4, 1827, paving the way for Diard's membership.[119]

Boie had been functioning as directing member of the Commissie. When he died, this position automatically went to his companion Macklot. When it then turned out that the Commissie was expected to accompany a military/exploratory expedition to New Guinea, Macklot petitioned Du Bus to appoint Diard to take care of the Java business of the Natuurkundige Commissie (mainly cataloguing and shipping their collections to Leiden) in their absence. As a fellow naturalist, Diard had already frequently been in touch with the members of the Commissie and was always a welcome guest. In Boie's words:

> Macklot and I felt very attracted to the person of Diard and we still miss his company, which had become a necessity, every day. He visited us regularly at the late hour of 10 PM, always happy, full of entertaining anecdotes and relevant remarks. He has been everywhere, has observed sharply, and gives the assurance that he really has not yet done anything.[120]

The existing good relations between Boie and Macklot and Diard inspired Macklot's request for Diard's involvement in the work of the Commissie. Both Diard's expertise and his pleasant character created a sense of trust between the naturalists. No doubt prompted by his good working relationship with Diard, Du Bus granted this request. Diard was made a full member of the Natuurkundige Commissie on January 31, 1828, and was given the same salary as Macklot (f500).[121] In the course of the next two years, Diard even became directing member, the same rank as Macklot, which gave each of them equal authority over the dealings of the Natuurkundige Commissie.[122] This situation existed until the death of Macklot on May 12, 1832, when Diard became the sole directing member of the Commissie.[123]

---

119 Veth, *Overzicht van hetgeen gedaan is*, 52.

120 'Macklot en ik gevoelden ons zeer tot den persoon van Diard aangetrokken en wij missen nog dagelijks zijn gezelschap, dat ons eene behoefte geworden was. Hij kwam dikwerf des avonds te 10 ure nog bij ons, altijd vroolijk, vol onderhoudende anecdoten en treffende opmerkingen. Hij is overal geweest, heeft scherpzinnig waargenomen, en geeft daarbij de verzekering van eigenlijk nog niets gedaan te hebben.' Letter from Boie, quoted in Veth, *Overzicht van hetgeen gedaan is*, 62–63.

121 Res.GG, February 4, 1828, no. 43. NL-HaNA, Koloniën, 1814–1849, 2.10.01, inv.no. 2513, quoting Res.CG, January 31, 1828, no. 22.

122 Veth, *Overzicht van hetgeen gedaan is*, 68.

123 Macklot was killed during the uprising of Chinese workmen at the agricultural establishment in Krawang.

When Temminck learnt that Diard had written to the Muséum to tell them they should not count on receiving any more collections from him, and when the collections that Diard had made during his Borneo expedition were promised to 's Rijks Museum in Leiden,[124] Temminck welcomed the steps taken by Diard and wrote to him on September 26, 1828:

> I do not want to finish this [letter] without showing you all the satisfaction I feel at having attached you to our Natuurkundige Commissie. I expect the most satisfactory results for the science we cultivate and for the museum to which you, like me, now have the same interest in devoting your zeal and your devotion.[125]

During the New Guinea expedition, Diard somehow became co-directing member with Macklot. Veth suggests this happened on January 15, 1830, when Macklot, Müller and Van Oort received a raise, but the resolution from January 8, 1830, that includes this raise, already identifies Diard as directing member.[126] When Macklot was killed on May 12, 1832, Diard took over as sole directing member and remained in this role until August 26, 1841.[127] During these years he combined his role as directing member with a plethora of agricultural experimental activities, as we will see in the next chapter.

During his years of membership, there is no record of Diard going on a collecting trip for the Natuurkundige Commissie himself. It was only during his 1858–1859 expedition to Ceylon and Mauritius that he gathered collections for 's Rijks Museum in Leiden again,[128] even though the main goal of this expedition was to inspect agricultural methods.[129] Diard's career came full-circle when he received an appointment to collect for 's Rijks Museum in Leiden in 1862 on the recommendation of its director Schlegel, who had succeeded Temminck in 1858.[130] However, that was not to be. Diard's zeal to provide

124 Veth, *Overzicht van hetgeen gedaan is*, 63–64.

125 'Je ne veux pas terminer celle-ci sans vous avoir témoigné toute la satisfaction que j'éprouve de vous avoir attaché a notre commission scientifique. J'en attends les résultats les plus satisfaisants pour la science que nous cultivons et pour le musée auquel vous avez dès-a-présent comme moi le même intérêt a consacrer votre zêle et votre dévouement.' Quoted in Veth, *Overzicht van hetgeen gedaan is*, 64–65.

126 Veth, *Overzicht van hetgeen gedaan is*, 68. Res.GG, January 8, 1830, no. 4. NL-HaNA, Koloniën, 1814–1849, 2.10.01, inv.no. 2833. Diard already appears as directing member ('chef') in Res.GG, October 23, 1829, no. 42. NL-HaNA, Koloniën, 1814–1849, 2.10.01, inv.no. 2830. It is unclear exactly when and if Diard was officially promoted to this role.

127 Res.GG, August 26, 1841, no. 1aa. NL-HaNA, Koloniën, 1814–1849, 2.10.01, inv.no. 2621.

128 Diard, "Pierre-Médard Diard," 208; Veth, *Overzicht van hetgeen gedaan is*, 141–142.

129 Peyssonnaux, "Vie, voyages et travaux," 65.

130 Veth, *Overzicht van hetgeen gedaan is*, 144, 148–149.

perfect specimens for the Museum led to his demise. According to this brother Hippolyte, Diard had developed an improved technique of preservation which involved the use of arsenic. Diard did not want his co-workers to be exposed to this toxic substance, and insisted on preparing himself. Eventually, this led to his death by arsenic poisoning on February 17, 1863.[131] When summarising the results of Diard's collecting efforts in the Indies, Hippolyte writes:

> As a result, the Leiden Museum became one of the richest museums in Europe, and no other could be compared to it, in all that concerned the natural history of Java.[132]

10 A Holistic Diard in a Transimperial Setting

Now that we have looked at the narratives of the three Diards and the chronological accounts of the lives of French, British and Dutch Diard have been harmonised, it is time to consider the complete Diard in a transimperial setting. It is important to first note that Diard was not a foreigner from an 'empireless' nation. As such, he lacked the neutrality that allowed many German and Swiss naturalists to work without any issues in the Netherlands Indies.[133] As a result of the conflicts that followed in his wake, he was never able to develop the unblemished reputation that allowed the Sarasin cousins to work peacefully in the Netherlands Indies.

Diard's career went from conflict to conflict and he was tossed about by the winds of distrust and suspicion. The clear and obvious external reason for this was the unrelenting competition between the three empires in Southeast Asia. Even though the Napoleonic wars in Europe had ended, Diard arrived at a time when much still needed to be settled in the colonial arena. The British had become masters of Southeast Asia in 1812, while the French empire had been reduced to tiny pockets in a vast British sea. A powerful French Indochina was still a couple of decades away, while the Dutch were just finding their feet again in the Netherlands Indies. In such an unsettled time, it's no surprise that a Frenchman travelling between empires would encounter conflict.

131 Diard, "Pierre-Médard Diard," 208–209.

132 'Il en résulta que le musée de Leyde devint un des plus riches musées d'Europe, et qu'aucun autre ne pouvait lui être comparé, pour tout ce qui concernait l'histoire naturelle de Java.' Diard, "Pierre-Médard Diard," 208.

133 See Teichfischer, "Transnational Entanglements," for foreign nationals in the military medical service in the Netherlands Indies.

However, there is more to it than just competition between empires in Southeast Asia. The fact that Diard was working as a naturalist actually even aggravated the situation for him. Natural history was used as a means to advertise empire. As we saw in chapter 1, Falck used the example of British and French explorers, and even the German financing of Von Humboldt's expeditions, as an incentive to coax Willem I into financing Kuhl's preparation, which prepared the way for the founding of the Natuurkundige Commissie. If even the 'empireless' Germans invested in natural history of what was not even their own empire, all the more so should the Dutch invest in researching their own empire.

It was embarrassing to note that even the British (William Marsden, Thomas Stamford Raffles, and soon after John Crawfurd) had produced more scholarship on the Netherlands Indies than the Dutch themselves had.[134] It had become a matter of national pride that any future travel histories of the Netherlands Indies should be penned under the direction and tutelage of the Dutch government. This was one of the main reasons for the existence of the Natuurkundige Commissie. Both British and French naturalists had published widely on the riches of the (natural) world. In fact, national publications on unknown parts of the world served as a catalyst for colonisation and consolidated empire at home. After the restoration of the monarchy, French naval exploration received a new impetus and both Louis de Freycinet and Jules Dumont d'Urville undertook grand expeditions, with Diard narrowly missing out on joining De Freycinet. And for the British, Darwin was still to come.

These nationalistic imperial motives put natural history at the centre of the public identity of the different empires. As a result, it became a much more sensitive issue for a French naturalist to collect for the French Muséum in the colonial realms of the British or the Dutch. Diard had been sent out by Cuvier to collect for the Muséum in Paris. When he transferred to British service, it became problematic for the British when collections created in service and in pay of the East India Company were about to be sent to the French national collection, allowing the French to undercut the British and steal some of their national pride. This is why the ownership of Diard and Duvaucel's collecting efforts for Raffles was so strongly contested.

Likewise, Temminck was not happy with the fact that Diard was known to have sent collections from the Indies to Paris. Cuvier's boast that the Muséum had managed to procure an impressive collection of specimens from the Netherlands Indies before the Dutch even had initiated any research there

134 Marsden, *History of Sumatra*; Raffles, *History of Java*; Crawfurd, *History of the Indian Archipelago*.

served as a further impetus to block off any future collection shipments to the Muséum. Temminck followed the strategy of neutralising foreign collectors by absorbing them into the Dutch infrastructure. This happened to both Diard and Horner, even though in Horner's case, his Swiss nationality made him a much smaller threat than Diard.

Diard's transimperial career followed a pinball trajectory precisely because of this strong interimperial competition in the realm of natural history. In addition, Diard was an entrepreneurial and versatile person, always following his instinct and jumping at opportunities. It was this character trait that allowed him to move effortlessly between empires. If one door closed, another one would open. The evaluations of Diard's character by Finlayson and Boie as quoted earlier provide us with a glimpse of Diard as a lively, loquacious and generally likeable person, even though Finlayson says (or is this Raffles speaking?) that he had the tendency to overestimate himself and exaggerate somewhat. While this made him quite successful in some respects, it also made it easier for those in authority to criticise him and could turn people against him.

Diard had taken the opportunity offered to him by Cuvier to travel to faraway lands. When he encountered Raffles, he jumped at the opportunity and joined Raffles on his trip to the Malaysian Peninsula and Sumatra. Unfortunately, Diard's opportunistic desire to serve two imperial masters at the same time, invoked Raffles' ire and forced Diard to leave the service of the British and move to the Netherlands Indies. There he was almost tempted into working for the Dutch, but when French royal favour was bestowed on him, he left on his adventure to Vietnam. He managed to manoeuvre himself in a position that would allow him to travel widely, but was blocked by the British to follow the itinerary of his choice and ended up in the Netherlands Indies again. There, he continued to collect for the French as a free agent, while simultaneously working on agricultural projects for the Dutch. Only by severing his ties with the French could Diard finally be appointed as member of the Natuurkundige Commissie and would henceforth exclusively work for the Dutch.

In the terms of Hedinger and Hée, Diard is a mobile subject that moved effortlessly between empires. The contested nature of his transimperial loyalties and movements makes it abundantly clear that the cooperation, competition and connectivity between empires were indeed entangled. In the realm of natural history, the connectivity can be found in the fact that all three empires awarded a similar value to the practice of natural history. Creating collections and publishing travel reports were valued by all three empires, even to such an extent that it almost cut off any cooperative effort in advance. Diard's desire to collect simultaneously for Cuvier and Raffles almost comes out as naïve in a world of interimperial competition. Such a concept was doomed to fail and

Diard and Duvaucel were confronted with the hard reality of this when Raffles seized their collections. The fallout of this conflict would spill over to Europe, where the attribution of the discovery of the tapir to Diard led to an extensive public brawl in the British press.

In the end, Diard's loyalties were not allowed to be layered. The fierce rivalry between the three empires entirely blocked the road for Diard to set up any type of transimperial cooperation and caused him to be bounced around like a pinball from empire to empire. The three empires that he served made sure that any notions of layered transimperial loyalties that Diard might have entertained were ruthlessly suppressed. Each empire demanded Diard's singular loyalty and made it impossible for him to simultaneously serve another imperial master. While Diard may have happily continued to divide his loyalties between the various empires, he repeatedly was blocked from doing this until he finally settled in the Netherlands Indies and cut off any remaining connections to the French. Layered loyalties in a transimperial setting were an unattainable utopia for Diard.

11 Conclusion

French Diard (the intrepid young traveller), British Diard (the unreliable Frenchman) and Dutch Diard (the faithful civil servant) turn out to have been the same person all along. In a period when empire in Southeast Asia was still somewhat in flux, Diard travelled between empires. His adventurous and opportunistic character allowed him to move from the service of one empire to the other, but each of the three empires would not allow Diard to hold a double loyalty. This was due to the importance that was attached to natural history in each of the three empires. It was this connecting factor between the three empires that triggered a fierce rivalry when it came to ownership of collections. Raffles could not allow collections made in the pay of the British to end up in a French museum or publication, while Temminck could not allow collections made in the Netherlands Indies to find their way to France. Both moved to cut off Diard's connections with Cuvier in order to appropriate Diard's collections for their own empires. While Diard may have happily continued to work transimperially, his nationality alone made this impossible. The rivalry between the three empires even made Diard's work a source of contention in European naturalist circles. It was this rivalry that broke the complete Diard and left us with a French, British and Dutch version of this talented and versatile naturalist.

CHAPTER 5

Rival Views of Natural History: the Versatile Career of Pierre-Médard Diard

Fame is a fickle thing. The French naturalist Pierre-Médard Diard, who spent most of his working life in the Netherlands Indies, was not unknown in his own days, but even though he was considered famous, it was certainly not always clear to the public why. When news of his death in the Indies filtered through to the Netherlands, the *Nieuwe Rotterdamsche Courant* wrote the following on April 13, 1863:

> The well-known naturalist Diard recently passed away here; his deeds as president of the Natuurkundige Commissie, etc., seem to have been forgotten to such an extent that we have not been able to collect any material in order to provide an account of his life and deeds.[1]

Diard served as (directing) member of the Natuurkundige Commissie from January 31, 1828, to August 26, 1841. He was obviously remembered for this service, but his actual deeds were not. Even worse, the 'etc.' refers to a career that stretched from 1820 to 1863 and included a wide variety of utilitarian activities that covered almost all facets of science, agriculture, and industry, ranging from tin to tea, from silkworms to sugar.[2] The fact that shortly after his death Diard's deeds in the Netherlands Indies were already forgotten, has perpetuated through the last 150 years. Even today, there is no comprehensive overview of Diard's career in the Indies, while his versatile scientific contributions to natural history, agriculture and industry would certainly merit this. It is time to shed some light on Diard's career in the Netherlands Indies and study his life and work in the context of the concept of layered loyalties.

1 'De bekende natuurkundige Diard is dezer dagen hier overleden; hetgeen hij vroeger als president der natuurkundige commissie enz. heeft verrigt, schijnt zoodanig in 't vergeetboek geraakt te zijn, dat wij tot heden niet geslaagd zijn, eenig materiaal te verzamelen tot het opstellen eener schets van zijn leven en verrigtingen.' *Nieuwe Rotterdamsche Courant*, no. 102 (April 13, 1863): 1.

2 The news item in the *Nieuwe Rotterdamsche Courant* is based on *Bataviaasch Handelsblad*, no. 15 (February 21, 1863): 4, which gives some more information, but where the public is still asked for information that would allow the newspaper to come up with a reliable obituary.

 | DOI:10.1163/9789004696877_007

In the previous chapter, we saw how in his early career in Southeast Asia Diard moved between empires from 1818 to 1824 and how he finally settled in the Netherlands Indies. Thanks to his good relations with both governor-general Van der Capellen and commissioner-general Du Bus, Diard was able to turn his interest in agricultural development into a career. Diard's transimperial knowledge was recognised by Van der Capellen and led him to appoint Diard as cultivation inspector. Even though Diard's position as cultivation inspector was dissolved during the reorganisations of Du Bus' strict economising regime, this did not mean the end of Diard's usefulness. Through his relationship with Du Bus, Diard was able to safeguard his own position and continued to use 's Land Plantentuin for the purpose of small-scale agricultural experiments. Eventually, Diard was given a plot of land in Wanayassa to conduct his own cultivation experiments. He even was appointed as director of an agricultural establishment in Krawang of his own design that was intended to facilitate agricultural experimentation, a position he would hold from the opening of the establishment in April 1828 until July 9, 1829.[3]

After the return of the Commissie from Timor in 1830, they worked on Java and were always within only a couple of days travel from Diard. This allowed Diard the opportunity to communicate regularly with the members without joining them on their expeditions. From 1833 onwards, the members went on expeditions, always without Diard, to Sumatra, Borneo, and Celebes. Boie and Macklot had been travelling directing members, but Diard turned it into an administrative post. His experience of working closely with Van der Capellen and Du Bus had provided him with a good understanding of the working of government, and as such, Diard administratively handled all proposals from the Commissie to the government. Diard was the essential office manager for the Commissie who knew the ways of the government and was able to make his voice heard. This role still gave ample freedom to concentrate on his agricultural projects. Even though he was not in charge of the establishment in Krawang anymore, he had not given up on agricultural development. Throughout his directing membership of the Commissie, Diard continued to work on agricultural matters, devoting years of his life to sericulture, sugar manufacturing and tea cultivation.

Given Temminck's earlier resistance to the inclusion of an agricultural component in Korthals' instructions, it can be safely said that he would not have condoned Diard's agricultural activities, if he would have known about it. What these three projects of Diard show is that natural history without agricultural

3 For a detailed narrative of Diard's early agricultural experiments in the Netherlands Indies, see chapter 6 of my original dissertation, available at https://hdl.handle.net/1887/3643457.

innovation was not an option for him. In fact, at one point, Diard even suggested to create a second Natuurkundige Commissie that could focus solely on agricultural innovation, because he had clearly understood that in Temminck's world there was no place for this. But let us first turn to Diard's involvement with sericulture, sugar manufacturing, and tea cultivation.

1 Silkworm Experiments

When Pieter Bleeker (1819–1878, Figure 38), a military physician who had just finished 18 years of service in the Netherlands Indies, travelled via Paris on his way home in 1860, he happened to meet with French entomologist Félix Édouard Guérin-Méneville (1799–1874).[4] Both naturalists, the men discussed Guérin-Méneville's recent introduction of sericulture in France and the report that he had written for the French emperor. Guérin-Méneville suggested to Bleeker that perhaps sericulture would be a possibility for the Netherlands Indies. Bleeker eventually edited three papers with reports on sericulture in the Netherlands-based *Tijdschrift voor Nederlandsch Indië*.[5] In the introduction to the first of these, Bleeker mentions that the colonial government was not unfamiliar with 'the normal sericulture, that is, […] that of the silk of the *Bombyx mori*, in the past attempted on Java against great costs, but as it seems, failed because of ineffective leadership.'[6] In the report that follows, it becomes clear that this early failure of sericulture on Java took place under the direction of Diard.

'S Lands Plantentuin in Buitenzorg had several thousand mulberry trees in the late 1820s, the leaves of which are the main food for the domesticated silkworm *Bombyx mori* (Figure 39). Diard had already been experimenting in the Plantentuin since 1826 when he had received a batch of silkworms that had been procured from the British empire.[7] His initial experiments were not unsuccessful, which becomes clear from the fact that his silk received a silver medal at a public exhibition of products of Indies agriculture and industry in 1829.[8] This exhibition had been organised by Du Bus as part of his policy of

4 For more on Bleeker, see Fransen, Holthuis, and Adema, "Type-catalogue of the Decapod Crustacea," 212–213.

5 Bleeker, "Invoering der volkszijdeteelt."

6 'De gewone zijdeteelt, dat is, […] die der zijde van de Bombyx mori, vroeger op Java met groote kosten beproefd, maar naar het schijnt, door ondoelmatige leiding mislukt.' Bleeker, "Invoering der volkszijdeteelt," 12.

7 *Javasche Courant*, no. 51 (April 26, 1828): 2.

8 *Javasche Courant*, no. 112 (September 24, 1829): 2.

FIGURE 38
Portrait of Pieter Bleeker, artist unknown, lithograph printed by C. W. Mieling, 1847–1865, RP-P-1905-4683, Rijksmuseum, Amsterdam. Public domain HTTP://HDL.HANDLE.NET/10934/RM0001.COLLECT.212361

stimulating agricultural experimentation and offered a platform for people to show their work to the public.[9] Diard's silk was evidently of good quality. He continued working on his silk experiments, and evidently managed to procure the necessary finances.[10]

Diard worked on sericulture development in a government context at the agricultural establishment in Krawang until July 1829, when he gave up his directorship of the establishment. Even without Krawang at his disposal, Diard continued his experiments and managed to procure an appointment as 'inspecteur over de zijde-kultuur' [sericulture inspector], on May 8, 1833, by governor-general Van den Bosch.[11] This post was made necessary by the growth of sericulture in the Indies. Diard received a monthly salary of f200, in addition to his salary as directing member of f500. As always, Diard came up with grand plans, but a proposal to start sericulture in nine residencies at a cost of f4,000 per residency was rejected by governor-general ad interim Baud.[12]

9 De Prins, *Voor keizer en koning*, 175; Groot, *Van Batavia naar Weltevreden*, 255.

10 Van Oort mentions an advance of 5000 guilders, given to Diard in 1830. Weber and Van Zanen, *Kunstenaar op Java*, 98.

11 Res.GG, May 8, 1833, no. 2. NL-HaNA, Koloniën, 1814–1849, 2.10.01, inv.no. 2847.

12 Res.GG, April 18, 1834, no. 3. NL-HaNA, Koloniën, 1814–1849, 2.10.01, inv.no. 2851.

FIGURE 39 '*Bombyx mori*. The Silk Moth.' Plate 13 from Harris and Westwood, *Aurelian*. Public domain. On October 12, 2024, Topaz Gigapixel ai was used to increase the resolution of this image.
HTTPS://WWW.BIODIVERSITYLIBRARY.ORG/PAGE/39060136

What was the nature of the sericulture experiments that Diard conducted? The main reason why these experiments needed to be conducted was that sericulture was an imported phenomenon. Due to a long history of domestication in China, silkworms were not able to survive and reproduce without human intervention. The aim of the experiments was to find out what the best ways were to develop a healthy and productive silkworm population and how to best manage this population.

There are several eyewitness reports that give us some insight in Diard's experiments. One such report is hidden away in a description of a volcano on Java by Commissie member Müller. Müller writes that, on the slopes of the volcano, Diard had built a brickwork basement, covered by a closed wooden shade. The aim of this refrigerator, invented by Diard, was to conserve the eggs of various breeds of silkworms for an extended period of three to nine months, or perhaps even longer, in order to slow down the development of the eggs. Slow development under stable and cool temperatures was meant to result in healthier caterpillars.[13] Diard worked on this experiment for a long time, because after Müller's departure for the Netherlands he asked for and received a sum of f400 to improve his refrigerator.[14]

A second report on Diard's experiments was written by Thomas Josephus Willer (1808–1865). The report was published as part of the Bleeker essays in 1863. Willer writes that Diard had been struggling with various diseases in his silkworm population, including grasserie and muscardine. He had failed to solve the issues by using smoke and various types of olfactory measures to clean the air in the worm hall inside the magnanery (the building in which sericulture takes place). Diard decided to experiment with moving the silkworms outside. He had heard that there were Chinese silkworm farms that raised their caterpillars in the open air. Diard writes, as quoted by Willer:

> I took some precautions against birds and ants, and placed on the mulberry trees a number of worms that were suffering most from grasserie. To my astonishment, I saw that they were healed overnight – and that they were soon spinning strong cocoons.[15]

13 Susanna, *Verhandelingen*, vol. 3, 451.

14 Res.GG, April 15, 1838, no. 20. NL-HaNA, Koloniën, 1814–1849, 2.10.01, inv.no. 2581.

15 'Ik nam [...] eenige voorzorgen tegen de vogels en de mieren, en plaatste op de moerbezieboomen een aantal wormen die het sterkst aan geelzucht leden. Ik zag tot mijne verwondering dat zij in één nacht waren genezen, – en dat zij weldra sterke tonnetjes sponnen.' Willer in Bleeker, "Invoering der volkszijdeteelt," 26.

Diard repeated the experiment with thousands of silkworms that he placed in special roofed racks in the mulberry plantations through which he led branches with leaves as food for the caterpillars. This combined the protection of indoors cultivation with the benefits of eating fresh mulberry leaves in the open air. Initial results were encouraging, but in the end, Diard concluded that it was impossible to apply this model on a large scale for several reasons: the protective nets were too expensive, there was no adequate protection against rain, the caterpillars would not be able to remount the racks or even the trees when they fell off, the caterpillars did not have the instinct to spread out, and the cocoons would be too exposed to burning sun by day and freezing cold by night.

A third eyewitness to Diard's sericulture experiments is Commissie artist Pieter van Oort. Van Oort had been with the Commissie since 1825 as artist-assistant to Boie and Macklot. From him we learn that Diard developed and built an almost industrial spinning wheel running on hydraulic power to unwind the silk cocoons. A large wheel was powered by water and set four spinning wheels in motion that were operated by Javanese boys.[16] While Van Oort praises this invention, his evaluation of Diard is generally negative. He makes it very clear that he believes that Diard is unfit for any type of commercial undertaking: 'Perhaps sericulture can bring some advantages to the country; but I think it should then be set up and executed by someone other than Mr. Diard. His Honour is unfortunate in all his undertakings.'[17]

In addition, Diard received enormous amounts of money for his agricultural projects but was not even capable of fulfilling the simple duties of a member of the Natuurkundige Commissie. He was supposed to collect, describe and dispatch, but simply did not do it. It was a waste of resources to give Diard, in addition to his 'small salary' ['bezoldigingje'] of 6,000 guilders per year, all this money for his agricultural establishment (80,000 guilders) and his silkworm experiments (5,000 guilders). Van Oort concluded that Diard was simply too well-liked by the people in power and because of that was able to easily pull the wool over their eyes so they did not see his failures: 'Oh! How my country is bamboozled!'[18]

---

16 Weber and Van Zanen, *Kunstenaar op Java*, 353, 359.

17 'Misschien dat de zijdeteelt eenig voordeel aan het land kan aanbrengen; maar dan geloof ik dat het door iemand anders moest als den heer Diard worden begonnen en uitgevoerd; Zijn Edele is in alle ondernemingen ongelukkig.' Weber and Van Zanen, *Kunstenaar op Java*, 157.

18 'O! Wat wordt mijn land gefopt!' Weber and Van Zanen, *Kunstenaar op Java*, 196.

These three eyewitness reports indicate the nature of Diard's experiments. Faced with the problems of transplanting a foreign industry into the Indies, he had to basically adapt most aspects of sericulture to the local conditions. Diard here serves as an inventor of small-scale innovations that he is experimenting with to see whether they are available for large-scale production processes. The individual experiments serve the single goal of turning sericulture into a sustainable and profitable industry. While the small-scale experiments seem to have been successful, or at least to have had some potential, from the reports by Bleeker and Willer it becomes clear that Diard's experiments with the *Bombyx mori* did not lead to a thriving sericulture industry. The conclusion of their research was that earlier attempts had failed, and that now a new attempt needed to be made with a different kind of silkworm. Cultivating *Bombyx mori* was a road best avoided.

Diard seems to have been rather well off in this period of his life. When he asked for his second sugar contract (see below), he even offered to continue to serve as sericulture inspector and directing member of the Natuurkundige Commissie in an honorary capacity, meaning that he would do the work but not receive the salary. This offer was eagerly accepted and from June 1, 1836, Diard was not paid anymore for his work in sericulture and with the Natuurkundige Commissie.[19] When it became clear that Diard intended to depart for Europe, he was honourably discharged with gratitude as sericulture inspector on July 15, 1841, and replaced by Louis Marie Rollin Couquerque (1791–1855).[20]

2 Diard as Sugar Baron

One of the main colonial cash crops in the Indies was sugar cane. According to Cees Fasseur, in his classic study on the cultivation system, after 1847 sugar became the second-most important cash crop after coffee.[21] However, the organisation of the sugar industry was somewhat unusual in the sense that it was not as strongly coordinated by the government as coffee and indigo for example were under the cultivation system as established by Van den Bosch in 1830. The industry was supported and stimulated by the government, but the sugar mills were private establishments that existed by the grace of

19 Res.GG, April 19, 1836, no. 17. NL-HaNA, Koloniën, 1814–1849, 2.10.01, inv.no. 2860.

20 Res.GG, July 18, 1841, no. 1. NL-HaNA, Koloniën, 1814–1849, 2.10.01, inv.no. 2620. Rollin Couquerque's registry card is available at NL-HaNA, Koloniën, 1814–1849, 2.10.01, inv.no. 3093, f. 213.

21 Fasseur, *Kultuurstelsel*, 63, translated as *Politics of Colonial Exploitation*.

FIGURE 40 'La fabrique de sucre située dans le domaine de Serpon, érigée pas Mr. J. J. van Braam.' ('The sugar factory located in the Serpon estate, erected by Mr. J. J. van Braam'), lithograph by H. J. Backer, 1842, KITLV 47B22, Leiden University Libraries, Digital Collections. Original plate from Braam, *Vues de Java*. Public domain
HTTP://HDL.HANDLE.NET/1887.1/ITEM:855369

a sugar contract with the government (Figure 40). The deal was simple: the government would make sure that local personnel was available to plant the government-provided cane for a fixed salary (*plantloon*). The private entrepreneur was expected to build a sugar factory in which they would process the harvested cane. An agreed amount of sugar would be sold to the government for a fixed price. Any surplus production could be sold by the owner of the mill on the open market. An extra advantage for the sugar contractors was that a new contract usually included an interest-free loan that could be used to finance the building of the mill. All in all, the system was very advantageous for the contractors who could make large amounts of money because the personnel salaries were artificially kept low and the production was always higher than the mandatory amounts.[22]

22 See Fasseur, *Kultuurstelsel*, 64–65. For more on sugar cultivation in the Netherlands Indies in the nineteenth century, see Knight, *Colonial Production*; Leidelmeijer, *Van suikermolen tot grootbedrijf*; Knight, *Sugar, Steam and Steel*; Bosma, Giusti-Cordero, and Knight, *Sugarlandia Revisited*; Bosma, *Sugar Plantation in India and Indonesia*.

Diard's first attempt to become involved in sugar cultivation dates from 1835, two years after his appointment as sericulture inspector, in the form of another grand plan. Diard requested to be allowed to start a sugar factory, either in cooperation with the government, or at his own expense. The plan was to use this factory to experiment with equipment to improve the manufacturing process. Diard indicated his desire to first travel throughout Java to inventory current techniques and then travel to Europe and the West Indies to familiarise himself with the most recent developments. This journey should result in a scientific report to the government with information on all useful branches of trade and agriculture and could result in the 'formation of a new committee of natural historical research.'[23] Baud wisely considered this plan to be too grand and put the sugar factory and the journey to Europe and the West Indies on hold.

However, the plan to explore agricultural practices on Java appealed to him. On June 1, 1835, Baud charged Diard with a geological-agronomical journey through the residencies of Java with a fourfold goal. Diard should 1) judge the suitability of each residency for the cultivation of crops destined for the European market, 2) find out why experiments with these crops had such widely different results, 3) suggest improvements for the cultivation of these crops and identify which residencies would be suitable for the wider introduction of certain crops, and, 4) identify the defects in mechanical equipment currently used.[24] An itinerary for Diard's inspections is not available, but snippets from the archives suggest that he kept working on this assignment on and off during the next six years.[25]

While this geological-agronomical journey on Java may have been much to Diard's liking, it did not give him the experimental sugar factory he had asked for. Three weeks after the assignments for the journey had been given to him, he sent in a much smaller proposal, asking permission to rent a plot of land of 250 *bouw* near the village of Karang Sambong in the residency of Cheribon

23 'Tot de zamenstelling eener nieuwe commissie voor natuurkundig onderzoek.' Nothing further is specified regarding this committee, but it seems that Diard intended this second Natuurkundige Commissie to focus on agricultural and industrial innovation. Perhaps Diard intended this committee to pick up some of the work of the Hoofdkommissie van Landbouw that had been dissolved in 1833. However, this second Natuurkundige Commissie never materialised.

24 Res.GG, June 1, 1835, no. 1. NL-HaNA, Koloniën, 1814–1849, 2.10.01, inv.no. 2855.

25 See Res.GG, June 4, 1836, no. 3. NL-HaNA, Koloniën, 1814–1849, 2.10.01, inv.no. 2862; Res.GG, March 11, 1837, no. 5. NL-HaNA, Koloniën, 1814–1849, 2.10.01, inv.no. 2568; Res.GG, August 12, 1839, no. 6. NL-HaNA, Koloniën, 1814–1849, 2.10.01, inv.no. 2597; Res.GG, October 23, 1840, no. 4. NL-HaNA, Koloniën, 1814–1849, 2.10.01, inv.no. 2611.

in order to build a sugar factory and ownership of the buildings of the former indigo factory Bandjar Kadipaten. Baud saw no problem with this and asked the resident of Cheribon for a draft sugar contract.[26] The final contract was approved on September 17, 1835, and stipulated that Diard would receive the buildings from the former indigo factory in ownership and the rent of a plot of land of 250 to 300 *bouw* for a period of 25 years.[27]

In 1836, Diard managed to add a second sugar contract in Cheribon to his business portfolio. This was transferred to Diard from a certain Tan Uko, who had been in possession of the contract since July 17, 1833. Perhaps it was Diard's expectation of the extra income that inspired him to offer to continue both his official positions (directing member and sericulture inspector) in an honorary capacity. After consulting with the resident of Cheribon, governor-general De Eerens approved the transfer of the contract on October 31, 1836,[28] and the final draft for the new contract was approved on December 17, 1836.[29] In 1838, Diard requested permission to sell both his sugar contracts in Cheribon, possibly with an impending departure to Europe in mind. His request was approved on July 10, 1838, ending Diard's first period as sugar manufacturer.[30]

Within a year of selling his contracts, Diard would again be involved in sugar manufacturing. Catalyst for this was his report on the mechanical equipment used in sugar manufacturing on Java, based on his geological-agronomical inspection of the various crop cultivations on Java. This report, given to the government on February 28, 1839, convinced governor-general De Eerens to grant Diard, by exception, a sugar contract in the residency of Besuki (East Java) on March 29, 1839.[31] The draft contract was finally approved on August 2, 1839.[32]

The goal of this factory was to experiment with the use of vacuum pans, which were used in the production of sugar. The advantage this brought to the production process was that in the low-pressure vacuum pan sugar had a lower boiling point, which meant that less fuel was needed. Following directly from

26 Res.GG, June 23, 1835, no. 2. NL-HaNA, Koloniën, 1814–1849, 2.10.01, inv.no. 2855.

27 Res.GG, September 17, 1835, no. 8. NL-HaNA, Koloniën, 1814–1849, 2.10.01, inv.no. 2856. A *bouw* was a Javanese measurement for the amount of land that can be cultivated by one person. It was subject to local variations but it is defined here as 500 *roeden*, which comes down to one *bouw* roughly equalling 7,096.5 m², or 0.7 hectare. 250 to 300 *bouw* roughly equals 175 to 210 hectare, or 430 to 520 acres.

28 Res.GG, October 31, 1826, no. 9. NL-HaNA, Koloniën, 1814–1849, 2.10.01, inv.no. 2565.

29 Res.GG, December 17, 1836, no. 20. NL-HaNA, Koloniën, 1814–1849, 2.10.01, inv.no. 2565.

30 Res.GG, July 10, 1838, no. 28. NL-HaNA, Koloniën, 1814–1849, 2.10.01, inv.no. 2584.

31 Leidelmeijer, *Van suikermolen tot grootbedrijf*, 343. See Res.GG, March 29, 1839, no. 3. NL-HaNA, Koloniën, 1814–1849, 2.10.01, inv.no. 2592.

32 Res.GG, August 2, 1839, no. 2. NL-HaNA, Koloniën, 1814–1849, 2.10.01, inv.no. 2597.

that, a lower temperature meant that less sugar would be lost to the chemical process of inversion.[33] There were several types of vacuum pans on the market and it was Diard's intention to perfect the production process with the use of vacuum pans. Along with the contract, Diard also received an interest-free loan of fl50,000 that would allow him to build the model factory. The model factory was located in the district of Sumber Karang in Probolinggo, at this time part of the residency of Besuki.

In an attempt to expand his sugar manufacturing activities, Diard purchased two more sugar contracts in Besuki in a public auction. The sale to Diard was confirmed by De Eerens on November 27, 1839.[34] When an issue came up regarding the financing of the purchase (Diard's bondsmen had a change of heart), the government cancelled the sale of both contracts on June 25, 1840.[35] The situation would get worse for Diard when he was warned because he was running the risk of not completing the construction of the factory in time for the harvest.[36] Shortly after, he fell gravely ill and had no choice but to transfer the contract for his model factory in Sumber Karang to Dirk van Aalst (1812–1867) on September 8, 1840.[37] Van Aalst would continue Diard's experiments with vacuum pans.[38]

In the space of just over five years, Diard had owned and lost five sugar contracts. The first two, in Cheribon, were voluntarily sold to new manufacturers, the third one, in Besuki, had to be sold due to ill health, and the approved sale of the last two, also in Besuki, fell through when his financial backers withdrew. Given the fact that many sugar manufacturers developed their factories over a long period of time, it is clear that Diard was not interested in sugar manufacturing as a long-term career. His sole goal was to improve the production processes, and all his requests for sugar contracts were always made in the context of a grand plan to travel the world in order to learn of the latest developments in sugar manufacturing (and other types of cultivation) and then return to Java to implement those findings in order to increase the yields.

This grand agricultural tour of the world never happened on the scale Diard had proposed.[39] His final proposal of this nature was denied on August 26, 1841, but it did leave him with an appointment as honorary cultivation

33 Leidelmeijer, *Van suikermolen tot grootbedrijf*, 52–53.
34 Res.GG, November 27, 1839, no. 2. NL-HaNA, Koloniën, 1814–1849, 2.10.01, inv.no. 2600.
35 Res.GG, June 25, 1840, no. 1. NL-HaNA, Koloniën, 1814–1849, 2.10.01, inv.no. 2607.
36 Res.GG, July 18, 1840, no. 9. NL-HaNA, Koloniën, 1814–1849, 2.10.01, inv.no. 2608.
37 Res.GG, September 8, 1840, no. 2. NL-HaNA, Koloniën, 1814–1849, 2.10.01, inv.no. 2610.
38 Fasseur, *Kultuurstelsel*, 245.
39 It cannot be excluded that Diard travelled the world on his own without any government support between 1842 and 1848. I have not seen any definite proof of Diard's whereabouts

inspector and a gratification of f2,000 to pay for his way back to Europe via Mauritius and Bourbon (La Réunion).[40] While this title was a recognition for Diard's career in agricultural development in the Indies, it simultaneously allowed the government to reap the benefits from Diard's further work without having to pay him. Diard studied agricultural methods on Mauritius and Bourbon from September 25, 1841, to February 24, 1842.[41] After his return to the Indies in early 1848, Diard's request for a new contract (made in 1847 in the Netherlands) was denied on January 6, 1849.[42] Another government-sponsored trip, from November 10–12, 1857,[43] to April 21–26, 1859,[44] allowed him to once more inspect the sugar industry on Mauritius, Bourbon and cinnamon and coffee cultivation on Ceylon.[45] Diard's sugar experiments even allowed him to develop a sugar cane variety, most commonly known as the Belloguet, that would be exported and used in various countries.[46]

The legacy of Diard's early work in sugar manufacturing on Java, however, was not an impressive one. Purchasing and selling or losing five contracts in just as many years does not look like a stable platform for innovation. The construction of his model sugar factory in Sumber Karang does not even seem to have been fully complete when Diard was forced to sell the contract to Van Aalst. As a result, the vacuum pan experiments were not conducted by Diard himself and any results that came from them must be credited to Van Aalst.

This rather unsuccessful outing in the sugar industry was not unremarked upon by Commissie member Forsten, who had arrived in Batavia in December 1838. He rather accusingly writes that Diard had left for his new sugar mill in September 1839, leaving Forsten to represent the Commissie with the Bataviaasch Genootschap.[47] This remark fits in with the general tone of the rest of the document (a summary of Forsen's activities from 1838 to 1840), in which Forsten is somewhat critical of and displeased with Diard.

---

in those years, but I also have not seen any mention of experience gained in this way after Diard's return to the Indies.

40 Res.GG, August 26, 1841, no. 1aa. NL-HaNA, Koloniën, 1814–1849, 2.10.01, inv.no. 2621.

41 Scalliet, *Antoine Payen*, 669.

42 Res.GG, January 6, 1849, no. 10. NL-HaNA, Koloniën, 1814–1849, 2.10.01, inv.no. 2714. See also Diard's registry card. NL-HaNA, Koloniën, 1814–1849, 2.10.01, inv.no. 3093, f. 486.

43 *Java-Bode*, no. 91 (November 14, 1857): 8. The precise date of Diard's departure is not given.

44 *Bataviaasch Handelsblad*, no. 61 (April 27, 1859): 8. The precise date of Diard's arrival is not given.

45 Registry card Pierre Diard. NL-HaNA, Koloniën, 1814–1849, 2.10.01, inv.no. 3093, f. 487.

46 Storey, *Science and Power in Colonial Mauritius*, 45, 62; Deerr, *Cane Sugar*, 27.

47 Kort verslag van het in Indië verrigte van 14 December 1838 to April 1840. NCAO, 1076_004.

Upon his arrival in the Indies, Forsten had immediately requested to start receiving his salary from the Indies government. When he still had not received a response by March 1839 and he was forced into inactivity because of a lack of funds, he enquired with the director of finances why he still had not received a response. He then was told that the documents needed for him to receive his funding had already been sent to Diard for his consideration on December 22, 1838. The director of finance immediately approved Forsten's salary request and Diard was reminded that he still needed to approve the documents. Clearly, Forsten blamed Diard's investment in his sugar factory for the problems that he himself experienced. However, as we have seen in chapter 4, the circumstances that led to delay for Forsten mostly were out of Diard's hands.

3 Tea Trouble

During his time as cultivation inspector, Diard had already been involved in tea experimentation (Figure 41). Diard himself was convinced that he had been instrumental in the introduction of tea to Java. In an unidentified archival document,[48] Diard mentions that he had already encouraged Van der Capellen in 1821 to experiment with tea cultivation. During Diard's absence, the government ordered tea seeds to be imported from Dejima (Japan) in 1822, 1823 and 1824, but all of these seeds arrived spoiled. Upon his return, Diard wrote instructions for safely transporting tea seeds and finally managed to receive seeds in good condition in the second half of 1826.[49] These seeds were planted in Buitenzorg[50] and eventually made their way to the agricultural establishment in Krawang.

During his years with the Natuurkundige Commissie, Diard continued to occupy himself with tea experiments. While his experiments with silkworms and sugar were government-sanctioned, in the form of the official post of

48 Van der Chijs, *Geschiedenis*, occasionally references a document by Diard from 1839. I assume this refers to some kind of tea cultivation report that Diard had written for the government, but the lack of a reference by Van der Chijs makes it impossible to identify this document. Van der Chijs adds that this document must be placed in the context of several people vying for the honour of being the one to have introduced tea cultivation to Java, so he is correct in approaching it with some caution.

49 Van der Chijs, *Geschiedenis*, 2–3. See *Gedenkboek der Nederlandsch Indische theecultuur*, for more on tea cultivation in the Netherlands Indies. Diard is mentioned in the contributions of Bernard and Cohen Stuart. See Ukers, *All about Tea*, for worldwide tea cultivation.

50 Van der Chijs, *Geschiedenis*, 3, quoting a report from the Hoofdkommissie van Landbouw from July 4, 1827.

FIGURE 41 'Thé, *Thea Sinensis, Rich.*' Plate 43 from Reveil and Dupuis, *Flore médicale*. Public domain. The tea shrub, *Thea sinensis*, is currently known as *Camellia sinensis*. On October 12, 2024, Topaz Gigapixel ai was used to increase the resolution of this image.
HTTPS://WWW.BIODIVERSITYLIBRARY.ORG/PAGE/12108560

sericulture inspector and the five sugar contracts that were transferred or sold to Diard, his tea experiments were mostly of a private nature. Diard never received any payment from the government for his involvement, even though his supervision of tea cultivation in Cheribon was officially recognised. This may have been because his role as directing member of the Natuurkundige Commissie was assumed to include an involvement in tea experiments. When Bürger sent a batch of 1,500,000 tea seeds from Japan to Java in 1832, they arrived at 's Land Plantentuin and Diard was the one to report this to the government. The seeds were destined by governor-general Van den Bosch for use in the agricultural establishment in Krawang.[51] The government was so pleased with this shipment that they awarded Bürger a gratification of fl,000.[52]

In addition to this, his involvement with sericulture also offered Diard the opportunity to work on tea experiments. In 1832, he received a plot of land in Tjitjoeroek (Priangan, Java) in use from the resident in order to conduct experiments with silkworms, tobacco and sheep, without any further government support. However, Diard used part of the land to plant tea shrubs. During the next ten years, this plantation would grow and produce yields that even exceeded the most optimistic estimates.[53] The result of Diard's experiments were much more successful than the results from the work of the official tea inspector.

The position of tea inspector, equivalent to Diard's position of sericulture inspector, was given to Jacobus Isidorus Lodewijk Levien Jacobson (1799–1848) from January 1, 1833.[54] Jacobson had arrived on September 27, 1827, in service to the Nederlandsche Handel-Maatschappij, an export company that had been founded by Willem I in 1824.[55] In the years leading up to his appointment, Jacobson had repeatedly travelled to China to discover the secrets of tea cultivation and import tea experts, tea seeds and everything else needed for the successful cultivation of Chinese tea on Java. Jacobson travelled back and forth to China six times in five years. His final journey was necessary because most of the Chinese teamakers he had brought with him on his previous journey had been killed during the uprising in Krawang in May 1832.[56]

51 Res.GG, January 14, 1832, no. 1. NL-HaNA, Koloniën, 1814–1849, 2.10.01, inv.no. 2543.

52 Res.GG, January 24, 1832, no. 9. NL-HaNA, Koloniën, 1814–1849, 2.10.01, inv.no. 2842.

53 Van der Chijs, *Geschiedenis*, 125–27.

54 Van der Chijs, *Geschiedenis*, 16. Jacobson's registry card is available at NL-HaNA, Koloniën, 1814–1849, 2.10.01, inv.no. 3094, f. 546.

55 For a history of the NHM, see De Graaf, *Voor handel en maatschappij*.

56 For a full account of Jacobson's involvement with tea cultivation on Java, see Van der Chijs, *Geschiedenis*, 1–33.

A government assessment on the state of tea cultivation on Java in 1834 concluded that the situation was not as desired. Both Jacobson and Diard had handed in reports on the state of the three tea plantations Wanayassa, Bodja Negara and Tjiseroepan (all in Priangan, Java), and the general state of tea production in Priangan, while additional information was provided by assistant-resident of Wanayassa Pierre François Clignett (1802–1841) and cultivation director Burchard Johan Elias (1799–1871).[57] The conclusion was that the tea produced by Jacobson was of inferior quality compared to the tea of Chinese teamakers what was produced without Jacobson's involvement.

As a result, governor-general Baud decided to limit Jacobson's responsibilities to Priangan, assign the planned tea cultivation in Cheribon to Diard, and order a committee of three to inspect the three plantations in Priangan for a further evaluation. This committee consisted of former cultivation director Van Sevenhoven, current cultivation director Elias, and the resident of Priangan, Otto Carel Holmberg de Beckfelt (1794–1857).[58] They were to be accompanied by Diard and Jacobson.[59] The final report of the committee was devastating for Jacobson. He was to blame for both the poor state of the tea shrubs and the bad relationships with the local officials and personnel. In an act of shrewdness from Baud, Jacobson was banned from any further involvement in tea cultivation in the residency of Priangan, but the district of Wanayassa was administratively removed from Priangan and added to Krawang. Jacobson was appointed assistant-resident of Wanayassa and allowed to continue to work as tea inspector only in that district.[60] And all this because of a very critical report by Diard![61] It is no surprise that there was no love lost between Jacobson and Diard and the former would continue to blame Diard for stealing the oversight of tea cultivation in Cheribon from him and maintained that 'through [his] quackery the government lost a lot of profit.'[62]

In the end it would be Jacobson's name that would be associated with the introduction of tea cultivation to Java and not Diard's. The tea plantations in Cheribon were a place where Diard would experiment, but only for a short

57 Clignett's registry card can be found at NL-HaNA, Koloniën, 1814–1849, 2.10.01, inv.no. 3093, f. 541. Elias' registry card is available at NL-HaNA, Koloniën, 1814–1849, 2.10.01, inv.no. 3094, f. 541.

58 Holmberg de Beckfelt's registry card can be found at NL-HaNA, Koloniën, 1814–1849, 2.10.01, inv.no. 3093, f. 45.

59 Res.GG, October 18, 1834, no. 6. NL-HaNA, Koloniën, 1814–1849, 2.10.01, inv.no. 2853.

60 Res.GG, January 12, 1835, no. 10. NL-HaNA, Koloniën, 1814–1849, 2.10.01, inv.no. 2854.

61 Van der Chijs, *Geschiedenis*, 27.

62 'Intusschen had het Gouvernment door deze kwakzalverij veel winstderving.' Van der Chijs, *Geschiedenis*, 259.

while.[63] He was assisted by Chinese teamaker Ahoey, who was seconded to him on February 23, 1836, for a period of two months.[64] When Diard reported that his experimental method was extremely successful, he received additional land and Ahoey was allowed to stay on in order to travel between the small tea plantations in Cheribon.[65] However, Jacobson's star had started to rise again and he was given permission to inspect several plantations on Cheribon in the context of his comparative research into tea cultivation. Diard was assured of the government's continued interest in and appreciation of his efforts.[66]

As a result of Jacobson's inspections, all tea cultivation was ordered to follow the general guidelines, with the exception of Diard's experimental tea plantations.[67] Finally, on June 26, 1838, Diard's request to be relieved of all responsibilities regarding tea cultivation in Cheribon was granted.[68] Van der Chijs points out that Diard's request was made shortly after he had received a report on the quality of his tea from Amsterdam.[69] What the verdict was is not entirely clear, but it must have been of such a nature that Diard believed it would not convince the government in the Indies to continue to support his experiments and rather than fight Jacobson to the bitter end, he decided to give up his involvement with tea, also with his planned worldwide cultivation inspection tour in mind.

4 Silkworms, Sugar, and Tea

Diard's experiments with sericulture and tea cultivation had already started in 1826 as part of his short-lived tenure as cultivation inspector. After (or perhaps despite) his appointment as directing member of the Natuurkundige Commissie in 1828, Diard continued with his silkworm and tea experiments. His involvement in sugar manufacturing started later, as a result of a grand plan to travel the world to visit and inspect cultivation sites in other colonial realms. This plan was deemed too grand and Diard was allowed to undertake

63 For the nature of Diard's experiments, see Van der Chijs, *Geschiedenis*, 254.
64 Res.GG, February 23, 1836, no. 1. NL-HaNA, Koloniën, 1814–1849, 2.10.01, inv.no. 2561.
65 Res.GG, June 17, 1836, no. 9. NL-HaNA, Koloniën, 1814–1849, 2.10.01, inv.no. 2862.
66 Res.GG, May 22, 1837, no. 1. NL-HaNA, Koloniën, 1814–1849, 2.10.01, inv.no. 2570.
67 Res.GG, September 16, 1837, no. 1. NL-HaNA, Koloniën, 1814–1849, 2.10.01, inv.no. 2574, confirmed in Res.GG, November 15, 1837, no. 20 and Res.GG, November 29, 1837, no. 1. NL-HaNA, Koloniën, 1814–1849, 2.10.01, inv.no. 2576.
68 Res.GG, June 26, 1838, no. 6. NL-HaNA, Koloniën, 1814–1849, 2.10.01, inv.no. 2583.
69 Van der Chijs, *Geschiedenis*, 258.

such an inspection tour on Java and start his own model sugar factory in order to experiment with the mechanical manufacturing processes.

Even though all his work took place in the developmental stages of what would become large agricultural sectors, Diard's contributions were never decisive enough to securely attach his name to any of these three cultivations. One of the main reasons for this is that Diard never specialised in one specific cultivation, but always worked on various projects simultaneously. He never devoted his undivided attention to one single crop, much in contrast with people like Jacobson.

In addition, his directing membership of the Natuurkundige Commissie was considered to be his main responsibility, especially because this was a well-paid position. It also did not help that none of his experiments proved decisive for the further course of the development of these cultivations. Bleeker's conclusion was that Diard's work had only shown that cultivating the silkworm species *Bombyx mori* on Java was not possible. When it came to tea, Diard was eclipsed by Jacobson, even though the quality of his work does not seem to have been the problem. Likewise, Diard's dabbling in sugar manufacturing is only a footnote in the history of sugar cultivation on Java, despite it resulting in a sugar cane variety that was exported and used worldwide.

For Diard himself, there never was a conflict between his agricultural experiments and his directing membership of the Natuurkundige Commissie. As we have seen, both Van Oort and Forsten were not happy with Diard's agricultural activities and believed that they distracted Diard to such an extent that his work for the Natuurkundige Commissie suffered from it. However, the question must be asked what Diard's own perception of his role in the Netherlands Indies and in the Natuurkundige Commissie was. We can find an answer when we look at the museum of natural history in Batavia that was conceived in 1829, founded in 1835, opened in 1837, and abolished in 1843. Diard was involved with this project from the beginning and his speech at the opening ceremony gives us a detailed insight into why this museum was such an important project for Diard. It is to the story of the museum that we now turn.

5 The Colonial Museum in Batavia

Diard was a member of the Bataviaasch Genootschap van Kunsten en Wetenschappen. The Genootschap was founded in 1778 and was the first society of its kind in Asia. It had weathered all the regime changes in the previous half century and was still fully functional by the time Diard arrived in the Indies. His first involvement with the Genootschap already took place in early

1825 when he held 'a very important lecture' on the deer of the Indies, within three months of his second arrival in the Indies.[70] The lecture was deemed of such importance that publication of the lecture with four accompanying plates of Javanese and Japanese deer in the *Verhandelingen*, the journal of the Genootschap, was announced, even though it never materialised.[71] The journal normally contained a good deal of natural historical content, with past contributions by Horsfield, Reinwardt and Blume, so Diard's lecture on deer would have been a good fit. On January 8, 1836, Diard would even become a directing member of the Genootschap.[72]

Already in 1828, plans had been developed to set up a natural history museum in Batavia under supervision of the Bataviaasch Genootschap, filled with objects collected by the Natuurkundige Commissie. The plan was initiated by commissioner-general Du Bus and further developed by a three-man committee of the Genootschap. Du Bus asked Diard to evaluate this plan for him and Diard's report is published as an appendix to Peyssonnaux' biography of Diard.[73] The plan included provisions for the renovation of an already existing government building and a strategy to build a collection: the Natuurkundige Commissie would have to donate part of their collections (both objects and drawings). In addition, a network of local leaders should be instructed by the government to collect objects of natural history according to guidelines that were to be set up by the members of the Natuurkundige Commissie. When Du Bus left his position, the plan lingered for a while, until it was finally picked up again by governor-general Baud who issued a decree on the founding of the Colonial Museum on August 13, 1835.[74] The full text of this decree was published in the front matter of the journal of the Genootschap.[75]

Baud started the decree by stating that the government was fully supportive of the founding of a museum of 'natural history, antiquities, rarities, etc.'[76] provided it was connected with the 'Royal Museum in Leiden'[77] as much as

70 'Eene zeer belangrijke voorleezing.' *Bataviasche Courant*, no. 6 (February 5, 1825): 6.

71 The and Van der Veur, *Verhandelingen van het Bataviaasch Genootschap*, includes a complete table of contents of the *Verhandelingen* and does not mention any article or lecture by Diard.

72 Fritze, "Voorberigt," xxv.

73 Peyssonnaux, "Vie, voyages et travaux," 79–84.

74 See Groot, *Van Batavia naar Weltevreden*, 231–288, for an extensive narrative of the development of the museum plans.

75 Fritze, "Voorberigt," xviii–xxii.

76 'Natuurlijk historie, oudheden, zeldzaamheden, enz.' Fritze, "Voorberigt," xviii.

77 'Koninklijk Museum te Leijden.' Fritze, "Voorberigt," xix. This is clearly a reference to 's Rijks Museum van Natuurlijke Historie.

possible. The government would support the exploitation of the Colonial Museum and its collection with f200 per month. The board of the Museum was to be formed by the board of the Bataviaasch Genootschap with a maximum of four additional members. There always needed to be a member of the Natuurkundige Commissie on the board. The Natuurkundige Commissie also had to donate the triplicates and quadruplicates from their collections to the Museum and was instructed to hire local personnel in the various residencies and outer districts to collect for the Museum (paid for from their existing budget). The goal of the Museum, as set out in Baud's decree, was 'to expand the scientific knowledge in the Netherlands Indies and to make contributions in the interest of scientific activities in the motherland.'[78]

During the first board meeting of the Bataviaasch Genootschap after this decree, on January 8, 1836, Diard was appointed as directing member of the Genootschap, which automatically made him a directing member of the new Museum. As such, he was involved in directing the involvement of the Natuurkundige Commissie with the Museum. Over the next three years, as Hans Groot describes, it would turn out to be a somewhat hit-and-miss affair. Commissie assistant Overdijk, who was appointed as curator after his companion and predecessor Van Gelder had left government service, proved to be extremely hard to manage and threatened to seriously disturb the organisation of the collections of the Museum.[79] Despite these challenges, the Museum finally opened to the general public in July 1839 (Figure 42).[80]

When news of the founding of the Colonial Museum in Batavia reached the Netherlands, it was not received well. Temminck complained about this new museum, in a letter dated December 14, 1836, stating that he considered its founding to be detrimental to the cause of 's Rijks Museum. Temminck argued that a rival museum in Batavia would severely reduce the number of specimens that were sent to Leiden and even suggested that this was perhaps even the main (hidden) reason for the founding of the Museum. Baud, who had in the meantime returned to the Netherlands, responded to Temminck's accusations by saying that the founding of the Museum was 'one of those measures, which are sometimes granted, not out of conviction of its use and sustainability, but as appeasement of a revealed public desire.'[81] There was no reason

78 'Om de wetenschappelijke kennis in Nederlandsen Indie te vermeerderen en bijdragen te leveren in het belang der wetenschappelijke verrigtingen in het Moederland.' Fritze, "Voorberigt," xxi–xxii.

79 Groot, *Van Batavia naar Weltevreden*, 265–266.

80 Groot, *Van Batavia naar Weltevreden*, 267.

81 'Een van die maatregelen, welke soms worden ingewilligd, niet zoo zeer uit overtuiging van derzelver nuttigheid en duurzaamheid als om toetegeven aan eenen algemeen

FIGURE 42 'Het Societeits-gebouw de Harmonie (Batavia).' ('The Society building the Harmonie (Batavia)'), lithograph by C. W. M. van de Velde. Plate 8 from Van de Velde, *Gezigten uit Neêrlands Indië*. Public domain. The Colonial Museum of Batavia was located in a renovated wing of the Harmonie. On October 12, 2024, Topaz Gigapixel ai was used to increase the resolution of this image. HTTP://HDL.HANDLE.NET/1887.1/ITEM:129862

to fear the Colonial Museum in Batavia because 'it probably already is, and may always be, a languishing institution, that will not cause any significant damage to the Leiden situation, and that, should the opposite turn out to be true, can be destroyed by the stroke of a pen.'[82] Baud also stressed that Diard's involvement with the Colonial Museum did not flow forth from any negative sentiments towards 's Rijks Museum.

But, as Baud had already predicted, the success of the Colonial Museum in Batavia as a natural history museum would be short-lived. Keeping the collections in the climatic conditions in Batavia in good order turned out to be

geopenbaarden wensch.' Verbaal, August 2, 1837. NL-HaNA, Koloniën, 1814–1849, 2.10.01, inv.no. 1109.

82 'Het is waarschijnlijk nu reeds, en zal wel altoos blijven, eene kwijnende inrigting, die aan het Leidsche geene noemenswaardige schade zal toebrengen, en die in het omgekeerd geval, met een pennestreek kan worden vernietigd.' Verbaal, August 2, 1837. NL-HaNA, Koloniën, 1814–1849, 2.10.01, inv.no. 1109. The quoted passages are both from a report by Baud to the minister of colonial affairs, written on May 24, 1837.

too difficult. Already in December 1839, an inspection of the collections of the Museum by Commissie member Forsten, who had become the representative of the Natuurkundige Commissie on the board while Diard occupied himself with his model sugar factory in Sumber Karang, had led to the conclusion that the collections were not impressive and partly in average condition. He advised the government to ship the collections to 's Rijks Museum in Leiden. The main reason, according to Forsten, of the sorry state of the collections was the fact that the directors were mostly only temporarily in Batavia and were occupied with other things, so that 'the one rejects and casts away what his predecessor has instituted.'[83] An additional report with the same conclusion by Commissie member Schwaner led to the decision, on July 20, 1843, to ship the entire natural history collection, excluding minerals and shells, to 's Rijks Museum in Leiden,[84] which meant the definitive reorientation of the Colonial Museum towards collecting antiquities and rarities instead of natural history.

6 Diard's View of Natural History

In 1837, the only visit of a member of the Dutch royal family to the Netherlands Indies took place. Young prince (Willem Frederik) Hendrik (1820–1879), grandson of king Willem I, spent seven months in the Indies, from February 9 to September 28. During his visit, prince Hendrik travelled the archipelago and ended with an extensive tour of Java.[85] Sunday, August 27, was a day with a busy programme.[86] The prince, accompanied throughout the day by governor-general De Eerens, attended the church service at the school building in Weltevreden (the European suburb of Batavia) and listened to a sermon by newly arrived reverend Wolter Robert van Hoëvell (1812–1879), founder of the *Tijdschrift voor Neêrlands Indië* in 1838. In the evening, he attended a formal session of the Bataviaasch Genootschap, where he listened to speeches by Van Sevenhoven, Diard and Bürger and was offered honorary membership of the Genootschap, which he graciously accepted. Van Sevenhoven was the chairman and spoke on the history of the Genootschap; Heinrich Bürger, whose connection with the Natuurkundige Commissie we have looked at earlier,

83 'Zoodat den eenen verwerpt en omverregooit wat zijn voorganger heeft daargesteld.' Kort verslag van het in Indië verrigte van 14 December 1838 to April 1840. NCAO, 1076_004.

84 Groot, *Van Batavia naar Weltevreden*, 307.

85 For more on prince Hendrik's visit, see Cornets de Groot, *Overzigt der reis*; Fasseur, *Indischgasten*, 231–247; Huyssen van Kattendijke, *Met Prins Hendrik naar de Oost*.

86 *Javasche Courant*, no. 69 (August 30, 1837) contains a report of the prince's activities during this weekend.

spoke on the nature of volcanoes.[87] The formal session was followed by a visit to the developing Colonial Museum of the Genootschap. Diard's speech was deemed to be of such relevance that it was printed in the *Javasche Courant* in its entirety in two instalments.[88] The full text of the speech runs to over 7,000 words and is printed in French.

Diard's speech was clearly meant as a celebration of the recently founded Colonial Museum, 'worthy of being opened to public curiosity and of taking rank among the most honorable establishments of the colony.'[89] Diard closed his speech by a rousing appeal to the prince to give 'an endowment which must, by an act of liberality worthy of the nation, raise at the same time a lasting monument to the noblest of sciences.'[90] This endowment was necessary because the current 'lasting monument' did not adequately serve its purpose and a new 'lasting monument' was needed to honour the worthy cause of the Museum. Finally, he invited the prince to add his name to the annals as 'the first protector of the Colonial Museum of Java.'[91] In order to back up his funding appeal, Diard had made a strong case for the value of the Museum by firmly placing it in the context of science and agriculture in the Indies. It is in this argument that we find strong traces of Diard's personal vision on the place of science and the relations between natural history and agricultural development.[92] Diard clearly saw the Natuurkundige Commissie in a different light than Temminck.

Diard hails 'the new era of prosperity and commercial splendour which is beginning for this colony.'[93] This commercial splendour goes hand in hand with 'brilliant cultural enterprises' that are financed by both 'public revenue' and 'private fortune.'[94] Trade and agriculture together will pave the way for a

87 Groot, *Van Batavia naar Weltevreden*, 266.

88 *Javasche Courant*, no. 74 (September 16, 1837), and no. 75 (September 20, 1837). All French quotations in this section are taken from the published speech. Van Sevenhoven's speech was published in Cornets de Groot, *Overzigt der reis*.

89 'Digne d'être ouvert à la curiosité publique et de prendre rang parmi les fondations les plus honorables de la colonie.'

90 'Une dotation qui doit, par un acte de libéralité digne de la nation élever tout à-la-fois un monument durable à la plus noble des sciences.'

91 'Le premier protecteur du Musée Colonial de Java.'

92 Even when one takes into account that a public lecture in the presence of prince Hendrik undoubtedly included passages that were more intended to gain the prince's interest rather than to express Diard's personal opinion, this speech is still a useful source in that it falls in line with the practical aspects of Diard's work that we have seen so far.

93 'L'ère nouvelle de prospérité et de splendeur commerciale qui commence pour cette colonie.'

94 'Brillantes entreprises de culture […] revenue publique […] fortune particulière.'

society where leisure time can be devoted to liberal studies. This 'happy inclination of spirits'[95] will only benefit from the founding of the Museum,

> intended not only to become the depository of all the objects whose research and observation can inspire the greatest taste for these same studies, but also to attract towards the Annals of the Society as towards a center of emulation and general correspondence the most interesting publications.[96]

The Museum will be a catalyst for further studies and will only give intellectual society a much-needed boost. In addition to these 'high hopes,'[97] the practical result of such a museum is also that agriculture, 'the basis of all prosperity,'[98] will profit from the results of the studies that flow forth from the presence of the Museum.

Diard explains that natural history should not only be aimed at the continuation of 'isolated research of some travellers,'[99] but also 'an investigation of the same facts, combined simultaneously in various places and in various circumstances on the same principles of observation.'[100] This position contains a strong echo of the contrast between the *naturaliste-sédentaire* and the *naturaliste-voyageur* as expounded by Georges Cuvier in his critique of a work by Von Humboldt.[101] It is no surprise that Diard, as a protégé of Cuvier, imbibed Cuvier's views of natural history so strongly that he would still hold on to them twenty years later. Diard's view on natural history faithfully echoes Cuvier's remark that 'the traveller can only travel one road; it is only really in one's study that one can roam freely throughout the universe.'[102] In the same way, the isolated expeditions of the members of the Natuurkundige Commissie can only bring in the objects and facts of one place. Only studying the collections and reports of several of these expeditions together, out of their original context,

95 'Heureuse inclination des esprits.'

96 'Destiné non pas seulement à devenir le dépôt de tous les objets dont la recherche et l'observation peuvent inspirer le plus de goût pour ces mêmes études, mais encore à attirer vers les annales de la société comme vers un centre d'émulation et de correspondance générale les publications les plus intéressantes.'

97 'Hautes espérances.'

98 'Base de toutes les prospérités.'

99 'Des perquisitions isolées de quelques voyageurs.'

100 'Une investigation des mêmes faits, combinée simultanément en divers lieux et en diverse circonstances sur les mêmes principes d'observation.'

101 See Outram, *Georges Cuvier*, 62–63, for the French text, and Outram, "New Spaces," 259–261, for an English translation.

102 Outram, "New Spaces," 261.

will allow the naturalist to 'reach reliable conclusions.'[103] And the Museum is the place to do that.

While from a European perspective Diard was a *naturaliste-voyageur*, this speech strongly suggests that he saw himself as a *naturaliste-sédentaire* whose 'reliable conclusions' based on the results of the expeditions of the Natuurkundige Commissie were meant to benefit society. In contrast to Temminck's desire to transport the natural riches of the Indies to Leiden and turn that into a 'sedentary' centre, Diard saw the Colonial Museum as a utilitarian centre in the colony. The Colonial Museum served as a rival centre to 's Rijks Museum, not only because the collection consisted of objects that were originally destined to go to Leiden, but also because the goal of the collection was in such marked contrast with Leiden. The Museum in Batavia was necessary, because studying the natural history objects in faraway Leiden would not lead to immediate benefits for the Indies, especially because the promised publications were still forthcoming. The founding of a rival museum in Batavia created a scientific centre where the results of research would immediately benefit the Indies.

In contrast to Temminck, Diard's interest was not in creating collections or publications.[104] The collections were a necessary means to enable useful research for the benefit of society. As such, the collections of the Natuurkundige Commissie in Batavia were intended primarily to serve the development of agriculture, which would lead to more prosperity for the Indies. Natural history in the colony 'can [...] bind itself by the most important relations to the development of trade and agricultural industry,'[105] and the observations of natural history 'can provide the most attractive and useful results.'[106]

7 The Layered Loyalties of Pierre-Médard Diard

His speech makes it abundantly clear that Diard was a naturalist driven by a desire to serve the common good. He saw himself as a naturalist in the broadest sense of the word and considered agricultural research to be a logical and necessary extension of natural history. Studying the natural world, either through reaching reliable conclusions based on objects and facts from various parts of the world, or through experimenting with crops and domesticated

103 Outram, "New Spaces," 260.

104 For Diard's publications, see Appendix 3.

105 'Peut [...] se lier par les rapports les plus importans au développement du commerce et de l'industrie Agricole.'

106 'Peuvent fournir les résultats les plus attrayans et les plus utiles.'

animals, was only as useful as it was beneficial to society. Herein lies the reason why Diard never joined one of the expeditions of the Natuurkundige Commissie. As directing member, he saw himself perfectly placed to take the next steps based on the collections of objects and facts that were gathered by the *naturalistes-voyageurs* of the Commissie.

The ultimate purpose of the Natuurkundige Commissie, in Diard's understanding of natural history, was to generate prosperity in the Indies. This concept of natural history research in service of a colonial society was not shared by the members of the Natuurkundige Commissie that were appointed in the Netherlands. Their first loyalty lay with Temminck and 's Rijks Museum, not with the Indies society. From their point of view, Diard was not committed to the goal of collecting objects and facts for 's Rijks Museum, while from Diard's point of view, the work of the members served an immediate purpose in the Indies which was not recognised by the members. Herein lies the root cause of the friction between Diard and the other members that sometimes surfaces in the writings of Van Oort and Forsten.

The Natuurkundige Commissie had originally been founded to collect specimens for 's Rijks Museum in Leiden, inspired by the vision of director Coenraad Jacob Temminck. All members appointed in the Netherlands had some form of loyalty to Temminck and 's Rijks Museum. This was in marked contrast with Diard, who became directing member of the Commissie in completely different circumstances. He had already spent several years in the Indies on agricultural projects. He had been involved in improving the production of indigo, tea and tin, while his agricultural establishment in Krawang, as carried forward by De Serière, was a significant element in the agricultural policies of Du Bus and Van den Bosch. Diard had received appointments or was granted significant requests by Van der Capellen (cultivation inspector), Du Bus (commissioner on an expedition to Borneo, the agricultural establishment in Krawang, directing member of the Natuurkundige Commissie), Van den Bosch (sericulture inspector) and De Eerens (the model sugar mill). The only dissonant voice in this impressive list of patrons is Baud, who negatively stresses 'the well-known versatile character of Mr. Diard' in a discussion with De Eerens on the granting of further sugar contracts.[107]

107 'Het bekende versatile karakter van den Heer Diard.' Quoted by Leidelmeijer, *Van suikermolen tot grootbedrijf*, 160. Baud's remark was clearly inspired by the fact that Diard's plans had not yet produced any tangible results. In the context of the discussion with De Eerens, it may also have been beneficial to Baud's case to exaggerate Diard's failure to produce results.

From Diard's speech on the occasion of prince Hendrik's visit to the Bataviaasch Genootschap and its developing Colonial Museum, it becomes clear what Diard's drive was. He believed that agriculture was the main driving force for prosperity. Natural history had a role to play in advancing agriculture for the benefit of society. His interests had a much wider orientation than Temminck's. Temminck was only looking for specimens for 's Rijks Museum, while Diard viewed natural history as a way to improve the prosperity of the Indies. Diard's loyalty lay with the society that he was a part of and the leaders of that society who had such a great appreciation of his contributions to the welfare of all. Temminck's narrower view of natural history did not leave much room for what Diard saw as the essence of natural history.

There was no reason for Diard to give priority to Temminck's view when he received his appointment to the Commissie. By paying his salary and continuing to involve him in agricultural projects, Diard's patrons in the Indies implicitly were in full agreement with Diard's view that agricultural experiments were of intrinsic value to society. The Natuurkundige Commissie during Diard's directing membership was the embodiment of two rival visions of natural history. The members continued to collect as *naturalistes-voyageurs*, while the directing member occupied himself with agricultural improvements and setting up a museum to allow the *naturaliste-sédentaire* to reap the benefits of the work of the *naturaliste-voyageur*.

Diard's agricultural activities were barely visible in the Netherlands. It was only when he seemed intent on diminishing the flow of specimens to 's Rijks Museum by founding the Colonial Museum in Batavia that tempers flared. Baud was able to quickly defuse the situation by stressing the fact that the Museum in Batavia was never going to be sustainable. Confirmation of this assumption by Forsten and Schwaner led to the demise of the natural historical focus of the collection. Diard had never intended the Museum to become a rival to 's Rijks Museum. The collection of the Museum, gathered by the *naturalistes-voyageurs* of the Commissie, was intended to be studied for the immediate benefit for local society. It was only Diard's recognition of the value of the work of the *naturaliste-voyageur* in his wider view of natural history that allowed him to let the *naturalistes-voyageurs* to do their work in peace. This rivalry of views could have seriously wreaked havoc in the Natuurkundige Commissie. Instead, only the Museum plans led to a short flare up of opposition by Temminck in the Netherlands, while Diard's perceived negligence of his duties as directing member was only revealed in private critiques by Van Oort and Forsten.

8 Conclusion

The story of Diard has shown that his versatility was what eventually condemned him to obscurity. As a transimperial traveller his public image was shattered into three hardly congruent national Diards. The British knew him as an imposter, the French as a *naturaliste-voyageur*, and the Dutch did not really know him at all. Upon his death, the *Nieuwe Rotterdamsche Courant* was only able to point to the fact that Diard had been the president of the Commissie. It was not even possible to write a proper obituary.

As we have seen throughout these two chapters, Diard was a man of grand plans and grand failures, but never a man of grand success stories. He was a man of initiative, brimming with ideas, never discouraged and he always cheerfully picked up the pieces to start afresh. His tireless work ethic allowed him to be active in so many different projects at once, but it was also this versality that prevented a legacy. Because he had to divide his attention over so many different projects, he never specialised in one topic so deeply that his contribution led to the association of his name with that particular topic. Diard was well liked by those in power, mainly because of his utilitarian streak. An experienced and talented naturalist with a strong tendency to use his (transimperial) knowledge to the benefit of society and who did not even always expect a salary was a rare commodity. As such, the governors-general had no hesitation to make use of Diard's services, even though his grand plans were often too grand (and expensive) for their tastes.

Because Diard worked for the prosperity of the Indies, reports of his tireless efforts never needed to cross the world seas. His results were practically applicable in the Indies which made it almost impossible to see from the Netherlands how versatile and influential Diard's work in the Indies actually had been. Even though he contributed greatly to the development of several important cash crops, his work was invisible in the Netherlands. He did not fit into the mould of a 'normal' Commissie member because he never went on expeditions and did not send collections home. As a result, Diard almost prefigured the direction that the Natuurkundige Commissie would take in the 1840s when Schwaner and Junghuhn were mostly used in utilitarian projects.

While Diard was in charge, the collecting expedition as conceived in the Netherlands continued to take place, but he was never the initiator of new expeditions. Diard's own initiative was limited to agricultural projects, which were only of importance to the governors-general in the Indies and not to 's Rijks Museum. Even Diard's royal decoration did little to heighten Diard's profile in the Netherlands beyond the fact that he was in charge of the Natuurkundige

Commissie. In the end, it is hardly a surprise that the *Nieuwe Rotterdamsche Courant* was not able to write a full report on Diard's career. Every element of it had been subservient to the greater good of Indies society and Diard had lost the ambition to make a name for himself as a naturalist due to the continuing conflicts surrounding his collecting activities. Now, more than 150 years after his death, these two chapters finally complete what the *Nieuwe Rotterdamsche Courant* was unable to write and restore Dutch Diard to his rightful place.

PART 3

'Hidden Histories' of Indigenous Participation

CHAPTER 6

Indigenous Employees as Natural History Assistants

> Wherever we went, we asked the local people for information about the animals that lived in their neighbourhood. Jack carried specially commissioned drawings of the creatures he was particularly keen to collect, including *Picathartes*, and showed them to local hunters to see if they recognised the animals represented and knew where we might find them. As the weeks went by, so we got to know some villages particularly well and became friends with their chiefs. One of them was exceptionally helpful and offered to stage a performance by his musicians and dancers in our honour.

Anyone who has read a nineteenth-century travel journal will immediately recognise the familiar patterns present in this passage. It clearly illustrates the roles that local people frequently had in supporting a natural history expedition. What we see in this passage is that the crew of the expedition uses the local population to gather information about what could be found in the present locality. They even call upon the help of local hunters to identify whether desired species were known to them and, if yes, where their habitats could be found. The relationship between the expedition and the local population became so intimate that a local chief decided to 'stage a performance' in honour of the expedition that had frequented their village for a while now. Both indigenous agency and indigenous knowledge were of tantamount importance to the expedition.

However, as the knowledgeable and observant reader may have noticed, the passage above is not from a nineteenth-century travel journal. In fact, is has nothing to do with the Natuurkundige Commissie at all. This quote is taken from the book *Life on Air* (2002) by naturalist extraordinaire Sir David Attenborough and describes the experience of the 4-man expedition crew while on the first *Zooquest* expedition in Sierra Leone in 1954.[1] The Jack mentioned in the passage is Jack Lester, the London Zoo Curator of Reptiles. *Picathartes* is the elusive bald crow that was first described by Temminck in

1 Attenborough, *Life On Air*, 40.

 | DOI:10.1163/9789004696877_008

his *Nouveau recueil de planches coloriées d'oiseaux*.[2] Currently, two species are recognised, both from West Africa.

In the more than one hundred and fifty years that have passed between the expeditions of the Natuurkundige Commissie and the publication of this travel narrative by Attenborough not much seems to have changed. As we shall see in this chapter, the quoted passage could have been a description of the daily experience of any of the expeditions of the Natuurkundige Commissie. The most significant difference between the two is a shift from collecting dead specimens for museum displays to collecting live specimens for zoo displays. However, collecting dead specimens for biological research is still common practice today, although it is not always undisputed.[3] But what has not changed at all is the fact that there is explicit but mostly anonymous reference to the assistance of the local population.

The narratives of the expeditions of the Natuurkundige Commissie that have reached us in the form of letters, diaries and travelogues are, without exception, written by Europeans. There simply are no extant reports of these expeditions in any form written from the perspective of the indigenous agents that worked around the clock to enable the naturalists to do their work. Unfortunately, this means that much of our knowledge of how indigenous people contributed to the work of the Natuurkundige Commissie remains obscure.

Felix Driver and Lowri Jones have identified the stories of indigenous contributions to exploration as 'hidden histories.'[4] They state that 'we need to make room in our histories for the local partners, guides, porters, fixers, interpreters, traders and officials who made journeys of exploration possible; and also the sponsors, patrons, publishers and editors who enabled accounts of these journeys to circulate more widely.'[5] So far, this book has told the story of the officials, the sponsors and patrons that made the work of the Natuurkundige

2 Temminck named it the *Corvus gymnocephalus*. Temminck and Laugier de Chartrouse, *Nouveau recueil*, plate 327 (vol. 2).

3 See Suarez and Tsutsui, "Value of Museum Collections," for the value of these collections of dead specimens. See Minteer et al., "(Re)extinction," for a strong stance against killing animals for science. For an equally strong response, see Wheeler, "Vital Bodies of Evidence." Available in open access under a different title here: https://www.newscientist.com/article/mg22229670-200-killing-animals-is-a-necessary-evil-for-natural-history/.

4 Driver and Jones, *Hidden Histories*. For Driver's reflections on the exhibition 'Hidden Histories of Exploration,' see Driver, "Hidden Histories Made Visible?" and Driver, "Exploration as Knowledge Transfer." For Jones' research, see Jones, "Local Knowledge and Indigenous Agency."

5 Driver and Jones, *Hidden Histories*, 5.

Commissie possible. It is time now to shed light on the 'hidden histories' of the indigenous people who contributed to the work of the Commissie.

Despite the fact that much of the stories of these indigenous contributors are 'hidden,' since we lack any medium carrying their own, unmediated voices, a significant part of these 'hidden histories' can be revealed through a careful reading of the European stories that we do have. Unlike Driver and Jones, who used extensive visual evidence for the exhibition 'Hidden Histories of Exploration,'[6] not only drawings and maps but also photographs and film, I mostly limit myself to the textual record of the expeditions of the Natuurkundige Commissie.[7] The corpus of texts from which these 'hidden histories' of indigenous agency in the expeditions of the Natuurkundige Commissie can be gleaned is vast. As has become clear from the previous chapters, there are countless letters, reports and narratives available from which it is possible to get a general idea of the nature of the indigenous support received by the Natuurkundige Commissie. I have chosen to concentrate on a representative selection from this material. For this chapter, I rely mostly on the recently published diary of Pieter van Oort,[8] the unpublished diary of Eltio Alegondus Forsten,[9] the published travel reports of Salomon Müller as published in the *Verhandelingen*,[10] and the published biography and letters of Heinrich Boie.[11] Combining these individual resources will allow me to reconstruct the manners in which the expeditions of the Natuurkundige Commissie were dependent on indigenous support in the realm of natural history.

---

6 This was held at the Royal Geographical Society (with IBG), from 15 October to 10 December 2009. Driver and Jones, *Hidden Histories*, was the companion volume to this exhibition.

7 Andreas Weber and Sylvia van Zanen have already unveiled the 'hidden history' of a Chinese artist employed by the Commissie, making use of the visual records (drawings) that are available. See Weber and Van Zanen, "Matters of Perspective." It is also possible to use the material remains of the Natuurkundige Commissie (both the natural history collections and the ethnographical collections) to study the 'hidden histories' of indigenous contributors. While visual evidence will be used on occasion to support the findings from textual evidence, this book does not offer the scope to systematically explore these 'hidden histories' from a visual or material point of view. See Ter Keurs, "Introduction"; Barnes, "Recording Cultures," and Ter Keurs, "Collecting in the Colony," for the ethnographical collections of Müller and Macklot in context.

8 Published by Weber and Van Zanen in 2021 in two volumes: *Kunstenaar op Java* and *Kunstenaar op Sumatra*.

9 Reisverhaal van Dr. Forsten naar Menado. NCAO, 1075.

10 Susanna, *Verhandelingen*.

11 Susanna, *Levensschets van Hendrik Boie*. I am using the Dutch version with added correspondence. There is an earlier German version available as well: Susanna, *Lebens-Abriss*.

As Dane Kennedy has shown, expeditions were virtually impossible undertakings without an enormous amount of indigenous support. As far as logistics were concerned, expeditions required transportation, supplies and labour.[12] In addition, indigenous actors served as 'guides, agents, interpreters, emissaries, and, above all, cultural brokers.'[13] This vast range of roles for indigenous actors is also present in the history of the Natuurkundige Commissie. In an earlier article, I identified 'organisational support, the provision of information, and the procurement of specimens' as the three main areas in which Forsten relied on indigenous agency.[14] Rather than taking the roles that indigenous actors could have had as my starting point, in this chapter and the next I investigate the functioning of indigenous agency according to the nature of the relationship of actors to the Natuurkundige Commissie.

Two main categories can be discerned. The first category, under scrutiny in this chapter, consists of a limited group of permanent indigenous employees of the Commissie. The common denominator for this group is that they received a monthly salary from the budget allocated to the Natuurkundige Commissie. I focus on how they practically contributed to the work of the Commissie by looking at their roles, their identity, the perception of their work in the Netherlands, and their place in the far-ranging expeditions of the Commissie. The second category, the subject of the next chapter, consists of a large number of indigenous people whose association with the Natuurkundige Commissie was much less formal and usually only temporary.

1 How to Thrive in the Promised Land

One of the members of the Natuurkundige Commissie was German zoologist Heinrich Boie (1794–1827). Boie left the Netherlands for the Indies on December 31, 1825, accompanied by geologist Heinrich Christian Macklot, preparator Salomon Müller and draughtsman Pieter van Oort.[15] The journey took over five months and the new recruits finally set foot on Indies soil on June 6, 1826. In a letter to his brother Friedrich, Boie expresses his elation at having finally reached his destination by crying out: 'Through all these misfortunes,

12 Kennedy, *Last Blank Spaces*, 131.

13 Kennedy, *Last Blank Spaces*, 163. See Konishi, Nugent, and Shellam, *Indigenous Intermediaries*, and Tiffany Shellam et al., *Brokers and Boundaries*, for a compilation of studies of indigenous agency in the context of exploration in Australia. See Schaffer et al., *Brokered World*, for a landmark study of brokerage.

14 Van Wingerden, "Science on the Edge of Empire," 816.

15 Susanna, *Levensschets van Hendrik Boie*, 40.

these dangerous times we have finally arrived in the promised land!'[16] For Boie, it had been a long road to the Indies indeed.

He had been a close friend of Heinrich Kuhl, who had left the Netherlands for the Indies as part of the original Natuurkundige Commissie on July 11, 1820, and ever since hearing of Kuhl's appointment, Boie has been eager to either accompany him or follow in his footsteps. A thoughtful review of Temminck's *Manuel d'Ornithologie* brought Boie to the attention of the museum director.[17] The resulting correspondence, combined with the deathbed request by Kuhl to appoint Boie as his successor, led to Boie's appointment at 's Rijks Museum in 1821 and the promise that he was to travel to the Indies on behalf of the Dutch government. Even though Boie was already appointed as member of the Natuurkundige Commissie on December 9, 1823, it would not be until the last day of 1825 that the second group of Natuurkundige Commissie members would finally leave the Netherlands.

During the long ocean passage, Boie had time to reflect on his future life in the Indies. His friendship with Kuhl and his stay at 's Rijks Museum had given him some familiarity with the situation in the Indies. He had been able to inspect and describe some of the collections resulting from the work of Kuhl and Van Hasselt and he had had ample opportunity to read their correspondence and reports. He was familiar with Temminck's belief that through the work of Kuhl and Van Hasselt 'one was as intimately familiar with the kingdoms of animals and plants of these regions [Java] as with the fauna and flora of our regions,'[18] which resulted in the assumption that it was now time to visit other islands in the archipelago.

However, on board the *Dijkzigt*, the ship that had already brought them safely from the Netherlands to the Cape and would also take them from the Cape to Batavia, he was met with laughter when he expressed this opinion to one of their fellow passengers, a Dutch military officer identified by Boie's biographer Susanna as Br.[19] Br., returning from temporary leave for health reasons in the Cape, had been part of the personal guard of governor-general Van der Capellen and was well-informed when it came to the previous natural history

16 'Per varios casus et tot discrimina rerum zijn wij dan eindelijk in het beloofde land aangekomen!' Susanna, *Levensschets van Hendrik Boie*, 187. The Latin quote is taken from Virgil, *Aeneid* I, 204.

17 Heinrich Boie, "Manuel d'ornithologie."

18 'Thans mag men zich vleien met het dieren- en plantenrijk dezer gewesten evengoed bekend te zijn als met de fauna en flora onzer streken.' Veth, *Overzicht van hetgeen gedaan is*, 38.

19 Susanna, *Levensschets van Hendrik Boie*, 172. The *Bataviasche Courant*, June 7, 1826, no. 23 suggests that Br. could stand for Brand.

missions of Reinwardt and Kuhl and Van Hasselt. He explains that Reinwardt was so overwhelmed with his administrative work that he had very limited time to work on his natural history collections. Kuhl and Van Hasselt had done admirable work, and it was only continued illness that prevented them from gathering truly staggering amounts of specimens.

However, it was not just continued ill health that hindered their work:

> Another hindrance also existed in the fact that the locals, who were not much accustomed to seeing Europeans, and were loath to spend time away from their families, only followed the orders of the residents to support the expedition until that time when they found a favourable opportunity to abscond, and that the small party soon found themselves in fully uninhabited forests without enough personnel.[20]

Fortunately, a solution to that problem was already available.

> The situation must be entirely different when one employs the required number of young Malayans, on whose excellent loyalty one can rely, when one treats them kindly, and also occasionally puts one's trust in them.[21]

Br. here identifies an issue that would continue to play a role in the interaction between the indigenous people and the Natuurkundige Commissie. The expedition members were foreigners in a strange world, in the days before wholesale colonialism had established itself. There were many blanks on the colonisers' (zoological and botanical) maps, and the members would need local assistance to help fill in the blanks. But who could they turn to? How could local support be organised in such a way that it would lead to a reliable infrastructure to collect as many specimens as possible? Br.'s advice indicates that there had been some failures in this regard and that local help was available but that the manner of relationship with the indigenous assistants to

20 'Eene hoofdzwarigheid bestond ook daarin, dat de inwoners, weinig aan het gezigt van *Europeërs* gewoon, en nergens liever zijnde dan in den kring hunner familie, het gebod der Residenten, om de reize te ondersteunen, slechts zoo lang opvolgden, tot zich eene gunstige gelegenheid, om te ontloopen, opdeed, en dat het kleine gezelschap zich daardoor weldra in de gansch onbewoonde woud-oorden, zonder genoegzame manschappen bevond.' Susanna, *Levensschets van Hendrik Boie*, 172.

21 'Geheel anders moet het daarentegen gelegen zijn, wanneer men het noodige aantal jonge Maleijers in eigene dienst neemt (gene wegloopers doen heerendiensten), op welke uitstekende trouw men zich gerust moet kunnen verlaten, wanneer men hen vriendelijk en ook somwijlen met vertrouwen bejegent.' Susanna, *Levensschets van Hendrik Boie*, 172.

the Commissie members mattered. Forced labour, as part of the set of obligations that indigenous rulers had towards the colonial government, clearly had its limits, so the ideal solution would be to create loyalty by making a small group of indigenous people morally and financially dependent on the work of the Commissie.

2 The Roles of the Indigenous Employees

That Boie took Br.'s advice to heart becomes clear from a letter to his mother on August 27, 1826, where he describes their living situation in idyllic Buitenzorg. Next to their living quarters were 'simple bamboo huts' where their servants lived with their families, 'ever vigilant to our needs, and without causing us any further trouble beyond paying them their modest monthly salary.'[22] Each expedition member had their own personal servant.[23] In addition to this, they would each employ a stable boy as soon as they would have received their horses. A cook had been employed who was also responsible for provisions and for candles and lighting. Müller employed an assistant in his laboratory (presumably to help him with preparing specimens, as this was Müller's main task), while Boie employed a hunter.

An earlier passage indicates that Boie's personal servant had been employed by Kuhl and Van Hasselt for his hunting skills and was also a skilled preparator.[24] Already on January 23, 1821, within a month after their arrival, Kuhl and Van Hasselt had been given permission to hire six indigenous assistants at a salary of f12 per month, to be expanded to twelve on June 11, 1821, in preparation for a planned expedition to Banten.[25] It is clear that some of these assistants had stayed on beyond the deaths of Kuhl and Van Hasselt.

As the comments by Br. to Boie show, it was generally thought best to employ a number of permanent servants, who could be entrusted with the essential tasks of the expedition. Paying them would create a level of loyalty towards their employers and prevent them from disappearing when other matters required their attention. These employees were used for very specific tasks.

22 'Eenvoudige bamboeshutten [...] steeds aandachtig op onze wenken, en zonder ons wijders eenigen last te veroorzaken, dan dat wij hun maandelijks hunne geringe soldij toetellen.' Susanna, *Levensschets van Hendrik Boie*, 208–209.

23 Van Oort mentions in his diary that he received his first Malay servant on June 29, 1826. See Weber and Van Zanen, *Kunstenaar op Java*, 58.

24 Susanna, *Levensschets van Hendrik Boie*, 199.

25 Res.GG, January 23, 1821, no. 12A. NL-HaNA, Koloniën, 1814–1849, 2.10.01, inv.no. 2774; Res.GG, June 11, 1821, no. 7. NL-HaNA, Koloniën, 1814–1849, 2.10.01, inv.no. 2460.

This practice was not unique to the Commissie.[26] Reinwardt, who immediately preceded the Natuurkundige Commissie in exploring the Indies, already took several employees with him on the expedition that took him on a grand tour throughout the archipelago visiting Timor, Ambon, Banda, Ternate and Manado between February 1821 and March 1822:

> On Monday, February 26 [1821], we boarded: besides myself, the gentlemen-artists A. J. and J. T. Bik and the gentlemen J. G. Landré and W. Kent, in addition to this my own assistants; my servant Philips, a small Dayak and three more men, mostly for the collection of objects of natural history, namely Unus for the plants, Pati for the insects, and a third one, Lisa, for preparing other specimens.[27]

The tasks allocated to Unus, Pati and Lisa were also specifically required for the natural history expeditions of the Natuurkundige Commissie. When the Boie party went on their first expedition, Pieter van Oort writes the following in his diary entry of December 28, 1826:

> On December 28 [1826] we left on our first Natuurkundige Commissie expedition to the interior. Our party existed of the gentlemen dr. H. Boie, dr. Macklot, G. van Raalten, S. Müller and myself. We had eighteen indigenous servants, consisting of five personal servants, six stable boys, one cook, one 'provisioneer,'[28] two hunters, two insect collectors, one preparator.[29]

---

26 See Camerini, "Wallace in the Field." Parallel examples of indigenous assistance on natural history expeditions in South America can be found in: Burnett, ""It Is Impossible to Make a Step without the Indians""; Roller, "Colonial Collecting Expeditions"; Antunes, Massarani, and Moreira, ""Practical Botanists"." For an example from a New Zealand context, see Kuster, "Global Commodity Chains." For general reflections on indigenous agency in colonial natural history expeditions, see Morris, "Non-western Collectors."

27 'Op den 26ste Februarij, des Maandags, konden wij ons aan boord begeven: te weten, benevens mij, de beide heeren teekenaren A. J. en J. T. Bik en de heeren J. G. Landré en W. Kent, waarbij tot mijne eigene hulp nog kwamen; mijn bediende Philips, een kleine Dajak en nog drie man, voornamelijk voor de inzameling van voorwerpen van natuurlijke historie bestemd, zijnde Unus voor de planten, Pati voor de insecten, en een derde, Lisa, voor het bereiden van andere voorwerpen.' De Vriese and Pijnappel, *Reis naar het oostelijk gedeelte van den Indischen Archipel*, 307.

28 Van Oort uses the word 'dispensier,' a person who is responsible for distributing and controlling provisions.

29 '28 december [1826]. Gingen wij onze eerste Natuurkundige Kommissie-onderzoeksreis naar de binnenlanden maken. Ons gezelschap bestond uit de heeren doctor H. Boie, doctor Macklot, G. van Raalten, S. Muller en mijn persoon. Wij hadden 18 inlandsche

The list of tasks that were allocated to the indigenous employees are a combination of the earlier tasks listed by Boie and the specific natural historical tasks listed by Reinwardt. The only role missing in Van Oort's overview is that of the plant collector, which may be explained by the fact that none of the members of this particular expedition was a botanist.[30] It also cannot be ruled out that the five employees tasked with collecting and preserving specimens additionally also collected plants.

In general, the three tasks of hunting, collecting plants and catching insects are the three most commonly mentioned natural historical tasks of the employees. Salomon Müller mentions in the *Verhandelingen* that 'none of our twelve Sundanese hunters, insect catchers, plant collectors and personal servants' had ever before seen or heard of it', with 'it' referring to the short-tailed gymnure (*Hylomys suillus*, Müller), an unusual mammal species that two of them had caught a specimen of on Java.[31] Müller also mentions that they used a small boat on the Borneo expedition 'to let several of our Javanese servants, hunters, insect catchers and plant collectors pass close to the riverside, in order to collect anything of interest that they would encounter.'[32]

Already on the first page of his travel diary, Commissie member Eltio Alegondus Forsten mentions that he will be accompanied by 'six Javanese, as hunters, animal and plant collectors, who are more or less skilled in this, as well as in preparing the collected objects.'[33] Forsten pays ample attention to his assistants. A survey of Forsten's terminology when he mentions his assistants, will tell us much about their roles. In his 233-page diary, Forsten uses the word 'jager' [hunter] including derivatives 70 times, mostly as a collective noun for his collecting assistants, sometimes referring to just the hunters. Another word he uses to refer to his servants is the word 'jongen' [boy], which appears 75 times (including derivatives) as a reference to his assistants. Of these 75 mentions, 15 refer to 'plantenjongens' [plant boys], 15 to 'insecte jongens' [insect boys],

---

bedienden, als 5 huisjongens, 6 staljongens, 1 kok, 1 dispensier, 2 jagers, 2 insektenvangers, een preperateur.' Weber and Van Zanen, *Kunstenaar op Java*, 59.

30 Zipelius and Korthals were the only specialised botanists connected to the Commissie.

31 'Dat geen onzer twaalf Sundanesche jagers, insektenvangers, plantenverzamelaars en huisbedienden, dit diertje ooit vroeger had gezien of van hetzelve hooren spreken.' Müller and Schlegel, "Beschrijving van een merkwaardig insektenetend zoogdier," 157.

32 'Om eenige onzer Javaansche bedienden, jagers, insektenvangers en plantenzoekers, steeds digt langs de rivierzoomen te voeren, ten einde te verzamelen, wat zich van belang aan hen opdeed.' Susanna, *Verhandelingen*, vol. 3, 342.

33 '6 javanen, als jagers, dieren en planten verzamelaars, welke reeds hierin meer of min bedreven waren, als mede het prepareren van de gevondene voorwerpen.' Reisverhaal van Dr. Forsten naar Menado. NCAO, 1075_003.

three to 'kommissie jongens' or 'jongens der kommissie' [Commissie boys], one to 'steenen jongen' [rock boy] and one to 'natuurjongens' [nature boys].

The word 'bediende' [servant] is only used once in reference to his assistants, while the ethnic collective of 'Javanen' [Javanese] is also used once in reference to his assistants. In total, Forsten refers 147 times to his employees. Given this abundance of references, it is all the more striking that Forsten only occasionally names them. He mentions Hendrik five times, Ammon four times, and Jeiler (or Jeilar) six times. Itjam only has one mention (and only because he dies on board the *Charles* on their way to Manado), while the others go unnamed.

In contrast to this, Pieter van Oort mentions the 'jagers' [hunters] and 'jongens' [boys] that are travelling with the expeditions on Java from 1826 to 1833 much less frequently. His published diary of 383 pages represents 627 hand-written pages,[34] almost triple the size of Forsten's modest 233-page diary. However, Van Oort only mentions the word 'jager' [hunter] and its derivatives 37 times, with only 21 of those referring to the Commissie employees. The word 'jongen' [boy] and its derivatives appears 54 times, with only 33 of those referring to the Commissie employees. Van Oort frequently uses the ethnic collective of 'Javanen' [Javanese] to refer to his assistants. He used the word 39 times to refer to them, often preceded by the possessive pronoun 'my' or 'our.' He also refers to the 'Boeginezen' [Buginese] ten times. In total, Van Oort refers to his assistants a modest 126 times, and names them just as rarely as Forsten. We learn of Asied (two mentions), Raden Waddir (or Wader) (two mentions), and Seiroen (one mention).

In Table 3, we see that Forsten, in his smaller diary, refers much more often to his assistants than Van Oort (147 vs. 126 times). Of Forsten's 147 mentions, in two thirds of the cases he uses a word that describes their natural history role (102), while of Van Oort's 126 mentions, the same happens in only a quarter of the cases (31). Forsten's diary was written as a scientific diary in preparation of an eventual publication, while Van Oort's diary was a much more personal affair, which could explain the disparity. Van Oort does not seem to feel the need to continually point out the importance of their assistants to the success of the mission of the Natuurkundige Commissie, while Forsten is interested in noting down the details surrounding the making of his collections.

34 11 pages for part 1, 30 pages for part 2, 21 pages for part 3, 411 pages for part 4, 55 pages for part 5, 99 pages for part 6, making a total of 627 hand-written pages. See Weber and Van Zanen, *Kunstenaar op Java*, 465–472.

TABLE 3 Role designations of indigenous employees in the diaries of Eltio Alegondus Forsten and Pieter van Oort

Designation	Forsten	Van Oort
NATURAL HISTORY ROLE		
'jager' (hunter)	70	21
'plantenjongen' (plant boy)	15	
'plantenverzamelaar' (plant collector)		1
'insecte jongen' (insect boys)	15	
'insektenvanger' (insect catcher)		1
'kapellenvanger' (insect catcher)		7
'steenen jongen' (rock boy)	1	
'natuurjongen' (nature boy)	1	
'preparateur' (preparator)		1
Subtotal	*102*	*31*
GENERAL OR OTHER TYPE OF ROLE		
'jongen' (boy)	40	30
'kommissie jongen' or 'jongen der kommissie' (Commissie boy)	3	
'bediende' (servant)	1	8
'dienaar' (servant)		1
'huisjongen' (personal servant)		1
'staljongen' (stable boy)		2
'staldienaar' (stable servant)		1
'begeleider' (companion)		3
Subtotal	*44*	*46*
ETHNICITY		
'Javanen' (Javanese)	1	39
Boeginezen' (Buginese)		10
Subtotal	*1*	*49*
Total	**147**	**126**

Two of the three common collective words used by Forsten and Van Oort to refer to the assistants do not indicate anything about the nature of their tasks. Both 'Javanen' [Javanese] and 'jongens' [boys] do not refer to a specific job. In modern times, the term 'jongen' [boy] sounds derogatory and perhaps even racist, but in those days, it was a normal word used to refer to servants. The only collective word that does inform us about the tasks of the indigenous assistants is the 'jagers' [hunters], but this does not necessarily mean that all they did was hunting. When 'jager' [hunter] is used on its own, it usually refers to the collective of hunters, plant collectors and insect catchers and as such covers the meaning of 'natural history assistants.'

In addition to shooting and collecting specimens, the 'natural history assistants' were also responsible for preparing the gathered specimens. Forsten writes the following in his diary:

> Close to five o'clock I saw the proa that I had sent out in the morning to pick up the two hunters that were left behind yesterday, and they brought O joy! a large male *sapi utan* with them.[35] It was one of the largest of its kind and measured 1,86 Dutch ell from the tip of its nose to the base of its tail. However tired the hunters were, they immediately had to start skinning. I had the skin preserved in arak, while the assembly of the skeleton was a nice job for the next day, during which they [the hunters] could rest.[36]

No rest for the weary! Another passage from Van Oort's diaries shows that the indigenous assistants had a large part to play in making camp as well:

> I had my tent pitched and my Javanese crawled into the leaf shelters that had been built there by our people on our second journey to the mountain. The dense forest on the mountain ridge where our tents and shelters were pitched, was soon turned into a lively village. It is nice to see how soon the Javanese have settled at such an arrival. When they have put down their goods, they all go into the dense forest. Soon one sees some of them return with thin tree trunks, others with forest rattan, while the rest bring an enormous amount of bracken, branches, and large leaves. In an instant a small space has been cleared from wood and plants, there

35 A proa (from Malay *perahu*) is the general term for a ship of indigenous origin. A *sapi utan* is a species of anoa or dwarf buffalo. Two species exist on Celebes: mountain anoa (*Bubalus quarlesi* Ouwens, 1910) and lowland anoa (*Bubabuls depressicornis* Smith, 1827).

36 'Tegen vijf uren zag ik de praauw welke ik des ochtends uitgestuurd had, om de twee achtergeblevene jagers van gisteren aftehalen terugkomen, en zij bragten O vreugde een groot sapi-outang van het mannelijk geslacht mede, het was een der grootsten in zijn soort. Deszelfs lengte van den snuit tot aan den wortel der staart, was 1.86 Ned. el. Hoe vermoeid de jagers ook waren moeste zij toch dadelijk aanhet villen, de huid liet ik in arak werpen terwijl het maken van het schelet juist en goed werkje voor den volgenden dag was, waarbij zij konden uitrusten.' Reisverhaal van Dr. Forsten naar Menado. NCAO, 1075_041.

> are holes in the ground and a couple of stiles are erected with horizontal and slanted pieces of wood tied to them and one quickly sees a skeleton village. The branches are laid over [the structures], a kind of benches of rough logs are placed inside, dry wood is being gathered, and by rubbing some bamboos together a fire is made, and the occupants sit down with some comfort at their little fires, whose smoke billows over the tops of the highest oaks.[37]

Van Oort's passage has some ambiguity in it regarding the status of the Javanese he mentions. He first mentions 'my Javanese,' clearly referencing his natural history assistants. When he moves on to the description of how camp was made, he suddenly uses 'the Javanese.' This might indicate that this group includes a number of Javanese that were not necessarily only their natural history assistants. In a sense, the two categories of permanent employees and temporarily hired indigenous people flow into each other and Van Oort only very subtly differentiates between the two. In addition, it seems that it was not unusual for the permanent natural history assistants to take part in the work of the temporary employees.

One task that has so far been conspicuously absent is that of the artist. There are well-documented cases of (unnamed) indigenous artists working with western naturalists,[38] but in the case of the Natuurkundige Commissie there is no mention at all of indigenous Indonesian artists.[39] In fact, most of the drawings were made by European artists. Keultjes and Van Oort were appointed in the Netherlands specifically for this purpose, while Fievez de

37 'Ik liet mijne tent opslaan, en mijn Javanen kropen in de bladerhutten die door ons volk, op onzen tweeden togt naar den berg, waren opgeslagen. Het dichte woud op het bergjuk waar onze tent en hutten waren opgeslagen, was aanstonds in een levendig dorp herschapen. Het is aardig om te zien hoe spoedig de Javanen bij zulk eene aankomst ingerigt zijn. Wanneer zij hunne goederen hebben neer gesteld, gaan zij allen het digte bosch in. Eenige ziet men daarna terugkomen met dunne boomstammen, andere met bosch-rottang, terwijl de overige een menigte vaar en boomtakken, en groote bladeren aanbrengen. In een ogenblik is er eene kleine vlakte van hout en planten gezuiverd, gaten in den grond en eenige stijlen, waarover dwarsche en schuine houten zijn gebonden en men ziet alras een dorp in geraamte staan. De takken worden er overheen gelegd, een soort van banken van ruwe boomstammen er ingesteld, droog hout gezocht, en in hetzelve met een paar bamboezen te wrijven vuur gemaakt, en de bewoners zetten zich met eenig genoegen bij hunne vuurtjes, waarvan de rookwolken boven de toppen der hooge eiken uitrijzen.' Weber and Van Zanen, *Kunstenaar op Java*, 313.

38 For examples from Asia, see Fan, *British Naturalists*, 40–57; Noltie, *Raffles' Ark Redrawn*, 163–166; Guan, "Drawing Nature"; Magee, *Chinese Art*; Magee, *Art of India*.

39 Curiously, there is also no mention at all of any indigenous involvement in the mapmaking exercises of the Natuurkundige Commissie.

Malines, Von Gaffron and Van Pers were hired in the Indies as artists. Only two artists of non-European origin are known to have participated in the work of the Natuurkundige Commissie. One was Antoine Maurevert, who came from India and had most likely arrived in the Indies with Diard. Maurevert was clearly a foreigner in the Indies and as such, his appointment was a matter for the governor-general to consider. Upon the request of Kuhl and Van Hasselt, Maurevert was appointed as their artist.[40]

The second non-European artist to work for the Commissie was Chinese artist Tsing Wang Ho, who worked with Van Oort for a brief period.[41] He first appears in Van Oort's diary on July 31, 1831, and was cast aside unceremoniously in March 1832 to make way for European artist Fievez de Malines, much to the chagrin of Van Oort.[42] Unlike Maurevert, Tsing Wang Ho's appointment had not been initiated by the governor-general, or, at least, I have found no record for this. He suddenly appears in Van Oort's diary, without any archival track record. Van Oort states that Fievez de Malines was given the same salary as Tsing Wang Ho (f60).[43] When governor-general Van den Bosch kindly requested the Commissie to hire Fievez de Malines, he asked the Commissie to give Fievez an appropriate 'daggeld' [daily allowance]. It seems that Tsing Wang Ho accordingly also received a 'daggeld' of around f2 per day.

At f60 per month, Tsing Wang Ho was paid much better than the indigenous assistants, but much less than Maurevert (f120, later raised to f150).[44] Apparently, there was a big difference in status between Maurevert and Tsing Wang Ho. One reason could be that Maurevert was directly in the employ of the government, while Tsing Wang Ho was in service of the Natuurkundige Commissie and therefore part of their budget. The most likely reason is that, as a Chinese, Tsing Wang Ho was considered to be on par with an indigenous person, and therefore of lower status. There had been Chinese communities in the Indies Archipelago for such a long time that they almost were considered a distinct indigenous people group. It was not rare to encounter Chinese merchants or labourers and many towns had Chinese quarters.[45] Even the fact that

40 Technically, Maurevert was appointed as artist at Reinwardt's department of agriculture, art and science, but in practice he worked for Kuhl and Van Hasselt. Res.GG, March 16, 1821, no. 12. NL-HaNA, Koloniën, 1814–1849, 2.10.01, inv.no. 2459.

41 Weber and Van Zanen, "Matters of Perspective."

42 Weber and Zanen, *Kunstenaar op Java*, 209.

43 Weber and Zanen, *Kunstenaar op Java*, 209.

44 Res.GG, March 16, 1821, no. 12. NL-HaNA, Koloniën, 1814–1849, 2.10.01, inv.no. 2459; Res.GG, July 30, 1821, no. 16. NL-HaNA, Koloniën, 1814–1849, 2.10.01, inv.no. 2461.

45 Taylor, *Social World of Batavia*, 19. For the economic role of the Chinese on Java, see Claver, *Dutch Commerce*; Tjiook-Liem, *Chinezen uit Indonesië*.

Tsing Wang Ho most likely immigrated from Canton to Java himself and was not a descendant of the resident Chinese population did nothing to put his status on the same level as Maurevert's.[46]

3 The Identity of the Indigenous Assistants

A question that can be asked is whether the same indigenous assistants were continually employed under different Commissie members throughout the thirty years of the existence of the Natuurkundige Commissie. As we have seen above, Kuhl and Van Hasselt were given permission to first hire six assistants, and later even twelve. Fairly quickly after his arrival, on February 27, 1827, Boie and his party were also given permission to hire six indigenous employees at a salary of f12.[47] After Boie's death, during the preparations for the New Guinea expedition, Macklot received permission on January 29, 1828, to hire fourteen indigenous assistants, eight of which were natural history assistants (three hunters, a preparator, two insect catchers and two for miscellaneous natural history tasks).[48] Shortly after their return to Java, on October 5, 1830, the Commissie was given ten Buginese recruits as assistants.[49]

When the Commissie left for Sumatra in June 1833, it is certain that they brought several Javanese assistants with them, but Van Oort does not specify their number nor their tasks.[50] They also hired several indigenous inhabitants of Nias, an island to the west of Sumatra, but it is unclear what their precise role or relationship with the Commissie was. In March 1836, we find a record of three hunters employed at the Colonial Museum.[51] Again, Horner was given permission to hire four indigenous hunters on May 3, 1837,[52] while Schwaner was allowed on July 28, 1843, to decide for himself whether he wanted to continue hiring Forsten's assistants, as long as he would pay for their salaries from his own budget for collecting costs.[53] In a letter to Temminck, Schwaner writes that he had taken three hunters, two insect and amphibian collectors and a

46 Weber and Van Zanen, "Matters of Perspective," 46.

47 Proposal of the governor-general to the commissioner-general, January 19, 1827, no. 33. NL-HaNA, Koloniën, 1814–1849, 2.10.01, inv.no. 2798; Res.GG, February 27, 1827, no. 35, quoting Res.CG, January 23, 1827, no. 18. NL-HaNA, Koloniën, 1814–1849, 2.10.01, inv.no. 2501.

48 Res.GG, January 29, 1828, no. 1. NL-HaNA, Koloniën, 1814–1849, 2.10.01, inv.no. 2809.

49 Res.GG, October 5, 1830, no. 2. NL-HaNA, Koloniën, 1814–1849, 2.10.01, inv.no. 2540.

50 Weber and Van Zanen, *Kunstenaar op Sumatra*, 45, 127, 137.

51 Groot, *Van Batavia naar Weltevreden*, 267.

52 Res.GG, May 3, 1837, no. 1. NL-HaNA, Koloniën, 1814–1849, 2.10.01, inv.no. 2570.

53 Res.GG, July 28, 1843, no. 15. NL-HaNA, Koloniën, 1814–1849, 2.10.01, inv.no. 2647.

plant collector. In addition, he had taken an indigenous child, who goes by the name of Piet Messert, with him as a factotum.[54]

It is clear that hiring indigenous assistants was a common practice during the entire history of the Natuurkundige Commissie. There are even two examples of indigenous assistants being hired to work independently for the Commissie on remote locations without accompanying a European member or assistant. Three indigenous plant collectors and insect catchers, selected by Diard, accompanied the 1835 expedition to New Guinea under command of Michiel Langenberg Kool.[55] On June 15, 1838, Diard received permission to send two indigenous preparators (at a salary of f25) to Japan to instruct the two Japanese collectors of specimens in preparing and caring for natural history specimens.[56] The two Japanese collectors, Magoeits and Foské, worked in the botanical garden in Dejima and had been appointed after Bürger's departure in 1835 in order to continue the natural history work in Japan.[57] The salaries of the two indigenous preparators were paid from the budget of the Natuurkundige Commissie.[58]

As we have seen above, Boie re-employed a hunter who had learned his skills under Kuhl and Van Hasselt. We also know that some of the assistants that accompanied Horner on his Sumatra expedition were subsequently employed by Forsten.[59] And Schwaner mentions that he continued to hire some of Forsten's assistants.[60] This shows that there was a practice of deliberately hiring experienced assistants. However, there are also examples of indigenous employees leaving the service of the Natuurkundige Commissie. When the Commissie was about to depart for New Guinea in 1828, many of their experienced hunters were not interested in travelling to such distant lands. Only one hunter stayed with the Commissie (only to fall overboard and drown during the journey) and the majority of the fourteen assistants were newly hired employees.[61] In addition, the three hunters that worked at the Colonial

54 Veth, *Overzicht van hetgeen gedaan is*, 108.

55 Res.GG, February 12, 1835, no. 1. NL-HaNA, Koloniën, 1814–1849, 2.10.01, inv.no. 2557; Mörzer Bruyns, *Met de Triton en Iris*, 64–65.

56 Res.GG, June 15, 1838, no. 1. NL-HaNA, Koloniën, 1814–1849, 2.10.01, inv.no. 2583.

57 For Bürger's departure from Japan, see Kernkamp, "Heinrich Bürger," 116. For the appointment of Magoeits and Foské (this is the Dutch spelling of their names), see Res.GG, February 1, 1836, no. 3. NL-HaNA, Koloniën, 1814–1849, 2.10.01, inv.no. 2858; Res.GG, May 26, 1837, no. 2. NL-HaNA, Koloniën, 1814–1849, 2.10.01, inv.no. 2570.

58 Res.GG, July 18, 1838, no. 7. NL-HaNA, Koloniën, 1814–1849, 2.10.01, inv.no. 2584.

59 Van Wingerden, "Science on the Edge of Empire," 816. See Kort verslag van het in Indië verrigte van 14 December 1838 to April 1840. NCAO, 1076_003.

60 Veth, *Overzicht van hetgeen gedaan is*, 108.

61 Mees, "Vogelkundig onderzoek," 6.

Museum in March 1836 had already left service in July because they had found better pay elsewhere. These three hunters, Amon, Goemok and Djamieng had enjoyed a salary of only f10.[62] Based on this information, we can conclude that a long employment did happen, but was never guaranteed. Some would stay around for a long time, while others tried their luck elsewhere, either because of personal reasons, or because of financial incentives.

In many cases, we don't know the names of many of these indigenous employees, which makes it hard to gauge how common it was for them to stay in service for a long time. The lists of the names of indigenous employees that I have been able to find will help us understand the extent of this practice.[63] In an extant booklet of fieldnotes of Van Oort, taken in January 1833, there is a page with what seems to be a list of employees in reference to salaries or rice allowances.[64] There are seven names on the list: Racim, Indjam, Asied, Lapir, Sipan, Siedin, Soernu. Except for Asied, none of these names appear in Van Oort's diaries.

Another set of lists can be found in Forsten's financial reports and contain salary information.[65] An overview of the information in these lists is available in Table 4. Forsten arrived in Batavia on December 14, 1838. From this overview, we can see that he already had three indigenous employees in February 1839, a hunter and two insect catchers. We know, from a decree by Du Bus from February 27, 1827, that the earlier indigenous employees from the Boie party received a maximum salary of f12 per month,[66] just as the assistants hired by Kuhl and Van Hasselt mentioned earlier.[67] The salaries given to Forsten's assistants do not seem to be exceptional compared to this figure of 12 years earlier. Already in April 1839, Forsten hired four additional employees. In his 'Kort verslag' Forsten indicates that he continued to hire the assistants who had just returned from the Sumatra expedition after Horner's death in April 1839,[68] so it is likely that all four new hires (Jeilar, Miding, Ammon and Marsinna) were experienced hunters and preparators.

62 Groot, *Van Batavia naar Weltevreden*, 266–267.

63 It is not at all inconceivable that the vast archives of the Natuurkundige Commissie contain additional such lists, but these three lists certainly illustrate the point.

64 Aantekenboekje No. 8, January 1833. NCAO, 1119_189.

65 NL-HaNA, Van Beresteyn, 2.21.018, inv.no. 1523.

66 Res.GG, February 27, 1827, no. 35, quoting Res.CG, January 23, 1827, no. 18. NL-HaNA, Koloniën, 1814–1849, 2.10.01, inv.no. 2501.

67 Res.GG, January 23, 1821, no. 12A. NL-HaNA, Koloniën, 1814–1849, 2.10.01, inv.no. 2774; Res.GG, June 11, 1821, no. 7. NL-HaNA, Koloniën, 1814–1849, 2.10.01, inv.no. 2460.

68 Kort verslag van het in Indië verrigte van 14 December 1838 to April 1840. NCAO, 1076_003.

TABLE 4 Salaries paid to Forsten's indigenous employees between February 1839 and December 1840. All names and numbers taken from NL-HaNA, Van Beresteyn, 2.21.018, inv.no. 1523

Salary payment dates → Names and job titles (when given) ↓	February 28, 1839	March 30, 1839	April 30, 1839	May 31, 1839	June, 1839	July 31, 1839	August 31, 1839	September 30, 1839	October 31, 1839	November 30, 1839	December 31, 1839	January 16, 1840	February 29, 1840	March–December, 1840
Reban, 'jager' (hunter)	16	16	14	14										
Leidan, 'insectejongen' (insect boy) Beo, 'insectejongen' (insect boy)	22	22	 10	10 										
Jeilar, 'jager' (hunter)			16	16	16	16	16	16	16	16	16	16	16	(16)
Midin(g)			15	15	15	15	15	15	15	15	15	15		
Ammon, 'jager' (hunter)			15	15	15	15	15	15	15	15	15	15	16	(16)
Marsinna			12											
Ledin				10										
Droes, 'insecte jongen' (insect boy)					10	10	10	10	10	10	10	10	10	(10)
Modair / Madani, 'plantenjongen' (plant boy)					10	10	10	10	10	10	10	10	10	(10)
Itjam					10	10	10	10	10	10	10	10		
Jeilin, 'insecte jongen' (insect boy)					10	10	10	10	10	10	10	10	10	(10)
Jamir, 'plantenjongen' (plant boy)					10	10	10	10	10	10	10	10	10	(10)
Sayin							10	10	10	10	10	10		
Hendrik, 'mandoor'													25	(25)
'jagers' (hunters)														97

The composition of Forsten's group of employed assistants was shuffled in June 1839, when he took on a group of five new employees (Droes, Modair, Itjam, Jeilin and Jamir) to replace the five employees that had left the service of the Commissie in the previous month (Reban, Leidan, Beo, Marsinna and Ledin). It is unfortunate that we are not aware of the reasons why these five employees left their position. Were they dissatisfied with their salary reduction and did they decide to leave? Reban had to take a pay cut from 16 to 14 guilders, Leidan and Beo both from 11 to 10. And perhaps Marsinna never even meant to stay on beyond Horner's Sumatra expedition. Or were they perhaps not to Forsten's liking and were they let go? There is also no indication that helps us understand where the five new assistants come from or how they were recruited. In August 1839, another new assistant (Sayin) was hired, bringing the total to nine indigenous assistants.

The next big change happened when Forsten finally departed for Celebes and was expected to be gone for a long period of time. Whether Forsten only had space and money for a smaller number of employees, or whether Miding and Sayin were not interested in joining another expedition far from home is not clear, but these two assistants did not join Forsten on his expedition. Itjam did join the expedition but died on the way to Celebes. The only new hire for the expedition, Hendrik, immediately receives the highest salary. From Forsten's diary it becomes clear that Hendrik served as his personal assistant. His role as *mandoor* [overseer] also is a good reason for a higher salary. It is apparent from Van Oort's diary that the *mandoor* played a significant role in coordinating the local porters that were hired continually when the expedition was on the move.[69] The fact that Forsten employed his own *mandoor* rather than rely on locally hired *mandoors* shows that he did not expect to find capable people to take on this role during his travels.

From these lists we can conclude that, even though some indigenous employees stayed on between different expeditions, most of them worked for the Commissie for only a limited period of time. When we compare the lists from Van Oort's notebook with the lists from Forsten's financial report, there is only one possible person that seems to have been continually employed from 1833 to 1840. Van Oort's Indjam is possibly the same person as Forsten's Itjam. When Itjam died on board the *Charles* on February 17, 1840, a sale of his possessions was held on board and on the official report of this sale (signed by the ship's captain Prosser and Forsten) his name is spelled as Idjam, so the spelling variation is not prohibitive against identifying Indjam with Itjam.[70] It is not

69 See Weber and Van Zanen, *Kunstenaar op Java*, 170, 171, 174, 221, 298.

70 NL-HaNA, Van Beresteyn, 2.21.018, inv.no. 1523.

clear what happened to the 28 guilders and 16 cents (copper) that resulted from the sale. Given the fact that in the document Itjam is identified as a Malayan from the residency of Bandong, kampong Ronga [?], it may be possible that the proceeds from the sale of his goods were sent back to his relatives.

4 Indigenous Assistants from a Dutch Perspective

Even though the indigenous hunters and preparators were skilled at their tasks, some doubts were expressed about the quality of their work back in the Netherlands. However, it is likely that these doubts were motivated by other reasons than the actual quality of the work. When news reached the Netherlands of the death of Commissie member and botanist Alexander Zipelius during the Timor expedition, Temminck proposed to the ministry of internal affairs to appoint Pieter Willem Korthals as his replacement and to let him be accompanied by two skilled employees of 's Rijks Museum as preparators.[71]

The response from the minister was that a staggering sum of more than f 600.000 had already been spent on the Natuurkundige Commissie in the period 1819–1829 and colonial affairs were only able to take on the appointment of three new members when internal affairs would take care of the costs incurred before their arrival to the Indies. This led to the suggestion to only send Korthals, because:

> [the] collection of zoological objects by Mr. Blume on Java that has been said to be well-maintained, and largely, if not entirely, prepared by Javanese. If this is true, then the Javanese have shown that they possess the required skills to prepare objects and one can surmise that it is not necessary to send people from here for those tasks.[72]

While Temminck may not have disagreed with the statement that the Javanese possessed the required skills, he nevertheless objected to this suggestion and argued as follows:

---

71 Veth, *Overzicht van hetgeen gedaan is*, 69.

72 'De verzameling aan zoölogische voorwerpen door den Heer Blume op Java bijeengebragt en welke gezegd is in allen opzichte zoo wel verzorgd te zijn, grootendeels, zoo niet geheel, door Javanen is bewerkt. Is dit zoo, dan hebben de Javanen getoond, dat zij voor het prepareren van voorwerpen de vereiste bedrevenheid bezitten en mag er verondersteld worden dat het niet nodig is voor die werkzaamheden personen van herwaarts te zenden.' Quoted by Den Ouden, "C. L. Blume," 30.

> Although one may praise the skills of the Javanese in preparing the objects, they will always be tools that are useless without leadership. Evidence of this are the collections of Mr. Blume that are so to say thrown together without insight or skill; of several, sometimes unimportant objects there are hundreds of specimens, while important animals that are not easy to find or procure have been neglected. By only using Javanese as assistants, one will collect, as Mr. Blume has done, a very large amount, but also a rather less insightful and less well-ordered stock of only averagely prepared objects, hardly of use to science.[73]

Den Ouden delicately comments that Temminck here contradicts his earlier judgement on the quality of the preservation of Blume's collection when he judged that these objects were better preserved than those in the shipments from Reinwardt and Kuhl and Van Hasselt.[74] Temminck obviously used all his rhetoric powers, even stooping down so low as to describe the indigenous assistants of the Commissie as 'werktuigen' [tools], to prevent the rejection of his proposal of sending two assistants with Korthals to the Indies. Temminck prevailed and on July 30, 1830, Korthals was appointed as member of the Natuurkundige Commissie and was to be accompanied by Van Gelder and Overdijk, who were appointed as preparators and assistants.

It seems obvious that Temminck's derogatory remarks on the quality of the work of the Javanese assistants were entirely beside the mark, even though he may have touched upon a real issue when he describes Blume's collections as being 'thrown together without insight or skill.' An example of how this might have happened in practice can be found in one of Boie's letters:

> In the coffee plantations that are close to our bamboo huts it is swarming, since the beginning of March, with natives that are occupied with harvesting coffee, and, because the goal of our stay here is known to all, they

73 'Want hoe men de behendigheid der Javanen in het bereiden der voorwerpen ook roemen moge, het blijven altijd werktuigen welke zonder leiding van geen het minste nut zijn. Getuigen hiervan de verzamelingen van den Heer Blume als het ware tezamen geraapt en zonder keur of kunde opgehoopt; van eenig, soms niets beduidend voorwerp een honderdtal dubbelden, terwijl belangrijke gedierten, wier verkrijging of ontmoeting zich niet voor den voet opdoen, veronachtzaamd worden. Door het gebruik alleen van Javanen, als handlangers, verzamelt men, zoals door den Heer Blume, een zeer talrijke hoeveelheid, maar ook tevens een min kiesch als goed geordenden en voor de wetenschappen iets afdoenden voorraad van slechts middelmatig bereide voorwerpen.' Quoted in Den Ouden, "C. L. Blume," 30.

74 Den Ouden, "C. L. Blume," 30.

> bring us, without distinguishment, all the animals they can find. Sitting at the edge of a window, with a wallet full of copper coins in front of me, surrounded by bottles and boxes, I have founded what my companions call my 'zoological booth,' and where I, especially when the weather is beautiful, receive so many visitors from early morning till evening that I am sometimes exhausted. Especially three or four little boys stand out in their enthusiasm; and, because I pay good money for rare objects, they will not stop looking for others. [...] It is true, that one, in order not to discourage these folks, often pays them for something that is thrown out as soon as they have turned their backs; but the same people, only moments later, often bring another object that may surpass the previous object in value perhaps a hundredfold, and in the meantime the expenses do not exceed the sum of a couple of cents. The skill of the natives in catching and tracking animals to their hideouts is indeed admirable; however great the zeal of the traveller may be, he will never make a good catch alone.[75]

It was common practice for the Natuurkundige Commissie to simply buy specimens. It is understandable that Blume's collections were of a different nature than Boie's collection. After all, Blume was a botanist, and may have simply kept everything that his Javanese assistants found and collected, while Boie was very familiar with the collections of 's Rijks Museum and could more easily evaluate the worth of certain objects.

Not all Commissie members succeeded as well as Boie in selecting the right objects for shipment back to the Museum. This becomes apparent from the

75 'In de koffijtuinen, welke in de nabijheid van onze bamboeswoningen gelegen zijn, wemelt het, sinds het begin der maand Maart, van inboorlingen, welke met het inoogsten van de koffij bezig zijn, en, daar het doel van ons verblijf alhier aan allen bekend is, brengen zij ons, zonder onderscheid, alle dieren, die hun in handen vallen. Aan den hoek van een venster gezeten, met eene wel gevulde beurs met koperen munt vóór mij, en rijkelijk omringd van flesschen en dozen, heb ik opgerigt, hetgeen mijne reisgezellen mijn *Zoologisch* kantoor heeten, en waar ik, vooral bij fraai weder, van den vroegen ochtend tot den avond, zoo druk bezoek ontvang, dat ik somtijds geheel ben afgetobd van vermoeijenis. Drie of vier kleine jongens vooral, onderscheiden zich door hunnen ijver; en, daar ik hun de zeldzame voorwerpen wel betaal, houden zij niet op, met anderen te zoeken. [...] Het is waar, dat men, om de lieden niet te ontmoedigen, hen dikwijls voor iets betaalt, hetwelk, zóódra zij den rug hebben gekeerd, wordt weggesmeten; maar die zelfde menschen brengen, eenige oogenblikken later, dikwerf een ander voorwerp, hetwelk het eerste veelligt honderdvoud in waarde overtreft, en alle de onkosten gaan ondertusschen het bedrag van eenige centen niet te boven. De vaardigheid der inboorlingen in het vangen en opsporen der dieren, in hunne schuil plaatsen, is inderdaad verwonderingswaardig; en hoe groot de ijver ook moge zijn van een' reiziger, alleen zal hij nimmer eene goede vangst doen.' Susanna, *Levensschets van Hendrik Boie*, 48–50.

instructions that were given to Forsten, when it is clearly stated that he should limit himself to collecting objects that are not present in Museum, or only present in small numbers, and that he should correspond with the director to make sure that he will not send objects back to the Museum that are not required there.[76]

These strict instructions even inspire Forsten to offer a reward for those people that can provide him with specimens that he is specifically looking for. On January 26, 1842, he pays out a reward of 50 guilders to a local fisherman who has been able to procure a dugong (*Dugong dugon*) for him, while on July 20, 1842, he pays out a reward of 5 guilders to a local fisherwoman to whom he had shown an empty nautilus shell. He had offered her 5 guilders if she could procure a nautilus (*Nautilus pompilius*) with the animal still in its shell for him, and already the next morning she brought him one.[77] Forsten's financial report over 1840 holds a post for 'gekochte dieren' [purchased animals] for almost every month, and some purchases are deemed important enough to get their own entry, like when he purchases a living 'babi-rousa' (*Babyrousa celebensis*, North Sulawesi babirusa) for 4 guilders in September 1840.[78]

5 Indigenous Assistants as Foreigners

As we have seen above, the hunting and gathering was mostly done by the permanent employees of the Commissie, while the members occupied themselves with preparing, describing, drawing, and cataloguing the collections. But sometimes the permanent employees lacked the right hunting skills or the required knowledge of the local landscape to be able to find the animals and plants from the environment they were in. After all, as far as we can tell they were from Java and of either Javanese or Sundanese ethnicity.[79] Van Oort also mentions a group of eight Buginese hunters[80] that were temporarily employed by the Commissie without much success:

76 Article 4 and 5, Provisionele Instructie voor den Heer Med Doctor E. A. Forsten. NCAO, 1079_006.

77 For the dugong, see Reisverhaal van Dr. Forsten naar Menado. NCAO, 1075_107. For the nautilus, see Reisverhaal van Dr. Forsten naar Menado. NCAO, 1075_118.

78 NL-HaNA, Van Beresteyn, 2.21.018, inv.no. 1523.

79 The Javanese and the Sundanese are the largest ethnic groups in modern-day Indonesia. While the Sundanese find their origin in West Java, the Javanese have their strongholds in Central and East Java.

80 The Buginese are a people group from southern Sulawesi, but due to their naval skills lived all across the archipelago in coastal areas.

> When I arrived back to Krawang, I forgot to mention that while we were travelling to Tjabang our eight Buginese left for Batavia. They were soldiers, recruited in Makassar, who were surplus to requirement when they arrived and were subsequently partly put on loan to us and partly to the agricultural establishment in Krawang. Because they wanted to use them as soldiers again, the government demanded their return from the Natuurkundige Commissie and from the director of the agricultural establishment in Krawang. This nation is good at soldiery, but for our activities they were less suitable. Mr. Müller is currently busy hiring Javanese again, just as we used to have. You can do much more with them than with the wilful Buginese. They are softer of heart, listen better to what they are told and fit in much better in the interior of Java with the inhabitants and ourselves.[81]

Originally, ten Buginese recruits were made available to the Commissie as assistants and the Commissie was given f72 per month to provide for them.[82] It seems that the Buginese were not entirely happy in their unanticipated role of assistants to a natural history expedition. They were obviously also not familiar with the customs and culture of the areas in which the Commissie was active. And it was not only Van Oort who was not entirely happy with their Buginese assistants. When Macklot complained, the ten Buginese recruits were returned to the army on December 16, 1831, and the Commissie was allowed to look for replacements.

However, not everyone was unhappy with their Buginese assistants. While Van Oort was glad to see the back of them, Korthals was positively annoyed when he lost his Buginese assistants. Given the fact that Van Oort only mentions eight Buginese, it is likely that two of the original ten recruits were

81 'Toen ik te Krawang weder terugkwam, heb ik nog vergeten te melden dat intusschen, dat wij de reis naar Tjabang deden, onze 8 Boeginezen naar Batavia zijn vertrokken. Het waren te Macasser aangeworvene jonge soldaten, die de regering bij hunnen aankomst niet nodig hadden en derhalven gedeeltelijk aan ons en gedeeltelijk aan de Stichting van Landbouw te Krawang ter leen afstonden. Thans echter dezelven als soldaten weder willende plaatsen, eiste de regering dezelven weder van de Commissie van Natuurkundig Onderzoek en van den Directeur der Stichting van Landbouw terug. Dit volk is zeer goed voor soldaten, maar voor onze werkzaamheden waren zij minder geschikt. Thans is de heer Muller bezig om weder Javanen, zoo als wij oudtijds hadden, aan te neemen. Met deze is veel meer als met de ondeugende Boeginezen uit te voeren. Zij zijn zachter van inborst, hooren beter naar het geene hun wordt gezegd en schikken zich in de binnenlanden van Java veel beter naar de bewoners en naar ons.' Weber and Van Zanen, *Kunstenaar op Java*, 198.

82 Res.GG, October 5, 1830, no. 2. NL-HaNA, Koloniën, 1814–1849, 2.10.01, inv.no. 2540.

working with Korthals as plant collectors. Korthals was only informed of the impending departure of his Buginese by a letter from Macklot on December 25, 1831. Immediately upon receipt of the letter, he replied to Macklot in order 'to let my feelings on this be known,'[83] suggesting that he was not happy at all. When Korthals reports on January 1, 1832, that his Buginese have left to rejoin their regiment, he adds that 'the new year did not find me in a joyous mood.'[84] It is obvious that the two Buginese that worked with Korthals much better answered to his needs than the eight Buginese working with Müller and Van Oort. Somehow, this fact eluded Macklot, or was not of enough weight to counteract the complaints of Van Oort and presumably Müller as well.

Just as the Buginese were foreigners on Java, so the Javanese assistants were foreigners on Celebes, New Guinea, Ambon, Timor, Sumatra and Borneo. Consistent reference to Javanese or Sundanese hunters in reports on all expeditions to the Outer Islands points to the direction that the indigenous assistants that joined the Commissie on their expeditions were all hired on Java. They were unfamiliar with the wildlife on the other islands, and while hunting is to some extent a universal skill, the Javanese hunters lacked knowledge of the local wildlife and of the required hunting methods for species that were less prevalent on Java or required special skills.

6 Conclusion

Despite the fact that we cannot hear the unmediated voices of the permanent indigenous employees of the Commissie, it has proven possible to reconstruct a consistent image of their contributions to the work of the Natuurkundige Commissie by carefully gleaning through the textual heritage of the Commissie. These traces of 'hidden histories' have revealed that paid employees were a normal feature in the daily practice of the Natuurkundige Commissie. From the earliest beginnings of Kuhl and Van Hasselt in 1821 until Schwaner's Borneo expedition in the mid-1840s, indigenous Javanese (and sometimes Buginese) were continually employed in the service of the Natuurkundige Commissie. We know some of their names, but many remain anonymous. It is obvious that without these indigenous hunters and collectors it would have been impossible for the Commissie to gather the amounts of specimens that they did.

83 'Om mijne gevoelen hierover te kennen te geven.' Dagboek Korthals, December 25, 1831, NL-LdnNAT, Korthals_1_084.

84 'Het nieuwe jaar vond mij in geene vrolijke stemming.' Dagboek Korthals, January 1, 1832. NL-LdnNAT, Korthals_1_086.

It seems that, when possible, preference was given to hiring experienced indigenous assistants who had worked with the Commissie previously, but it is also clear that the indigenous natural history assistants were free to resign from their position whenever they wanted to. They seem to have been free labourers, who were never forced to accompany the Commissie to places they did not want to go. The fact that only one of the existing indigenous employees was willing to accompany the Commissie to New Guinea is a strong indication for this. In the next chapter, I will take a look at the category of the temporarily employed indigenous actors and we will see that this category included a number of actors whose involvement with the Commissie may not always have been of a voluntary nature.

CHAPTER 7

Porters, Knowledge Brokers and the Mediation of Indigenous Knowledge

In the previous chapter I have considered the roles played by the formally employed indigenous assistants of the Natuurkundige Commissie. The common denominator for this group is that they were hired permanently and received a monthly salary. We have seen that this group of assistants served in the roles of personal servants, cooks, stable boys, but mostly as natural history assistants. While this specific group of indigenous assistants was of tantamount importance to the success of the natural history mission of the Natuurkundige Commissie, they were not the only indigenous agents. While the Javanese assistants were essential to the success of the mission, their numbers were not adequate for the logistics required for the large-scale expeditions of the Natuurkundige Commissie. Even if they played an important role in making camp and the daily inner workings of the expedition, when the expedition was on the move it was a different story. Enormous amounts of equipment, collections and personal belongings needed to be transported and a dozen or so specialised assistants would not do. Another problem was that when the Commissie travelled to the other islands in the Indies Archipelago, much of the knowledge of the Javanese assistants became void, because they had also become foreigners.

In this chapter, I first consider the two groups of indigenous agents that solved the issues of transport and knowledge for the Commissie. The first group are the indigenous people that formed the logistics crew of each of the expeditions of the Commissie, namely the porters. The second group are the indigenous people that took over the role of knowledge brokers from the Javanese assistants, either when their knowledge was so specific that the Javanese did not possess it, or when the Commissie travelled to areas where the Javanese assistants had also become foreigners. I specifically look at interpreters and hunting guides. In the third section of this chapter, I look at how the knowledge received through indigenous brokerage was mediated by the Commissie in such a way that the indigenous agents providing the original knowledge gradually disappeared from the European view. To conclude, I look at how the theme of layered loyalties can be considered to function in the various indigenous actors that formed the support structure for the expeditions of the Natuurkundige Commissie.

 | DOI:10.1163/9789004696877_009

1 The Indispensable Services of Porters

No expedition of the Natuurkundige Commissie was possible without logistical support. Whenever an expedition was on the move, large amounts of goods needed to be transported. There is not much reference to the exact nature of these goods, but we can glean some information here and there. One of the heaviest goods that were taken along were the books of the Commissie. Every member needed a working library of European natural history works. In chapter 2, we have seen Forsten's regular requests for the latest zoological literature. Often, the members took these libraries along with them to be able to determine new specimens in the field. A second bulky commodity were the collections themselves and the required material for packaging and preservation. This included empty glass bottles for wet specimens and enough arak to preserve them with. After a successful expedition, large amounts of crates and boxes with specimens needed to be transported back. When Forsten sent his first shipment to Leiden, it consisted of five crates and five barrels and that was only after he had prepared them for shipment and packaged them as carefully as he could. These crates and barrels all came from the same areas, so it must have been quite the job to transport all these specimens.

But the members also took their personal luggage with them. Müller mentions that on a specific trip on Borneo they were unable to take matrasses with them, suggesting that it was common practice to transport matrasses through the tropical forests.[1] Müller describes them as 'a small, thin matrass [...] filled with the soft silky cotton from the kapokh-tree.'[2] One of the plates (Figure 43) in the *Verhandelingen* shows the large tents that the Commissie members used during their expedition on Timor. Given the fact that as Europeans, the members and assistants themselves hardly carried anything themselves, all this weight came down on the shoulders of indigenous porters. The European members mostly travelled on horseback (if the terrain allowed it) or let themselves be transported on palanquins. Forsten regularly suffered from ill-health and often used a palanquin for transportation during these times. When mountains needed to be climbed, or one needed to travel through forests following rhinoceros paths, the members themselves also travelled on foot, but wherever the infrastructure allowed it, horses, ox-drawn carts and porters were used to transport all the necessities of a travelling expedition.

---

1 Susanna, *Verhandelingen*, vol. 3, 419.

2 'Eene kleine, dunne matras [...] met het zachte, zijachtige katoen van den kapokhboom gevuld.' Susanna, *Verhandelingen*, vol. 3, 183.

FIGURE 43 'Onze legerplaats in de vlakte Wienoto' ('Our camp on the plain of Winoto'), by Pieter van Oort. Plate 29 (Timor) from the volume *Land- en volkenkunde* of the *Verhandelingen*, NCAO, 871_029. Public domain. An indigenous man shows a snake to two members of the Natuurkundige Commissie that can possibly be identified as Salomon Müller on the left and Pieter van Oort on the right. HTTPS://DH.BRILL.COM/NCO/VIEW/NCO_NNM001000871_029/MAKINGSENSE

Transporting all these goods sometimes required enormous amounts of porters. Both Van Oort and Forsten regularly mention the hiring of *koelies*. The Dutch word *koelie* is the equivalent of the English word coolie, from the Hindi word for 'day labourer.' The use of the word *koelie* was widespread in the Netherlands Indies. Nowadays, the word coolie has decidedly negative connotations. In the late nineteenth century, the word *koelie* in the Netherlands Indies would also come to be associated with slavery and abuse.[3] At the time, the word did not yet fully encompass a negative meaning as is made clear by the following passage, taken from the caption accompanying the plate of the coolie in Auguste van Pers' *Oost-Indische Typen* (Figure 44).[4] This book collects a series of stereotypical people that one could encounter in daily life in the Netherlands Indies and gives a short description of their role in society:

---

3 See Breman, *Koelies, planters en koloniale politiek*; Houben et al., *Coolie Labour*; Jan Breman, "Colonialism and Its Racial Imprint."

4 Van Pers and Hageman, *Nederlandsch Oost-Indische typen*. Most of the plates were engraved by Carl Wilhelm Mieling (1815–1903) based on drawings by Commissie assistant Auguste van Pers with accompanying text mostly written by Johannes Hageman (1817–1871). For Mieling, see http://www.biografischportaal.nl/persoon/11165346. For Hageman, see Muller, "Hageman (Johannes)."

FIGURE 44 'De lastdrager (kouli)' ('The porter (coolie)'). Lithograph after a drawing by Auguste van Pers, 1853–1856, KITLV 51R1, Leiden University Libraries, Digital Collections. Original plate from Van Pers and Hageman, *Nederlandsch Oost-Indische typen*. Public domain
HTTP://HDL.HANDLE.NET/1887.1/ITEM:856868

> Do you see that porter over there, who is more than half-naked, and perhaps carries all his wealth with him? To earn some doits, he brings an entire fortune of his fellow humans to a different place. It is the coolie, the porterman. [...] However couth and unkempt they may look, nonetheless coolies are largely honest. And their use? That is irrefutable. The newly arrived, the traveller, the merchant, the auction visitor, how unlucky would they be if there were no coolies to be hired for a small sum?[5]

This is a somewhat derogatory description of the role of the coolie: they are incredibly useful, and though they look entirely untrustworthy, you can trust their honesty. This sentiment is echoed by Van Oort when he writes that 'one can confidently trust the common porters with one's goods without anything being stolen.'[6]

This picture of the rough but honest coolie, who is always ready to assist the traveller in carrying his goods for a few coins, quietly glosses over the actual situation in most parts of the Indies Archipelago. In his diary entry on the government of Ternate, Forsten explains how the coolie-system worked. The sultan of Ternate is nominally in control, but is embedded in the Dutch powers structure as a colonel. The island has a small Dutch enclave from which the resident rules. The resident appoints the sultan and his family in their positions and upon their appointment they are required to swear loyalty to the Dutch government.

> Our government is the protector of the sultan, who is not allowed to grow any spices. The sultan is obliged to assist the government by lending proas and men and these men will be paid by the government, every coolie receives 20 doits per day with the necessary rice.[7]

5 'Ziet gij daar dien sjouwer, die meer dan half naakt is, en welligt zijn ganschen rijkdom met zich draagt? Om ettelijke duiten te winnen, brengt hij een geheel vermogen van zijn evenmensen naar eene andere plaats. Het is de *Koeli*, de sjouwerman. [...] Hoe woest en schooijerig ook van uitzien, zijn de *Koeli's* niettemin meerendeels eerlijk. En hun nut? dit is onwedersprekelijk. De nieuw aangekomene, de reiziger, de koopman, de venduganger, wat waren zij ongelukkig, indien er gene *Koeli's* voor weinig geld ten hunnen dienste waren?' This text was written by Johannes Hageman.

6 'Men kan de gemeene dragers gerust zijne goederen aanvertrouwen, zonder dat er iets van wordt gestolen.' Weber and Van Zanen, *Kunstenaar op Sumatra*, 51. This is only one of six times that Van Oort uses the Dutch word 'drager' (which is equivalent to the English 'porter') in his entire diary.

7 'Ons Gouvernement heeft den Sultan in zijne bescherming, er mogen geene specerijen aangekweekt worden. Den sultan daarentegen is verpligt het Gouvernement hulp te verleenen door het leenen van praauwen en manschappen waarvoor deze laatste betaling van het

From the diaries of Van Oort and Forsten it becomes clear that such arrangements existed on a lower level all across Java, Sumatra and Celebes as well. Wherever the Dutch government was established there were either treatises with local sultans that obliged them to supply coolies at prearranged prices, or general arrangements with local village and district heads. The numbers of coolies used by the Commissie ranged from a dozen or so to several hundreds. Large numbers of coolies had to be ordered in advance to give the local leaders time to gather the required numbers, but smaller amounts of a couple of dozens could often be supplied on the same day.

However, organising coolies was not always a smooth process. Van Oort writes: 'Only with a lot of effort the 56 coolies that we required were gathered.'[8] And it did not always work out ('We just could not get any coolies.'),[9] sometimes to such an extent that travel plans needed to be amended: 'I wanted to return in the evening, but was unable to because there were no coolies to be had.'[10] On Sumatra, the situation was particularly difficult. When the Commissie travelled there in 1833, they did so in the wake of the colonial army that had been sent there to quench a 'rebellion' by the indigenous Padri.[11] This meant that the coolie arrangements did not work as smoothly as on Java or Celebes. Shortly after their arrival, Van Oort complains that 'we can each only get six coolies which is not enough to take all our necessary baggage.'[12] In order to solve this problem, the Commissie was supplied with six additional Nias. These Nias had been sold by their village heads to the government into debt slavery for only f76 and had to work off their debts for a duration of ten years. Van Oort does not often mention them, so it is unclear how long they stayed in the service of the Commissie.

Even though most of the thousands of coolies who travelled with the Commissie were paid for their services, it must be noted that they were working under duress. The case of the Nias sold into debt slavery may look exceptional

---

Gouvernement geniet, zijnde voor iedere koeli 20 duiten per dag met de benoodigde rijst.' Reisverhaal van Dr. Forsten naar Menado. NCAO, 1075_090.

8 'Met veel moeite kwamen de 56 koelies die wij nodig hadden bij elkander.' Weber and Van Zanen, *Kunstenaar op Java*, 221.

9 'Wij konden maar geen koelies krijgen.' Weber and Van Zanen, *Kunstenaar op Java*, 273.

10 "s Avonds wilde ik weer teruggaan, maar kon niet omdat er geen koelies te krijgen waren.' Weber and Van Zanen, *Kunstenaar op Java*, 287.

11 Van Zanen and Weber, "Schreeuwende apen"; for the Padri War, see De Moor, "Warmakers in the Archipelago"; Hagen, *Koloniale oorlogen*, 421–424; Groen et al., *Krijgsgeweld en kolonie*, 61–69.

12 'Wij kunnen ieder maar zes koelies bekomen en dus niet genoeg om al onze nodige zaken mee te nemen.' Weber and Van Zanen, *Kunstenaar op Sumatra*, 44n27.

as a very rare example of the use of unpaid forced labour in the history of the Commissie, but the coolies often did not have a choice.[13] Forsten describes what happened if a village head really could not supply any coolies:

> It is the custom over here that, if one wants to travel from one place to another, or from the Minahasa to the neighbouring country, the inhabitants of one place should bring you to the closest main village and hand you over to the people there who in their turn should serve you and lead you further. If there is a shortage of people or something like that, then you can demand that the ones who have assisted you first, continue to assist you and lead you further, but then they will receive extra payment, over your payment, from the foreign village itself where they assist you.[14]

It is possible that some of the coolies volunteered for this task, but it can certainly not be ruled out that they did not have a choice in the matter when selected by their village head. In addition, the coolies could never be certain that their service really ended with the next village reached, since the 'guest' they were assisting could always demand further service from them if the next village could not or refused to supply new coolies. In extreme situations, the government could also use the coolie system as a punitive measure as Van Oort signals when he writes that several villages on Sumatra 'are now punished with occupation and supply of coolies without pay.'[15]

It is hard to estimate how many coolies were actually involved in the work of the Natuurkundige Commissie. From the diaries of Van Oort and Forsten it becomes clear that there must have been thousands of them. Van Oort reports they needed one hundred coolies to carry their baggage on their first

---

13 It must be noted that in several projects that Commissie members were involved in punitive forced labour did play a large role. Many of the workers in Diard's agricultural establishment were convicted to forced labour. The mining exploits of Schwaner, Von Gaffron and Comblen on Borneo also involved forced labour (even though payment was sometimes given).

14 'Het eigenlijke gebruik, alhier is dat zoo men van de eene plaats na de andere, of uit de Menahassa naar het naburige land gaat, de inwoners van de eene plaats u tot de naaste hoofdnegorij brengen en u aldaar aan de lieden overgeven, die nu op hunnen beurt u moeten dienen en verder brengen, komen er hier nu menschen te kort of dergelijke, dan kunt bij vergen dat die geene die u het eerste geholpen hebben, u verder helpen en verder brengen, doch dan krijgen zij boven de betaling van u nog extra betaling van die vreemde negorij zelve, in welke zij u adsisteren.' Reisverhaal van Dr. Forsten naar Menado. NCAO, 1075_048.

15 'En werden thans met bezetting en koelielevering zonder betaling gestraft.' Weber and Van Zanen, *Kunstenaar op Sumatra*, 123.

expedition,[16] while Forsten mentions the use of three hundred coolies when transporting his goods to his expedition headquarters in Tondano.[17] Forsten's assistant, Heinrich von Gaffron, has left a detailed financial documentation of his use of coolies during two short surveying expeditions along the coast of Ambon (Table 5).[18] The first expedition lasted eighteen days, from August 19 to September 5, 1842, consisting of 25 travel stages, while the second expedition lasted only five days, from October 19 to 23, 1842, consisting of twelve travel stages. For the first expedition, Von Gaffron hired between four and fourteen coolies per travel stage, and for the second, between four and seventeen.

TABLE 5 Coolie and proa payments by Heinrich von Gaffron. All information taken from taken from NL-HaNA, Van Beresteyn, 2.21.018, inv.no. 1523. All mathematical inaccuracies present in the original. Numbers in brackets added by the author

Date	Stage	Proa price	Coolies	Payment	Total
August 19, 1842	1	1,00			1,00
August 21, 1842	2		4	5	0,20
August 22, 1842	3	0,30	6	7,5	0,75
August 22, 1842	4	0,30	6	7,5	0,75
August 22, 1842	5		14	7,5	1,05
August 23, 1842	6	1,125	9	22,5	3,15
August 23, 1842	7		12	2,5	0,30
August 25, 1842	8		14	22,5	3,15
August 25, 1842	9		10	5	0,50
August 25, 1842	10		10	12,5	1,25
August 25, 1842	11	0,50	6	12,5	1,25
August 25, 1842	12	0,60	8	15	1,80
September 1, 1842	13		12	12,5	1,50
September 1, 1842	14		10	2,5	0,25
September 1, 1842	15	0,60	12	25	3,00
Proa back to Ambon	16	0,60			0,60
September 3, 1842	17		10	7,5	0,75
September 3, 1842	18	3,75			3,75
September 4, 8142	19	2,70			2,70
September 4, 8142	20	0,30	6	7,5	0,75

16 Weber and Van Zanen, *Kunstenaar op Java*, 59.
17 Reisverhaal van Dr. Forsten naar Menado. NCAO, 1075_014.
18 NL-HaNA, Van Beresteyn, 2.21.018, inv.no. 1523.

TABLE 5 Coolie and proa payments by Heinrich von Gaffron (*cont.*)

Date	Stage	Proa price	Coolies	Payment	Total
September 4, 8142	21	0,75			0,75
September 4, 8142	22		10	12	1,20
September 5, 1842	23		10	7,5	0,75
September 5, 1842	24		10	12	1,20
September 5, 1842	25	1,00			1,00
Total			(179)		34,15
Date	**Stage**	**Proas**	**Coolies**	**Payment**	**Total**
October 19, 1842	1		8	5	0,40
October 19, 1842	2		4	5	0,20
October 20, 1842	3		12	5	0,60
October 20, 1842	4		12	5	0,60
October 20, 1842	5		11	5	0,55
October 20, 1842	6		12	5	0,60
October 20, 1842	7		12	5	0,60
October 21, 1842	8		12	15	1,80
October 21, 1842	9		6	12	0,75
October 22, 1842	10		12	5	0,60
October 22, 1842	11	1	17	24	4,08
October 23, 1842	12	1	17	24	4,08
Total			(135)		14,66

Von Gaffron paid between five and 24 doits per coolie,[19] with an average of roughly 9½ doits per coolie per travel stage.[20] On one occasion the price of a proa was included in the price per coolie, but proas mostly needed to be rented separately at a price of between 30 doits and f3 and 75 doits. For the first expedition, Von Gaffron spent a total of f20 and 42½ doits on 179 coolie payments,

19 A doit was a low-value copper coin in the Netherlands Indies (with 120 doits making up a Netherlands Indies guilder) that was 'the preferred means of payment among the indigenous population for small-scale transactions.' Wolters, "Doit Infestation in Java," 112.

20 A note by Van Oort suggests that payment for coolies was dependant on the distance travelled. Von Gaffron does not record the distances for each stage. See Weber and Van Zanen, *Kunstenaar op Java*, 37n83.

for the second f14 and 66 doits on 135 coolie payments, leading to an average of just over 11 doits per coolie per travel stage. Compared to the f12 per month salary of the permanent assistants, coolie wages come out as incredibly low, so it financially made a lot of sense for Europeans to hire coolies wherever and whenever they were needed instead of hiring a permanent group of coolies.

It is likely that a large percentage of these individual payments represented individual people, as it is possible that individual coolies received multiple payments for repeated service. If the coolie-system functioned as Forsten described, then every travel stage (which went from village to village) would have been accompanied by a different group of coolies. This leads to a potential number of 314 coolies hired in the span of only 23 expedition days. One has to take into account that this was only a small expedition (Von Gaffron went on his own, without Forsten). The longer an expedition lasted and the larger it was, the larger was the number of coolies hired to make such an expedition possible.

If this pattern can be assumed for many of the expeditions of the Natuurkundige Commissie, and I believe it can, the total number of coolies that worked for the Commissie must have run into staggering numbers. During the entire period of thirty years in which the Commissie was active it is likely that over ten thousand individual coolies were hired, possibly even tens of thousands. Unfortunately, it is not possible to be more precise than this.

2 Knowledge Brokerage: Language Skills

Even though the coolies supplied essential services, without the right knowledge an expedition still could not be successful. As we have seen in the previous chapter, the Javanese assistants that were paid on a monthly basis provided the Commissie with specific knowledge and skills, mostly in the realm of natural history. However, when the Commissie travelled to other islands in the Indies Archipelago, much of the local knowledge of the Javanese assistants became obsolete. Different people groups had different customs, different islands had different flora and fauna, different languages were spoken. The Commissie was very much in need of local indigenous knowledge brokers that could fill the gap that on Java had been adequately filled by the Javanese assistants. From the writings of the members of the Natuurkundige Commissie two quintessential roles emerge: interpreters and hunting guides.

For the interpreters, we do not see a lot of examples. One reason for this is that most members and assistants of the Commissie spoke Malay themselves.

At the time, this was the lingua franca used in trade and international contacts in the Indies Archipelago.[21] Müller seems to have been a gifted linguist and regularly mentions speaking to the indigenous population in their own language. In his travel reports in the *Verhandelingen* Müller mentions the use of interpreters on Timor (but only because the indigenous leader they wanted to speak to did not know a word of Malay),[22] but a very obvious example of where interpreters were absolutely required is during the 1828 expedition to New Guinea.

In 1828, the *Triton* and *Iris* were sent on an expedition to New Guinea in order to found a colony on the west coast. Commissioner-general Du Bus had ordered Macklot to join the expedition and upon his request had allowed Macklot to take the entire Commissie along with them. So Macklot, Van Raalten, Müller, Van Oort, Zipelius and newly recruited artist (seconded by the army) Latour travelled along on the *Triton* and *Iris*. The expedition was commanded by Jan Jacob Steenboom (1786–1828).[23] On Ambon, the expedition was joined by government official Arnoldus Johannes van Delden (1804–1885), and three interpreters.[24] They are mentioned by name: Abdul Kadir from the island of Ceram, and Palukan and Pattybarombang from the island of Ambon (Figure 45).[25] These interpreters were necessary because the languages spoken in New Guinea were unknown to Europeans, even to many of the indigenous people of the Moluccas. As it turned out, many of the languages spoken by the Papuans were also unknown to the three interpreters. When the interpreters were able to understand the local population, relations were usually rather amicable and it was possible to have a decent conversation. However, it was when communication broke down that the situation spiralled out of control.

There are three major and two minor accounts of the expedition. The major accounts were written by Van Delden, military officer Justin Modera (1803–1866), and Müller. The accounts by Van Delden and Modera have recently been published by Willem Mörzer Bruyns, while Müller's account is part of the geographical/ethnographical volume of the *Verhandelingen*.[26] The minor accounts were both written by Macklot and contain a summary of his correspondence and his official report.[27] While much of these reports contain

21 Van Ronkel, "Maleisch."
22 Müller, "Bijdragen tot de kennis van Timor," 205.
23 Mörzer Bruyns, *Met de Triton en Iris*, 33.
24 Mörzer Bruyns, *Met de Triton en Iris*, 35.
25 Mörzer Bruyns, *Met de Triton en Iris*, 233.
26 Susanna, *Verhandelingen*, vol. 3, 1–80.
27 Macklot, "Uittreksels uit brieven van Dr. H. C. Macklot"; Macklot, "Verslag van het Land."

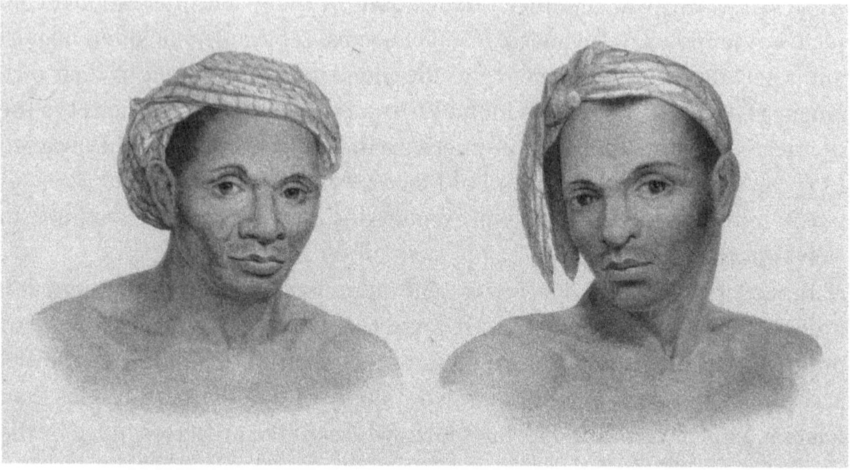

FIGURE 45 'Kustbewoners van het eiland Ceram' ('Coastal residents of the island Ceram'), by Pieter van Oort. Plate 21 (Ceram) from the volume *Land- en volkenkunde* of the *Verhandelingen*, NCAO, 871_021. Public domain. According to Müller, these two portraits are likenesses of two of the interpreters that joined the Commissie on the expedition to New Guinea. The original drawings are held at Wereldmuseum Leiden, RV-4537-89 and RV-4537-90.
HTTPS://DH.BRILL.COM/NCO/VIEW/NCO_NNM001000871_021/MAKINGSENSE
HTTPS://HDL.HANDLE.NET/20.500.11840/792147
HTTPS://HDL.HANDLE.NET/20.500.11840/792148

the type of observations that are usual for travelogues (a travel event journal interwoven with descriptions of lands, peoples, and customs encountered on the way), there is one striking passage where the communication with an indigenous people group breaks down and chaos ensues. Even the interpreters are not able to speak to this group of people. The story can be found in all three major accounts by Modera, Van Delden and Müller. Modera witnessed the event from his ship, while Van Delden and Müller lived through the event. Van Oort actually also lived through the event and immortalised it in a drawing that became a plate in the *Verhandelingen*. I will use this plate (Figure 46) in combination with Müller's narrative to unpack the event. An alternative depiction of the same event by Van Raalten was recently discovered (Figure 47).

The event took place on May 22, 1828. In the morning of that day, the expedition had an encounter with a group of indigenous people, with whom it was hard to get contact. They had approached them in two sloops. The interpreters were not able to speak to them and it had been hard to earn some trust. Some overtures were made to trade, but when the crew wanted to return the peaceful encounter suddenly exploded into violence, leaving three crew members wounded. In the confusion shots were fired, but none of the indigenous people

FIGURE 46 'Ontmoeting met de inboorlingen in de Prinses Marianne Straat' ('Meeting with the natives in the Princess Marianne Strait'), by Pieter van Oort. Plate 4 (New Guinea) from the volume *Land- en volkenkunde* of the *Verhandelingen*, NCAO, 871_004. Public domain
HTTPS://DH.BRILL.COM/NCO/VIEW/NCO_NNM001000871_004/MAKINGSENSE

FIGURE 47 Meeting with the indigenous people at the Princess Marianne Strait, drawing by Gerrit van Raalten, 1828, 42.5 × 37 cm, RV-4537-124, Collectie Wereldmuseum (Leiden). Public domain
HTTPS://HDL.HANDLE.NET/20.500.11840/792182

were killed. At around four o'clock in the afternoon, movement was seen in the treeline. Full of curiosity and interested to know how they would respond after shots had been fired, a sloop was sent to shore with Van Delden, Macklot, Müller and Van Oort on board.

In the plate, we can see the sloop approach the treeline. The interpreter is standing up and waving with a piece of cloth while trying to make contact with the indigenous people. Van Oort portrays himself, as always, with his hat on at work, while one of the military men joins the interpreter in holding out a necklace of beads. The men in top hats undoubtedly include Macklot and Müller. The sloop is an image of order, peace, and civilisation. In the treeline, upon first inspection, we see only a few human figures stand out from the trees. The longer one looks, however, the more human figures one distinguishes. Only careful observation leads to a final tally of nineteen human figures. Most of them hold back, while only a couple of them move towards the people in the sloop. One of these is waving a branch. In contrast to the sloop, the treeline is the image of chaos, disorder, and wildness.

The meeting lasted about half an hour but did not result in any meaningful contact. Müller describes his experience of seeing these indigenous people in the following terms:

> When we approached the shore, we noticed that the treeline was full of natives. They made a terrible noise, jumped, beckoned, nodded, and signified through a hundred other movements and gestures that we should land. Our interpreter from Ceram, in his turn, did not stop calling to them as loud and well as he could that we expected them with us, showing them white pieces of linen, bead necklaces and other such gifts. Some of them climbed down and appeared with green branches in their arms in front of the forest, where they stood up until their armpits and some even up until their neck in the water. There was no end to their beckoning, their waving with branches, and their strong barking calls: *kaka, kaka, – djewa, djewa, – njieuba, njieuba*, etc. All of them were shouting at different tones over each other, and the movements of one were even wilder and more ferocious than those of the other. Their brightly painted faces and unkempt hair were clearly visible to us. [....] When after about half an hour, we rowed westward along the edge of the forest to the mouth of the creek, the savages followed us there, just like large monkeys, climbing through the canopy of the trees and all over the protruding roots of the mangrove trees with their naked, dark bodies.[28]

28 'Den wal naderende, merkten wij op, dat het geboomte vol zat van inboorlingen. Zij maakten een vreeselijk misbaar, sprongen, wenkten, knikten en gaven ons door honderd

While the civilised sloop is all peace and quiet, the group of indigenous people explodes into chaos and a cacophony of noise. When reading through this passage the image is conveyed of a group of senseless monkeys jumping up and down the trees and shouting insults at the innocent travellers. What starts out as a peaceful group of 'natives' ends in a group of 'savages' who are 'like large monkeys.' The inability to communicate with these people turns them in the eye of the European beholder (in this case Müller) into senseless savages.

This is the only passage in the entire *Verhandelingen* where communication has broken down to such an extent that the indigenous people lose their humanity in the description. This only goes to show how important the language skills of the interpreters were. When these skills were sufficient and communication was made possible, understanding started to grow and indigenous people were approached on the level of fellow humans. Without those language skills, or when those language skills were insufficient, the indigenous people lost their humanity. Without the successful brokerage of interpreters, contact with the indigenous people groups was nearly impossible. When that communality of language is lost, humanity is also lost and people become part of the natural world. The loss of language seems to turn humans into objects of natural history rather than conversation partners.

3 Knowledge Brokerage: Hunting Guides

According to Peter Boomgaard, in nineteenth-century Indonesia 'hunting and trapping do not seem to have been major subsistence activities, although they were certainly not absent.' Most people were dependent on 'agriculture, fishing, and livestock rearing' for their subsistence, but that does not mean that

andere bewegingen en gebaren te kennen, dat wij toch aan land zouden komen. Onze Ceramsche tolk hield niet op, hun op zijne beurt, zoo luid en zoo goed hij kon, toe te roepen, dat wij hen bij ons verwachtten, hun tevens witte lappen linnen, snoeren koralen en andere soortgelijke geschenken vertoonende. Eenigen van hen klauterden af en verschenen met groene takken in de hand, voor het bosch, waar zij gedeeltelijk tot onder de armen, en enkelen zelfs tot aan den hals in het water stonden. Het wenken en zwaaijen met de takken, en hun sterk, blaffend geroep: *kaka, kaka, – djewa, djewa, – njieuba, njieuba* enzv., nam geen einde. Allen schreeuwden uit verschillende toonen, even geweldig door elkander, en de bewegingen van den eenen waren al wilder en woester, dan van den anderen. Hunne bont geschilderde gezigten en verwilderde haren waren ons duidelijk zigtbaar. [...] Toen wij, na omtrent een half uur toevens, westwaarts langs het bosch, naar den mond der kreek roeiden, volgden ons de wilden derwaarts, even als groote apen, met hun naakt, donkerkleurig ligchaam, vol behendigheid door de kruinen der boomen en over de uitstekende wortelen der mangle-stammen heen klauterende.' Susanna, *Verhandelingen*, vol. 3, 41.

hunting played no role in this.[29] As we have seen, purchasing animals was a common occurrence in the daily practice of the Natuurkundige Commissie. On December 29, 1831, Van Oort writes that he 'bought all kinds of animals from the people there, who gathered a nice bunch together.'[30]

Another recurring phenomenon was the extent to which the expeditions relied on indigenous knowledge when they were looking for specimens. This becomes apparent from Forsten's diary. We read of indigenous people giving him information on the number of villages and their makeup in the area,[31] giving him the local names of species,[32] explaining the nesting behaviour of the maleo (*Macrocephalon maleo*), a large bird endemic to Sulawesi,[33] and revealing the location of a prime pigeon-hunting spot (even though the ensuing hunting expedition did not result in any pigeons shot).[34] Forsten even complains when the indigenous people are close-lipped[35] and less forthcoming on where best to hunt deer and pigs. According to Forsten, they protect their own interests,[36] are unwilling to share the location of a boiling spring because they are afraid to go there,[37] and refuse to confirm or deny rumours that diamonds are to be found in a certain area (according to Forsten due to their 'enormous indifference').[38] Below we will see that Salomon Müller also extensively solicited indigenous knowledge during his travels.

The indigenous assistants of the Commissie and the members themselves were expert hunters and made use of all this indigenous information to gather large amounts of specimens. However, that does not mean to say that hunting was easy. Hunting was a dangerous job, especially when large mammals were involved. Gerrit van Raalten, who had been sent to the Indies as assistant to Kuhl and Van Hasselt, lived to tell the tale of how dangerous large mammals actually could be. While following a rhinoceros trail during a hunt, he surprised the animal and was badly wounded in the ensuing encounter. He was

29 Boomgaard, "Hunting and Trapping," 190–191.

30 'Ik kocht intusschen allerhande dieren van de menschen daar, die spoedig een aardig boeltje bij elkander bragten.' Weber and Van Zanen, *Kunstenaar op Java*, 186.

31 Reisverhaal van Dr. Forsten naar Menado. NCAO, 1075_007.

32 Reisverhaal van Dr. Forsten naar Menado. NCAO, 1075_018.

33 Reisverhaal van Dr. Forsten naar Menado. NCAO, 1075_040 and 1075_041.

34 Reisverhaal van Dr. Forsten naar Menado. NCAO, 1075_096.

35 Reisverhaal van Dr. Forsten naar Menado. NCAO, 1075_009.

36 Reisverhaal van Dr. Forsten naar Menado. NCAO, 1075_010.

37 Reisverhaal van Dr. Forsten naar Menado. NCAO, 1075_071.

38 Reisverhaal van Dr. Forsten naar Menado. NCAO, 1075_106.

saved from a certain death by one of the accompanying indigenous hunters.[39] Tigers were also a source of anxiety. Van Oort writes:

> I still shiver at the thought of how easily from the high reeds that surrounded us a tiger could have jumped forward and could have snatched and mangled one of us in the wilderness, while we were following each other helplessly on our horses.[40]

As Van Raalten's unfortunate accident and Van Oort's anxiety show, there was dangerous wildlife around against which the hunting party of the Commissie had no defence, let alone that they would have known how to actively hunt these animals. Fortunately, there were specialist hunters available for wildlife like tigers, rhinoceros, and crocodiles. These hunters, such as the *orang malim* that I will come back to later, 'were regarded as spiritual specialists, capable of dealing with animal and forest spirits which threatened lesser mortals,' and even 'tiger or crocodile charmers, who were supposed to be able to let these animals do their bidding.'[41]

In the official reports of the Natuurkundige Commissie, Salomon Müller writes about these specialised hunters and explains how their hunting techniques have benefitted the work of the Commissie. He confesses that the specimen of the *Rhinoceros sondaicus* (Javan rhinoceros, Figure 48) that he uses to describe the species was actually procured by an 'old experienced hunter,' who was a professional rhinoceros hunter and regularly accompanied them (for a fee). Müller describes the hunting techniques that the old hunter shared with them, and which would have prevented Van Raalten's unfortunate accident. The old hunter advised them to always make a lot of noise, thus eliminating the risk of surprising the rhinoceros, and to always try to shoot the rhinoceros from an elevated position, either driving it to the open, or climbing in a tree to create elevation. Rhinoceros were known to blindly charge directly towards a perceived danger, so waving a piece of clothing below you while sitting in

39 The story can be found in Van Oort's diary, Weber and Van Zanen, *Kunstenaar op Java*, 65–66. An extended version is available in Susanna, *Levensschets van Hendrik Boie*, 258–261, and Müller and Schlegel, "Bijdragen tot de natuurlijke historie der rhinocerossen," 186–187. Just like Alfred Duvaucel, Diard's early travel companion, Van Raalten survived the attack to succumb to fevers a couple of years after his ordeal.

40 'Ik ijs er nog van als ik herdenk, hoe ligt er uit het hooge riet dat ons rondom omgaf, een tijger ten voorschijn had kunnen springen, die een van ons, welke hulpeloos te paard, achter elkander voortstapten, in de wildernis had kunnen rukken en verscheuren.' Weber and Van Zanen, *Kunstenaar op Java*, 223nI.

41 Boomgaard, "Hunting and Trapping," 192–193.

FIGURE 48 *Rhinoceros sondaicus*, by Hermann Schlegel. Plate 33 (Mammalia) from the volume *Zoologie* of the *Verhandelingen*, NCAO, 864_298. Public domain HTTPS://DH.BRILL.COM/NCO/VIEW/NCO_NNM001000864_298/MAKINGSENSE

a tree would lure the rhinoceros close enough to shoot. To this hunter, the Natuurkundige Commissie owed at least two specimens.[42]

The hunting techniques on Sumatra to catch the *Rhinoceros sumatranus* (Sumatran rhinoceros, *Dicerorhinus sumatrensis*, Figure 49) were quite different. Müller describes how the indigenous people used to dig holes of roughly one by two meters wide and two meters deep right in the middle of an existing rhinoceros path in the forest. The edges would be lined with poles and sticks driven into the ground to prevent the rhinoceros from climbing out, and then all of it would be covered with leaves. Müller writes that all three specimens of the Sumatran rhinoceros that the Commissie had obtained, bought from the locals for 60 guilders each, were caught using this technique. In addition, seven tapirs that had also fallen into this trap were purchased for initially 48 guilders and later 36 guilders each. The Commissie had been able to eliminate

42 Müller and Schlegel, "Bijdragen tot de natuurlijke historie der rhinocerossen," 187–188. Van Oort also tell the story of 'an old hunter from Tjikao' ('een oude jager van Tjikao') and describes his hunting techniques as 'unusual' ('merkwaardig'). See Weber and Van Zanen, *Kunstenaar op Java*, 62n1.

FIGURE 49 *Rhinoceros sumatranus*, by Hermann Schlegel. Plate 34 (Mammalia) from the volume *Zoologie* of the *Verhandelingen*, NCAO, 864_299. Public domain
HTTPS://DH.BRILL.COM/NCO/VIEW/NCO_NNM001000864_299/MAKINGSENSE

the dangers of hunting rhinoceros by making full use of indigenous hunting techniques. Without these indigenous skills, it would have been much harder to procure all the rhinoceros specimens that they did.[43]

'Without a doubt, crocodiles are one of the most dangerous and most feared predators of the Eastern Archipelago. We think it is possible that in the Indies more people lose their lives to crocodiles than to tigers.'[44] This is how Müller introduces the treatise on the crocodiles of the Indies. And while they were extremely dangerous, they were also incredibly difficult to catch. Müller complains that crocodiles are 'often lost to the hunter.'[45] When shot, there is almost always enough life in them to flee to the water and only if they happen to get

43 Müller and Schlegel, "Bijdragen tot de natuurlijke historie der rhinocerossen," 192.

44 'Tot de gevaarlijkste en meest gevreesde roofdieren van den Oosterschen Archipel, behooren ongetwijfeld de Krokodillen. Wij houden het voor mogelijk, dat er in Indië niet meer menschen door de tijgers, dan wel door de krokodillen het leven verliezen.' Müller and Schlegel, "Over de krokodillen," 1.

45 'Voor den jager verloren.' Müller and Schlegel, "Over de krokodillen," 4.

FIGURE 50 *Tomistoma schlegelii* Müller, 1838. Skull, collected by Salomon Müller on Borneo, RMNH.RENA.35444, Naturalis Biodiversity Center, Leiden. Photograph in the public domain
HTTPS://DATA.BIODIVERSITYDATA.NL/NATURALIS/SPECIMEN/RMNH.RENA.35444

stuck in some reeds is there a chance of retrieving them. A series of gruesome anecdotes then illustrate the dangers that crocodiles pose to the Dayak people on Borneo.

But fortunately, there are also specialised crocodile hunters. These people are called *malim* on Java, and *pangereran* on Borneo. Müller describes how they obtained a skull from a 'skull-display.' Hunting crocodiles was a ritualised event which required extensive and careful preparations from the *pangereran*. The hunting technique used by the *pangereran* was fairly simple: a hook with meat was suspended on a rope above the water. In order not to anger the crocodiles, there were only ever as many hooks used as there had been animals or people taken by the crocodile that was the object of the hunt. When the crocodile was caught on the hook, it was killed, and its head was cut off and displayed on a wooden frame in order to warn and scare off other crocodiles. It was from such a 'skull-display' that Müller received a large skull for their collections from a group of *pangererans* who had been hunting crocodiles in a certain area for a period of seven months (Figure 50).

In areas where there was plenty of food available for the crocodiles a different hunting technique was used. A wooden plank was smeared with resin.

When a crocodile would leave the river to rest on the riverbank, it would look for a flat surface. The unfortunate crocodile that chose the plank as its resting place would get stuck to it and would be unable to dive, making it an easy target.[46]

In the cases of the rhinoceros and the crocodile, the Natuurkundige Commissie made extensive use of indigenous hunters. Specialised hunters were a normal occurrence in the Indies and whenever the Commissie encountered them, they made use of their services, often for a fee, or bought specimens from them. This allowed them to keep themselves and their indigenous assistants safe. Van Raalten's accident had shown that it was extremely dangerous to hunt the large forms of local wildlife without knowledge of the required hunting techniques. Fortunately, Van Raalten was the only Commissie member to ever fall victim to an angry rhinoceros, but Van Oort's palpable fear was not imagined. Relying on indigenous skill and knowledge kept the Commissie members and their indigenous assistants safe from harm.

4 Validating Indigenous Knowledge

The ultimate goal of the collecting mission of the Natuurkundige Commissie, at least from the point of view of 's Rijks Museum and Temminck, was to enrich the taxonomical knowledge of the Dutch colonies and bring the riches of the Indies into the European scientific discourse through Dutch mediation. It would not do if either the French or the English would publish zoological novelties from the Dutch colonial realm. That had to be done by scientists in Dutch service. The Dutch government was fully in agreement with this point of view. This becomes clear from the recurring article in the appointment decrees of the members that states that all zoological finds needed to be published in Dutch first before they could be translated or published in other languages. Article 3 of the decree in which Kuhl and Van Hasselt are appointed almost literally returns in the decree in which Boie, Macklot and Müller are appointed:

> No description of the discovered or transferred [to the Netherlands] objects, nor of the observed natural phenomena, shall be allowed to be published in a foreign language or to be sent to foreign societies and academics without permission of our Minister, and only after

46 Müller and Schlegel, "Over de krokodillen," 1–13. See Weber, "Over krokodillenvangers," for a popular account.

> earlier publication through one of the learned societies or journals of this realm.[47]

Even though this was worded in the form of a rather strict prohibition, Temminck did not seem to take this very seriously.

In a letter from December 3, 1824, he informs the minister about how the Kuhl and Van Hasselt collections will be published. Originally, the idea was that the members themselves, upon their return, would describe and publish their own collections, as stated in article 2 of the same royal decree. However, their early deaths prevented Kuhl and Van Hasselt from publishing their own findings. As a result, Temminck had to be creative and chose to use a variety of outlets to publish their collections. In the letter, he describes how he himself will publish the mammals in his *Monographies de mammologie*,[48] the birds in his *Planches coloriées d'oiseaux*.[49] Heinrich Boie will publish the reptiles and amphibians, while the fish will be published in the large work that Georges Cuvier (1769–1832) and Achille Valenciennes (1794–1865) are working on.[50] Wilhem de Haan (1801–1855), curator at 's Rijks Museum,[51] will publish the molluscs in cooperation with André Étienne d'Audebert de Férussac (1786–1836), while Jacob Sturm (1771–1848) from Nuremberg was set to take care of the insects.[52] When the minister objected to this on the grounds of article 3 from the appointment decree, Temminck indicated that a financial investment from the side of the government was required to make such a publication possible.

Two themes from how Temminck dealt with the Kuhl and Van Hasselt collection kept returning. The first one was the problem that hardly any of the members of the Commissie returned to the Netherlands. Only Salomon Müller and Pieter Willem Korthals ever tasted the pleasure of seeing their (in Müller's case adopted) homeland again. The second problem lay with the finances needed to make a publication possible. Eventually, the *Verhandelingen* would

---

47 'Geene beschrijving der ontdekte of overgezonden voorwerpen, noch der waargenomene natuurverschijnselen, zal in een vreemde taal mogen geschieden of aan Buitenlandsche Genootschappen en geleerden mogen toegezonden worden, dan met toestemming van onzen Minister voornoemd, en na de vroegere bekendmaking door een der geleerde genootschappen of tijdschriften van dit Rijk.' KB, May 2, 1820, no. 10. NL-HaNA, Staatssecretarie, 2.02.01, inv.no. 988.

48 Temminck, *Monographies de mammalogie.*

49 Temminck and Laugier de Chartrouse, *Nouveau recueil.*

50 Cuvier and Valenciennes, *Histoire naturelle des poissons.*

51 See Holthuis, *Rijksmuseum van Natuurlijke Historie*, 32–34.

52 Veth, *Overzicht van hetgeen gedaan is*, 37–38.

be published in 29 parts between 1839 and 1847 by Müller and Korthals, with contributions by Hermann Schlegel, Wilhem de Haan and Gerard Sandifort. Together these parts form three volumes. The first one deals with zoology, the second one with botany, and the third one with geography and ethnography in the form of several travelogues. Most papers in the zoological volume were written by Müller or Müller and Schlegel, with the paper on insects being written by De Haan. The botanical volume was written by Korthals, while the travelogues of the third volume were written by Müller.

As we have seen above, gathering the collections of the Natuurkundige Commissie had only been possible because of an extensive support structure of indigenous assistants. From a perusal of the *Verhandelingen* it also becomes obvious that indigenous support was not only indispensable for gathering the objects, but also for providing much of the factual knowledge that was required to publish the objects. Species descriptions were not complete without an evaluation of behaviour and habitat of the animal in question. Given the cursory nature of the expeditions of the Natuurkundige Commissie, much of this behaviour had not been observed by the Commissie members themselves. Indigenous knowledge was an important factor in the creation of new knowledge for a European context. Müller confirms the value of indigenous knowledge explicitly in his general introductory paper on the mammals of the Indies Archipelago when he writes:

> After long continued investigations of so many industrious men, and the manifold explorations undertaken by myself in the various regions of the archipelago with careful attention and questions directed to the natives, it seems to us that the Indies islands possess no or only very little large mammals of which we have no clue yet.[53]

Müller indicates here that it is not only his own research, but also his careful interrogation of the indigenous population that is the foundation of his publications. He even goes so far as to sometimes use the indigenous voice to contest with European scientific voices when it comes to the distribution of the Malayan weasel (*Mustela nudipes*) in the archipelago:

---

53 'Na de lang voortgezette onderzoekingen van zoo vele ijverige mannen, en de veelvuldige door mij zelven in verscheidene streken des Archipels, met zorgvuldige oplettendheid gedane navorschingen en aan de inboorlingen gerigte vragen, schijnt het ons toe, dat de Indische eilanden geene of slechts zeer weinige groote zoogdieren zullen bezitten, van welke wij nog geene aanduiding hebben.' Müller, "Over de zoogdieren," 3.

> According to French writers this species also occurs on Java; if that is the case, she has not been collected there by Mr. Reinwardt, Kuhl and Van Hasselt, nor observed in the wild by Mr. Boie of myself. Her Javanese origin must for that reason be questioned, all the more so because in the western part of the island no native knows anything about this animal.[54]

The testimony of the indigenous people in the area, or better, the lack thereof, supports Müller in his (correct) belief that the French are mistaken in stating that the Malayan weasel occurs on Java. The fact that Müller feels free to mention this in a scientific publication reveals the confidence that he places in indigenous knowledge. In addition, he has absolutely no problem with crediting the indigenous population for this knowledge.

Müller mentions two reasons why European writers, as in the case of the Malayan weasel, can be mistaken in their descriptions of species from the Indies Archipelago. He is singling out the French writers,

> who seem to have assigned to the animals that are sent to them from foreign regions that place from which they received them as region of origin, while they often originate from various islands and various other faraway regions. This has, among others, been the case with various skins that were send to Paris by Mr. Diard that he, according to his own statement to me, collected on Sumatra, in Siam and in Cochinchina; but partly only sent to Europe at a later stage from Java. This explains how, despite so many famous French explorers who have visited the Indies and its islands at various stages, in many French zoological publications so many incorrect statements on the region of origin of many species can be found.[55]

54 'Volgens Fransche schrijvers zoude deze soort ook op Java voorkomen; dan zij is aldaar noch door de Heeren Reinwardt, Kuhl en van Hasselt verzameld, noch door den Heer Boie of mij ooit in vrijen staat waargenomen. Hare Javaansche afkomst zal uit dien hoofde wel in twijfel getrokken mogen worden, te meer, daar ten minste in het westelijke gedeelte des eilands, geenen inlander iets hoegenaamd van dat dier bekend is.' Müller, "Over de zoogdieren," 30.

55 'Welke aan de hun uit vreemde oorden toegezondene dieren, zonder verder onderzoek, dat oord als vaderland schijnen toegekend te hebben, vanwaar zij die ontvingen; terwijl dezelve intusschen dikwerf van onderscheidene eilanden en uit verschillende andere verre streken afkomstig waren. Dit is, onder anderen, het geval geweest met verscheidene, door den Heer Diard naar Parijs gezondene dierhuiden, welke hij, volgens zijne eigene aan mij gedane verklaring, op Sumatra, Siam en Cochinchina bijeenverzamelde; doch gedeeltelijk eerst later van Java naar Europa overmaakte. Hierdoor laat het zich verklaren hoe, ondanks de vele zoo roemvol bekende, Fransche reizigers, welke Indie en de

This is also the reason why the French have attributed an incorrect region of origin to the Malayan weasel that Müller had been able to correct based on his own research that included indigenous testimony. An element of imperial rivalry undoubtedly also played a role in Müller's critique of the French.

But Müller warns that indigenous testimony can be a fickle thing and often gets lost in translation, leading to incorrect statements in zoological publications. He contests the supposition made by René Primevère Lesson, in his *Voyage autour du monde sur La Coquille*, that the North Sulawesi babirusa (*Babyrousa celebensis*) may also live on New Guinea.[56] Lesson had based his supposition partly on indigenous testimony. Müller (correctly) contests Lesson's assumption, based on his own research, and goes on to describe how he believes Lesson may have misunderstood the indigenous reports. That this can lead to serious misconceptions, especially in the case of the babirusa, is a historical fact:

> The uncivilised and gullible natives often give miraculous and exaggerated reports of the most ordinary creatures, and when one is not able to personally have a clear and concise conversation with them in their own language on such occasions, everything becomes even more complicated and, in many cases, absurd. In this regard I can speak from my own experience, and we even possess proof of stories misunderstood through an overactive imagination: for example, the unusual depiction provided by Nieuhof, identified as the sokotyro or the sukoterio.[57]

Lesson may have only misunderstood the location of where the babirusa can be found, but Johan Nieuhof (1618–1672) has made the bigger mistake of creating a misconstrued image of the babirusa based on the indigenous descriptions.[58]

Indische eilanden, op onderscheidene tijden en aan velerlei oorden bezocht hebben, in vele Fransche dierkundige werken zoo vele onjuiste opgaven ten opzigte des vaderlands van vele diersoorten gevonden worden.' Müller, "Over de zoogdieren," 13.

56 Lesson, *Voyage autour du monde*, vol. 3, 55.

57 'De onbeschaafde en ligtgeloovige inlanders geven dikwerf van de gewoonste wezens, zeer wonderbaarlijke en overdrevene voorstellingen, en wanneer men zich bij zoodanige gelegenheden niet zelven met hen en in hunne eigene taal, klaar en duidelijk kan onderhouden, wordt de zaak nog meer verwikkeld en daardoor dikwerf in de hoogste mate ongerijmd. Ik kan in dit opzegt bij eigene ondervinding spreken, en wij bezitten zelfs de bewijsstukken van diergelijke verkeerd begrepene verhalen eener verhitte verbeelding: zoo als, onder anderen, de door Nieuhof, onder den naam van Sokotyro or Sukoterio geleverde vreemdsoortige afbeelding.' Müller, "Over de zoogdieren," 41.

58 Nieuhof, *Zee en lantreize*, 294.

Müller here indicates that his proficiency in Malay allows him to carefully converse with the indigenous people and ask follow-up questions that will help him to clarify what they actually mean. While Müller clearly gives credence to indigenous testimony, he also imposes himself in between the indigenous informers and the audience of his publication and asserts that indigenous testimony is only of scientific value when it has been verified and confirmed by someone who speaks their language, placing himself in a unique position. Because his proficiency in Malay and his personal research have allowed him to question and verify the testimony of the indigenous people, there is no one in Europe who can rightfully contest his scientific conclusions. The fact that Müller here qualifies his indigenous informers as 'uncivilised and gullible' only adds credence to the importance of his mediation, because he is able to turn the 'miraculous and exaggerated reports' that were misunderstood by well-respected travellers like Nieuhof and Lesson into reliable scientific fact.

Another example of where Müller critically engages with indigenous reports is in his treatment of the 'big cat' species on Java. In the main text of his general paper on mammals in the archipelago he states that two species, Felis tigris (tiger, *Panthera tigris*) and Felis pardus (panther, *Panthera pardus*) appear on Java.[59] This statement is further clarified by an extensive note of four full pages that discusses the indigenous perception of tigers.[60] According to the note, the indigenous people distinguish six species of tiger, but Müller declares that their species perception is rather muddled.[61] Even the specialised tiger hunters (*orang malim*),

> who we consider to be the best informed when it comes to these animals, display, when one interrogates them closely regarding the distinguishing features of each species, not only deficient and wavering knowledge, but their reports often deviate considerably. From the joint reports that we have gained on this subject, we believe that we can conclude that the pretended six species in any case should be reduced to half, if not even less, true species.[62]

59 Müller, "Over de zoogdieren," 29.
60 Müller, "Over de zoogdieren," 51–54.
61 For an in-depth historical study of human-tiger relations on Sumatra, see Bakels, *Het verbond met de tijger*.
62 Müller, "Over de zoogdieren," 52.

Müller then enumerates all six species and lists the descriptions that he has been able to piece together based on the testimonies of all the *orang malim* that he has spoken to. He mentions that, rather than just basing their speciation on visible traits, the *orang malim* also consider food preferences, behavioural patterns and interaction with humans as decisive species characteristics. Müller believes that all these characteristics are due to differences in age, habitat, and other circumstances. After carefully considering all six species, Müller concludes that two of them refer to the felis tigris (tiger, *Panthera tigris*), three of them refer to the felis pardus (panther, *Panthera pardus*), while the sixth one could potentially describe a new species. This is Müller's mediation in action. He has been able to talk to the *orang malim* in their own languages, he has unravelled their 'confusing' and 'conflicting' reports, turned it into a methodical overview, compared that to his own observations and then comes to the conclusion that there are two species of 'big cats' on Java, while a third one could possibly exist but cannot be confirmed because he does not have any personal observation of it. The extensive note is meant to prove the validity of how Müller interprets and mediates indigenous knowledge in such a way that it can contribute to truly scientific European knowledge. In reality, the indigenous knowledge system is invalidated and overruled because it does not fit the European mode of thinking.

5 The Gradual Disappearance of Indigenous Knowledge

The examples above each come from the zoological volume of the *Verhandelingen*. This is not the only part of the *Verhandelingen* where we see traces of indigenous information. When it comes to species description in the *Verhandelingen*, two different levels can be distinguished. A practical example will illustrate the point. The zoological volume contains a paper on the sunbirds of the Indies Archipelago. One of the species described in it is the *Nectarinia solaris* (flame-breasted sunbird, *Cinnyris solaris*, Figures 51 and 52). This bird was encountered by Müller on Timor. The description of this bird consists of two paragraphs, as in most systematic treatments of species in the zoological volume. The first paragraph is a detailed physical description of first the male bird and then the female bird, consisting of exact measurements of the body and precise descriptions of body parts and colours. The second paragraph reads as follows:

The Nect. Solaris is fairly common in the gardens and villages of Timor and Semau. In Koepang one often sees her fluttering over the red flowers of the garnet bush, which have been planted there as enclosures around many houses, like a butterfly, flitting from one flower to the next and probing them with her pointed beak. It is because of her habit to preferably find food in these beautiful flowers, and also because the colour of these flowers is similar to the lower body colour of the male that this bird is labelled by the inhabitants of Koepang with the name *Boerang daliema*, which means garnet bird. One usually sees two, a male and a female, together.[63]

FIGURE 51 *Nectarinia solaris*. Figure 5 of plate 8 (Aves) from the volume *Zoologie* of the *Verhandelingen*. NCAO, 865_081. Public domain
HTTPS://DH.BRILL.COM/NCO/VIEW/NCO_NNM001000865_081/MAKINGSENSE

63 'De Nect. Solaris is in de tuinen en dorpen op Timor en Poeloe Samauw vrij gemeen. Te Koepang ziet men haar vooral dikwerf over de roode bloesems der granaatheesters, welke daar als heiningen rond vele huizen geplant zijn, gelijk een vlinder heenfladderen, zich nu eens op dezen, dan op genen nederzettende en hen met het spitse nebje doorzoekende. Het is uithoofde harer gewoonte, om bij voorkeur in die fraaije bloesems voedsel op te sporen, en ook om de overeenkomst der kleur van die bloesems met die des onderlijfs van het mannetje, dat dit vogeltje door de bewoners van Koepang met den naam van *Boerang daliema*, d.i. granaatvogel, bestempeld wordt. Men ziet er meestentijds twee, een mannetje en een wijfje, bij elkander.' Müller, "Over de zoogdieren," 60.

FIGURE 52 *Nectarinia solaris* Temminck, 1825. Male, skin, collected by Salomon Müller on Timor, RMNH.AVES.134970, Naturalis Biodiversity Center, Leiden. Photograph in the public domain
HTTPS://DATA.BIODIVERSITYDATA.NL/NATURALIS/SPECIMEN/RMNH.AVES.134970

The two different descriptions require different types of interaction with the indigenous population. The first type description is a piece of laboratory work that can be done by any competent European scientist. The only necessary but indispensable form of indigenous agency here lies in the procurement of the specimen. No indigenous knowledge is required for the scientist to write this type of description, but it would not be possible to describe the specimens if it had not been procured through indigenous agency first.

The second type of description is partly dependent on indigenous knowledge. Even though this example does not explicitly mention Müller's habit of questioning the local people, it is clear that he had a conversation with at least one local person who was able to tell him the local name of the bird and perhaps even explain to him why the bird had received this name. Indigenous testimony is seamlessly woven into a valid European scientific description. An additional description of the same bird can be found in the Timor travelogue in the third volume of the *Verhandelingen*:

> The Nect. solaris is very common in all gardens and even in the enclosures of garnet bushes, close to the houses. There one often sees this little

> bird fluttering around like a butterfly and penetrating from time to time with its pointed beak the open flowers, whose beautiful red colour seems to compete in splendour with the fiery red lower body of the male. This together results in such a lovely spectacle that the eye cannot get enough of it. Greedily, it preys, not only on the nectars from the flowers, but also on small beetles, especially from the garnet bush, and is it because of the similarity of its colour to the flowers of this bush that it has been labelled by the inhabitants of Koepang with the name of *Boerong daliema*,[64] which means garnet bird.[65]

Both passages are very similar. Both passages mention the local name, so Müller's interaction with the indigenous people in the area is visible in both types of texts.

According to Husson and Holthuis, the travelogue passage was published on June 12, 1843, while the species description was published on June 26, 1845, making my second quote the earlier published version.[66] For the species description, Müller deleted the reference to the bird feeding on beetles and the reference to the beauty of the scene, while he added an additional reason for the local name of the bird, and the observation that they are often encountered in pairs. It is easy to see why the reference to the beauty of the scene fits in the travel description but is removed from the scientific description. What is more difficult to understand is why the local name in the scientific description suddenly gets an extra explanation. Was it because Müller was re-examining the specimen and read from his notes or remembered that the name could also be explained from the feeding habit of the bird? Or was it because he was more focused on describing the feeding habits of the bird and then realised that the name could also be explained because of the association of the bird with red flowers? Here, the strict division between indigenous knowledge and

64 Note Müller's inconsistent spelling. In the previous passage this was the *Boerang daliema*.

65 'De Nect. solaris is in alle tuinen en zelfs in de heiningen der granaatheesters, digt bij de woningen, zeer gemeen. Vaak ziet men daar dit kleine vogeltje, als een' vlinder rondfladderen en het spitse nebje van tijd tot tijd diep binnendringen in de geopende bloesems, wier fraai roode kleur, met het vuurroode onderlijf van het mannetje, in pracht schijnt te wedijveren, en die te zamen zulk een liefelijk schouwspel opleveren, dat het oog zich moeijelijk aan hetzelve kan verzadigen. Gretig aast het, niet alleen op den nectar der bloemen, maar ook op [174] kleine kevers, vooral van den granaatheester, en het is uithoofde der overeenkomst zijner kleur met de bloesems van dezen heester, dat het door de bewoners van Koepang met den naam van *Boerong Daliema*, d.i. granaatvogel, bestempeld wordt.' Susanna, *Verhandelingen*, vol. 3, 173.

66 Husson and Holthuis, "Dates of Publication," 23.

European knowledge fades and indigenous knowledge morphs into European knowledge in such a way that it is extremely difficult to determine which part of European knowledge originally used to be indigenous knowledge.

This example was chosen to illustrate the gradual disappearance of indigenous knowledge into the European scientific discourse. As a result of this obscuring of the indigenous voice, there will undoubtedly be a large part of knowledge in the *Verhandelingen* that is not directly attributed to indigenous knowledge but would never have been known if Müller would not have extensively spoken with the indigenous people that he encountered all over the archipelago. While Müller generously acknowledges his indigenous informants, and sometimes even explicitly lends his ear to them in opposition to European scientists, it is obvious that Müller understands himself to function as an essential scientific filter between unmediated indigenous knowledge and validated European knowledge. His language skills and his own observations act as safeguards against fanciful notions. In the case of the tiger species this mechanism is out in the open and Müller explicitly lays his mediation bare for all to see, but in the case of the flame-breasted sunbird the situation starts to become diffuse, leading to the conclusion that much of Müller mediatorial work remains unseen and thus much of the indigenous knowledge that makes up the body of scientific European knowledge disappears from view through this.

6 Conclusion

The contrast could not be bigger between the studious assistants who collected and prepared specimens for the Natuurkundige Commissie and savage people who clambered around in the trees, 'just like large monkeys.' What this chapter has shown is that the role of indigenous people in the natural historical research of the Commissie is incredibly diverse and surprisingly well-documented. In fact, indigenous agency and indigenous knowledge were two major factors in the success of the collecting mission of the Natuurkundige Commissie. Without indigenous agency, there would have been no collections; without indigenous knowledge, there would only have been limited material to write on these collections. Both the European zoological press and 's Rijks Museum in Leiden gained a wealth of information and a treasure trove of specimens from these largely unnamed indigenous contributors. It remains unfortunate that we do not have any written sources available from the hands of the indigenous agents that made the work of the Commissie possible. That makes it difficult to analyse the motives for their participation and the layering of their loyalties. However, it is possible to come to some tentative conclusions on the nature of their motives and loyalties on a collective level.

When we consider the category of the employed indigenous assistants, a primary motive may well have been simply of an economical nature. A paid position with a decent salary would certainly have been desirable. This created a direct loyalty to the members of the Commissie, and an indirect loyalty to the goals of the Commissie. The interpretation of the goals of the Natuurkundige Commissie by the Commissie member who happened to direct their daily practice would of course have been leading for what the indigenous assistants would try to achieve in their work. It is likely that none of the indigenous assistants were very much aware of the conflicting loyalties that existed within the Commissie when their tasks were of a very practical nature. Perhaps a secondary motive for the indigenous assistants could have been a desire for adventure. Mobility in the archipelago for agricultural Javanese was limited and a job with the Commissie would offer an unprecedented chance to travel and discover life in the Outer Islands. What also cannot be discounted as a tertiary motive for the indigenous assistants is exactly what had brought most members to the Commissie: an intrinsic love of natural history or hunting. The indigenous assistants may not have been aware of the European scientific discourse, but they shared in an indigenous knowledge universe that allowed for the accumulation of a vast amount of knowledge of the natural world. Like Salomon Müller, who drew the attention of Boie and Macklot by his excellent preparation skills, several of the indigenous assistants may have had a strong personal affinity for nature.

The category of the casually encountered indigenous people is much harder to fathom. Not having a formal relationship with the Commissie also meant that their obligations were of a much different nature and much less clear-cut. Sometimes, their motives would have been economical, and they may have seen the Commissie only as a source of revenue. If those foreign *totoks* wanted to buy animals, then they were happy to sell these to them, but that did not create any loyalty beyond the single economic transaction. Sometimes, the relationship would extend beyond the simple one-time economic transaction and a hunter or guide would attach themselves to an expedition for a short while. This almost always involved a fee and did create some form of temporary loyalty that was over as soon as the attachment was finished.

However, the large majority of indigenous actors in the story of the Natuurkundige Commissie were coolies. While for some it may have been a voluntary exercise, the existence of the coolie-system indicates that very often coolie labour was not voluntary, even though it was paid. What is clear is that the task of a coolie was usually a job for a day, or at most a couple of days. In this case, the loyalty that forced them to take on these tasks was not directed at the government official (which is what the members of the Commissie were) asking for the service, but the indigenous leader selecting them to perform this task. This type of labour was much more prevalent in the practical organisation of an expedition but does not seem to have extended much to specific tasks relating to natural history. Expert skills and knowledge were recognised and valued for what they were by the Commissie and seems to have mostly been hired outside of the context of government-required labour.

All in all, Dutch military officer Br., who warned Boie against relying too much on forced labour, correctly judged the situation. The experience of Reinwardt, Kuhl and Van Hasselt, as described by Br., instilled in Boie the desirability of hiring a group of indigenous assistants whose loyalty could be counted on. This was a continued practice throughout the history of the Natuurkundige Commissie, even to such an extent that it is more than fair to say that indigenous agency and indigenous knowledge were the keys to success and that without indigenous support the Natuurkundige Commissie would have failed in its mission.

Conclusion

Jean Chrétien Baud served as minister of colonial affairs from 1840 to 1848. During those years, Baud faithfully visited king Willem II every Tuesday in order to advise him on governing the colonies. At one particular meeting with the king, on August 9, 1842, Baud 'demonstrated how favourably the work of our Natuurkundige Commissie was judged' with an issue of the French journal *Revue zoologique* in hand.[1] Of course, Baud had an ulterior motive in showing this to the king. Along with the favourable judgment of the work of the Commissie, he also handed the king a draft of a royal decree that contained the stipulations for a pension arrangement for the current members of the Natuurkundige Commissie. The ruse worked, and on August 14, 1842, the king signed the royal decree.

Baud was able to show to the king that the work of the Natuurkundige Commissie was favourably judged by contemporaries. In previous studies of the Natuurkundige Commissie, this contemporary judgement was entirely forgotten and the work of the Commissie was evaluated from modern perspectives. The high mortality rate of the members was interpreted as a failure, the collections were deemed of insufficient quantity, the publications were considered to be underwhelming. Part of the problem was that the Natuurkundige Commissie as a whole was mostly only studied with the desire to clear up the chronology and describe the sequence of events. In addition, much attention was paid to individual members, which made it easy to lose sight of the full project. Moreover, the narrative of the Commissie was firmly based in a European context and it was not noted that the reality in the field was an entirely different one from the one that entered the public domain in the Netherlands.

By shifting the perspective and bridging the geographical gap between the Netherlands and the Netherlands Indies, it becomes clear that the Natuurkundige Commissie was much more than a series of heroic explorers who died in their brave attempts to gather knowledge. It was a complex conglomerate, stretching half the globe and a period of thirty years, with more participants and stakeholders than was ever imagined. Approaching the Natuurkundige Commissie as such a complex reality has opened up the perspective and allowed us to see that the Commissie was much more than the sum of its members, much more than the collections they gathered, and much

1 'Aangetoond hoe gunstig beoordeeld wordt het werk onzer Natuurkundige Commissie,' Hugenholtz, *Geheim van Paleis Kneuterdijk*, 85.

 | DOI:10.1163/9789004696877_010

more than the publications that in later years came to be considered too little too late.

The issue of the *Revue zoologique* that Baud took with him to his meeting with the king contained such a positive review of the *Verhandelingen* that Baud believed it would help him to convince the king of the continued value of the Natuurkundige Commissie:

> The friends of the sciences will be grateful to the king of Holland for the services that he renders them by introducing the natural products of his rich colonies to Europe. [...] Attaching his name to such an enterprise is, for a sovereign, a title of glory so beautiful, and certainly more durable, than the conquest of a province and even a kingdom.[2]

Stronger praise was hardly possible. And this was precisely what king Willem I had set out to achieve with the Natuurkundige Commissie. He was in need of international prestige and hoped to gain that by showing the riches of the Dutch colonies to the world. The investment that the first stakeholder had made in the Natuurkundige Commissie had paid its dividends.

What the reviewer did not mention, was that the king was not the instigator of the Natuurkundige Commissie. It had been conceived by Temminck, who was in need of crates full of objects of natural history to fill 's Rijks Museum van Natuurlijke Historie in Leiden. The staggering amounts of specimens that made their way to Leiden were proof of Temminck's clever political manoeuvring through which he had managed to convince the king to become his benefactor. They were united by a single goal with mutual benefits. As this study has shown, this convergence of interests by the two major Dutch stakeholders in the story of the Natuurkundige Commissie gave it a flying start and ensured its continuation as soon as it had landed in dire straits.

However, as much as its crowing glory, the *Verhandelingen* were also a nail in the coffin of the Commissie. Through it, as far as the king was concerned, the Commissie had achieved the goal it existed for. In addition, the reality in the field had proved to be a totally different one from the one imagined behind Dutch desks. The building of a new colonial state demanded the full attention of the governors- and commissioners-general in the Netherlands Indies.

2 'Les amis des sciences seront reconnaissants envers le roi de Hollande pour les services qu'il leur rend, en faisant connaître à l'Europe les productions naturelles de ses riches colonies. [...] Attacher son nom à une pareille entreprise est, pour un souverain, un titre de gloire aussi beau, et certainement plus durable, que la conquête d'une province et même d'un royaume.' G. M., "Mémoires pour servir à l'histoire naturelle," 154.

If they were going to invest in an expensive project like the Natuurkundige Commissie, they might as well steer it in such a direction that they would be able to reap some benefits too. Information from the fringes of the fledgling empire would serve them well in consolidating and expanding their hold on the archipelago. When interest in the Commissie from the Netherlands waned, it waxed in the Indies. From a vehicle to generate international prestige and collections, it slowly turned into a fountain of knowledge of the colonial realm that could serve to further exploit the archipelago.

In the end, the Commissie outlived its usefulness. The king had his prestigious publication, Temminck had his massive collections, the governors-general had their information. The king had no further need of new publications, while Temminck and the governors-general had found other (cheaper) ways of continuing to receive what they needed. Interpreting the story of the Natuurkundige Commissie in the light of the direct interests of its stakeholders has supplied us with an explanation of how and why it became surplus to requirements. In fact, it is nothing short of a miracle that the Natuurkundige Commissie continued to function for such a long time.

To keep functioning for such a long time, it needed a veritable cast of characters. I have sketched the broad spectrum of participants in the Commissie, from university-educated Europeans to seafaring Ambonese interpreters. The usual cast of the fourteen royal appointees has been expanded by another dozen or so European participants that already happened to be in the Indies, several dozens of indigenous employees, and countless of temporary indigenous participants. This has given rise to a rich tapestry of agency that eclipses the usual storyline of a handful of tragic hero-explorers who died in the name of science.

The more or less well-known figures in the Commissie have been put in a different light by looking at them through the collective lens of the Natuurkundige Commissie. Using a geographical perspective, I have distinguished between the members and assistants that were appointed by the king in the Netherlands and those that only were added to the roster in the Netherlands Indies. While the Netherlands appointees have long dominated the narrative, there is a wholly different and mostly disregarded story that unfolded by looking at directing member Diard, one of the European members that was appointed in the Netherlands Indies. Hardly known in the Netherlands, he turns out to have been of considerable influence in more than one domain.

Diard's career embodies the rivalry that existed between the colonial powers in the realm of natural history. A Frenchman in British service, whose brief Vietnamese excursion was sandwiched between a life-long career in Dutch service, Diard is a prime example of a transimperial naturalist whose body of work became a problem precisely because he repeatedly crossed imperial borders. The interimperial rivalries even fractured the afterlife of Diard into three 'separate persons' whose lives and careers have been finally united again in this study.

Studying Diard's life has drawn the Natuurkundige Commissie into an international arena and has shown that natural history was uncontested as long as empires stayed within their own borders. Especially when borders became muddled, or had not yet been agreed upon, natural history could create legitimacy. Diard as an individual player could not achieve much in this regard, but the Natuurkundige Commissie as a whole could. It served as the backbone of the military expedition that was to stake the Dutch claim on the western parts of New Guinea for the next century and more. It surveyed mining opportunities and drew maps. It paved the way for the consolidation of Dutch power.

In addition, Diard's life has also shown that natural history and agricultural innovation where close friends. And because Diard saw both as sides of a single coin, he was able to combine the two and galvanise the development of various cash crops. Diard's career encourages us to look for other examples of transimperial naturalists to further tell the story of interimperial rivalry as well as examples of dual careers in natural history and agriculture to further elucidate the story of that single coin and its adherents.

The third category of participants, next to the Netherlands and the Netherlands Indies appointees of European heritage, were the indigenous participants in the story of the Natuurkundige Commissie. This group of actors has long been neglected or overlooked, while this study makes it clear that it is not extremely complicated to unveil some of their 'hidden histories.' The common denominator in these 'hidden histories' is that they have consistently played a more important role that they have been given credit for. From hunters and collectors, to preparators and personal assistants, from interpreters and guides, to porters and informants, without indigenous agency the Natuurkundige Commissie could not have functioned as it did. These 'hidden histories' provide a depth to the story of the Natuurkundige Commissie that has not been recognised before.

By looking through a lens of layered loyalties at the Natuurkundige Commissie as a whole, at its participants as a collective and as individuals, I have painted a picture of a lively and flexible conglomerate that consisted of far more strands than recognised before and developed along with the waxing and waning of the loyalties of its individual members to either of the three stakeholders. I have shown that a strong loyalty to the Dutch stakeholders and a strong commitment to their goals gave the Commissie a strong start. This was visible in the way the members were appointed and the way they understood their mandate. They saw themselves as collectors of objects of natural history for 's Rijks Museum, even though in practice they stood under the orders of the Indies government. Through the years, a shifting balance of these loyalties to the Netherlands Indies stakeholder changed the practice in the field. Collecting for the Museum almost became an afterthought, and mineralogical and agricultural research took precedent. An eventual loss of commitment from the Netherlands finally led to the Commissie's demise, but only after it had 'promote[d] the knowledge of the natural condition and the products [of nature] of our possessions in the East Indies,'[3] during a period of thirty years.

There is much more to learn about the Natuurkundige Commissie. This study has offered glimpses into other areas of research that would help to complete the picture that I have begun to sketch. While I have looked at the interests of the Dutch stakeholders, there is much left to be said on the political developments in the Netherlands and research into the ministerial decision-making process would help to understand how the Natuurkundige Commissie was embedded in the Dutch political and societal context. For example, the central role that Baud, De Kock, and Van den Bosch played in the history of the Commissie has been touched upon here, but there is much to be elaborated upon.

In addition, I have hardly touched the material heritage of the Commissie. Naturalis Biodiversity Center has a tower full of objects that can be used to tell the story of the significance of the collections of the Commissie to the developments of the biological sciences and our understanding of the natural world. So too with the ethnological collections at Wereldmuseum Leiden and the drawings and engravings for and in the *Verhandelingen*. These are full of stories that can continue to deepen and expand the stories of colonialism and natural history that this book has touched upon.

And finally, the 'hidden histories' uncovered in this study expose a great lack in our study of colonial natural history. While much work has been done in recent years in this regard, there still remains a world to be won. There are plenty of travel reports, letters, and diaries that I have not used for this study, so who knows what kind of 'hidden histories' of indigenous participation are still waiting to be uncovered.

---

3 'Bevorderen de kennis van de natuurlijke gesteldheid en van de voortbrengsels onzer bezittingen in de Oost-Indiën.' KB, May 2, 1820, no. 10. NL-HaNA, Staatssecretarie,2.02.01, inv.no. 988.

Afterthought

In 2020, Naturalis Biodiversity Center in Leiden had good reason for a celebration. It was 200 years ago that its direct predecessor, 's Rijks Museum van Natuurlijke Historie, had been founded by king Willem I. In honour of this anniversary a special exhibition was organised. It was titled 'Van onschatbare waarde' [Priceless]. The exhibition consisted of 25 objects of historical significance that were taken from the climate-controlled collection tower and shown to the general public in the Treasure Chamber on the top floor of the museum area. Each object was accompanied by a small text card explaining the significance of the object. The exhibition was supposed to have lasted for 200 days, but due to the pandemic lockdowns it was extended several times and eventually ran for almost 400 days (including lockdown closures) from September 30, 2020 to October 31, 2021.

When I visited the exhibition, I was, of course, eager to find out whether any objects from the collection of the Natuurkundige Commissie were featured. I was curious to see how Naturalis would tell the story I was researching for my PhD. I was very glad to find out that three objects with a connection to the Natuurkundige Commissie were present among the 25 priceless objects: a glass sponge collected by Reinwardt,[1] a false gharial collected by Müller,[2] and a whole drawer full of geological specimens collected by Junghuhn.[3] I was less thrilled to note the size of the text card accompanying the objects, too small to tell a story and drowned out by the extra-large photographs of some of the objects with public figures. I thought it was a shame that none of the three objects that could be used to tell the story of the Commissie were deemed worthy of a portrait with a public figure.

I was left with the sinking feeling that Naturalis had left much unsaid. Modern-day issues of climate change, human-induced mass extinction, and colonialism were mentioned, but hardly elaborated on. 'Though Naturalis appears willing to confront some of the more painful aspects of its own 'indelible colonial history', […] the design of the exhibit aims to accentuate visitors' sense of wonder, progress, and pride.'[4] Several of the 'showpieces' could have been used to unveil the hidden stories behind these objects, but Naturalis

1 *Sclerothamnus clausi* (Marshall, 1875), RMNH.POR.29.
2 *Tomistoma schlegelii* (Müller, 1838), RMNH.RENA.7935.
3 Collection drawer BE.500122.
4 Bertens and Wilson, "Wonder, Empire, Science," 44.

 | DOI:10.1163/9789004696877_011

decided to tell a different story. And this is exactly the problem in the wider world of natural history museums.

Subhadra Das and Miranda Lowe have convincingly shown that natural history museums in general, and the Natural History Museum in London in particular, are having a hard time dealing with the colonial origins of the larger percentage of their collections. According to Das and Lowe, 'the current absence of decolonial interpretation in contemporary natural history museums is problematic. At best, it misrepresents historical fact; at worst, it alienates audiences.'[5] In the case of Naturalis, Caroline Drieënhuizen en Fenneke Sysling have already identified the 'endurance of Eurocentric perspectives' in their object biography of Java Man.[6]

Even the general exhibition displays of Naturalis allow the visitor to stay oblivious of the fact that 'natural history collections are a central part of the history of imperial appropriation of the world,' and that 'collecting relied largely on colonial infrastructures [and] profited from the asymmetrical power structures and the exploitation of colonised territories and peoples.'[7] In Naturalis' displays, the importance of indigenous contributions, both in terms of agency and knowledge brokerage, is largely disregarded.

I believe that the time is ripe for Naturalis to start addressing the 'covert racism [that] exists in the gaps between the displays.'[8] In my own reflection, I can see two ways forward for Naturalis. First of all, it needs to become obvious to visitors that the construct of European knowledge rests on a foundation of indigenous agency and knowledge. Every object has a backstory and many of these backstories are steeped in imperialism and colonialism. Naturalis needs to take responsibility by publicly recognising this and telling these stories in their permanent displays.

Secondly, these gaps can only be filled if there is a solid foundation of research to stand on. While this study already provides a decent stepping-stone, there is much more work to be done in uncovering the 'hidden histories' behind the scientific material guarded in the collection tower. At the moment, research of this nature is only sparsely initiated and financed by Naturalis. I would like to invite Naturalis to show more initiative in initiating and financing research focused on the colonial provenance of the collection.

This study allows Naturalis to begin to tell the story of how one of their foundational collections was gathered in the field and how indigenous knowledge

5 Das and Lowe, "Nature Read in Black and White," 11.
6 Drieënhuizen and Sysling, "Java Man and the Politics of Natural History."
7 Kaiser, "Coloniality of Natural History Collections," 246.
8 Das and Lowe, "Nature Read in Black and White," 8.

is the foundation for much of the European knowledge that was generated from these collections. It opens up ways for Naturalis to address and research issues of colonialism and imperialism in their collections in a comprehensive way and invites Naturalis to tell their visitors the story of the Natuurkundige Commissie and the indigenous agency and knowledge that was its cornerstone.

Appendix 1: a Short Historical Narrative

Kuhl and Van Hasselt 1820–1821

When Kuhl and Van Hasselt arrived on Java with their assistants Van Raalten and Keultjes on December 22, 1820, they immediately set about collecting in the environment of Buitenzorg. On March 16, 1821, Maurevert was appointed as draughtsman on Kuhl and Van Hasselt's recommendation. After a couple of months, Kuhl and van Hasselt decided to prepare for an expedition to Banten (West Java), but a cholera outbreak prevented this. Instead, they decided to explore the mountains around Buitenzorg first, which happened in May, June and July 1821. Kuhl fell ill in August 1821 and died on September 14, followed by Keultjes one day later.

Banten Expedition 1822–1823

After having recovered from the shock of losing Kuhl and Keultjes, Van Hasselt and Van Raalten resumed their work and departed for the planned Banten expedition in August 1822. Van Raalten fell ill and was replaced by Bik in March 1823. Maurevert was also part of the expedition. Four months after Bik had joined the expedition, in August 1823, Van Hasselt himself also fell ill and was transported back to Buitenzorg where he died on September 8, 1823, practically ending the Banten expedition.

Processing the Kuhl/Van Hasselt Papers 1823–1825

With the deaths of Kuhl, Van Hasselt and Keultjes, the Natuurkundige Commissie was significantly reduced. Van Raalten was formally the only one left and set himself to the task of preparing the Kuhl/Van Hasselt papers and collections for shipment to Europe, assisted by Fievez de Malines and Kollmann. During these years, the Natuurkundige Commissie was in no state to even think about mounting a new expedition. Fortunately, the appointment in the Netherlands of Boie, Macklot and Müller on December 5, 1823, gave hope that their speedy arrival would soon revive the exploration of the natural world on Java. However, they would not arrive until June 1826.

Buitenzorg and environment 1826–1828

When Boie, Macklot, Müller and Van Oort, who had been appointed as artist on April 16, 1825, arrived in Batavia on June 6, 1826, they settled in the Buitenzorg area, awaiting their official appointment and further instructions (which would only happen on February 27, 1827). They were immediately joined by veteran Van Raalten and started exploring the area around Buitenzorg (mostly within one or two days travel and mostly to the east). On May 25, 1827, Zipelius was appointed as botanical assistant. The form of the work of the Commissie in these years consisted of small exploratory trips made

 | DOI:10.1163/9789004696877_012

up of one or a couple of individuals from the larger group. One of the more substantial trips was a month-long coal exploration expedition by Macklot to Lebak (Banten) in August 1827. On the day of Macklot's return, September 4, 1827, Boie died, leaving Macklot in sole charge of the Commissie. Upon Macklot's request, Diard was appointed to replace Boie on January 31, 1828, a role that, within two years' time, evolved into the status of directing member.

Triton Expedition 1828

When the government decided to found a (military) trading post on the island of New Guinea in order to formally take possession of the island before the British would do so, the Natuurkundige Commissie was attached to this mission in its entirety (except Diard). Macklot, Müller, Van Oort, Van Raalten, and Zipelius travelled to New Guinea. Latour was seconded to the Commissie as draughtsman. The party left Makassar on March 15, 1828, and travelled via Ambon and Banda to New Guinea where they arrived on May 21, 1828. After six weeks they found a suitable place for a settlement (Merkusoord) and stayed there for another six weeks.

Timor Expedition 1828–1829

On August 30, 1828, the entire Commissie, probably minus Latour, left New Guinea to travel via Ambon to Timor, where they stayed for a year. Both Zipelius (December 31, 1828) and Van Raalten (April 17, 1829) died on Timor. During their stay on Timor the Commissie undertook several smaller and larger expeditions. Macklot, Müller, and Van Oort arrived safely in Batavia on December 11, 1829.

Java Exploration 1830–1833

Spanoghe was appointed on April 3, 1830 in order to copy Zipelius' manuscripts. Macklot, Van Oort and Müller spent the next three years travelling and collecting extensively on West Java, joined on April 13, 1831, by new member Korthals with assistants Van Gelder and Overdijk. On April 9, 1831, a new expedition proposal was approved. The plan was to travel along the northern coast of Java to Madura and Bali and then return along the southern coast. They took it slowly and spent most of 1831 and 1832 in the Priangan and Krawang residencies. On May 12, 1832, Macklot found himself at the wrong place at the wrong time and was killed in an uprising of Chinese labourers. Many of the papers of the Commissie were lost to the flames during this uprising.

Sumatra Expedition 1833–1836

On June 4, 1833, Müller, Korthals and Van Oort travelled to West Sumatra in the wake of a military expedition. They were accompanied by Bürger, who was neither a member nor an assistant, but attached to them with his own specific task. Because of the ongoing Padri war, independent expeditions were out of the question during most of 1833. Müller travelled to Java in October 1833 to pick up some books and instruments, and returned on January 9, 1834, with Van Gelder and Overdijk. In the meantime, Bürger had finished his task alongside the Commissie and left Sumatra two days before Müller returned, on January 7, 1834. Most of 1834 and 1835 was spent travelling and collecting in West Sumatra. Van Oort died on September 2, 1834. Müller, Korthals, and Van Gelder left Sumatra on January 9, 1836, following Overdijk who had returned to Java on health leave in the last quarter of 1835.

Borneo Expedition 1836

Meanwhile, new member Horner had arrived in Batavia on July 8, 1835. He spent his first couple of months on Java and received his orders to join the Commissie on Sumatra on October 1, 1835, but during his preparations the entire Commissie was recalled to Java. An expedition to Borneo was planned and Müller, Korthals, and Horner departed for Borneo on July 11, 1836. After two expeditions, they left Banjarmasin on December 17, 1836. They arrived back on Java on January 9, 1837.

The Colonial Museum in Batavia 1835–1838

During the Borneo expedition Van Gelder stayed on Java where he spent some time working on the collections of the newly founded Colonial Museum until he resigned from the Commissie on March 5, 1838. The Colonial Museum was founded by the Bataviaasch Genootschap with Diard as one of its main propagators and its director. Diard had stayed on Java the entire time working on his agricultural experiments (sericulture, tea cultivation, sugar manufacturing) and never joined the Commissie on any of their expeditions.

Returning to Europe 1837–1841

On April 12, 1837, the first two members of the Commissie finally commenced their return journey to the Netherlands. Both Müller and Korthals were sent home to work on the publication of their collections. They arrived on August 21, 1837, with Korthals retiring in on February 11, 1842, and Müller on July 19, 1850, after the formal dissolution of the Commissie. They both worked on the *Verhandelingen*, and published additional material as well. Diard resigned from the Commissie on August 26, 1841, and temporarily returned to Europe. He would come back to the Indies in 1848.

Sumatra expedition 1837–1838

Back in the Netherlands Indies, on June 13, 1837, Horner travelled to Sumatra with Overdijk, in order to inspect coal layers there. They spent some time travelling around.

Horner died on Sumatra on December 7, 1838. Shortly afterwards, Overdijk was recalled to Java where he arrived on March 5, 1839.

Java 1838–1839

Meanwhile, Junghuhn had been provisionally appointed as assistant on July 1, 1838 for the duration of one year. He spent this year on Java and worked for the Commissie until November 1, 1839, when he returned to his post in the military medical service. On December 14, 1838, new member Forsten had arrived in Batavia. Through circumstances his planned expedition did not immediately materialise and he spent most of his time in the Buitenzorg area, occasionally joining Junghuhn.

Celebes Expedition 1840–1843

Overdijk was supposed to join Forsten on his expedition to Celebes. However, when Overdijk demanded more salary, he was taken up on his threat and was refused a higher salary and let go. Forsten's dislike of Overdijk played a role in this. On January 19, 1840, Forsten got the governor-general to appoint Von Gaffron as his assistant and they departed for Celebes on January 29, 1840, where they arrived on March 22, 1840. They travelled and collected on Celebes until April 15, 1842, with a brief stay on Ternate from June to September 1841. In April 1842, they travelled to Ambon. Forsten died on the neighbouring island of Ceram on January 2, 1843. Von Gaffron returned with Forsten's collections to Batavia where he arrived on June 25, 1843.

Borneo Expedition 1843–1848

New member Schwaner arrived in Batavia on August 6, 1842, in order to continue Horner's geological research. On May 30, 1843, he received instructions for a coal exploration expedition to Borneo. Von Gaffron was appointed to take part in this expedition on July 28, 1843, while on August 2, 1843, mining engineer Comblen was added to the party. They left for Borneo on August 14, 1843. Schwaner would spend several years on Borneo, with occasional shorter trips back to Java. Comblen (February 2, 1846) and Von Gaffron (May 19, 1846) left the service of the Commissie and took on positions in the developing coal mining industry on Borneo. Von Gaffron was replaced by Van Pers, who worked with Schwaner for just a couple of months until April 27, 1847. From November 2, 1847 to February 2, 1848, Schwaner travelled inland from Banjarmasin to Pontianak, a feat that no European traveller on Borneo had completed before him. After this expedition, Schwaner returned to Batavia.

Java 1845–1848

Due to an extensive lobby in the Netherlands, Junghuhn was reappointed, this time as a full member, on April 1, 1845. In the period between November 1839 and his reappointment he had travelled and researched extensively on Sumatra and East Java. From April 1845 to August 1848, he spent all his time travelling around on Java. He left for the Netherlands on August 27, 1848 in order to write up his reports and was allowed to continue working on his reports after the dissolution of the Commissie in 1850, publishing his monumental work on Java between 1850 and 1854. Junghuhn returned to Java in 1855.

The Final Years 1848–1850

After his return from Borneo, Schwaner worked on his reports with the assistance of François (June 14 to December 21, 1848), Strung (from October 30, 1848), Brendonck (October 30 to December 14, 1848) and Oehler (April 1, 1848 to March 1, 1851). He even temporarily rehired Overdijk to work on preparing his collections from December 12, 1848 to August 23, 1849. After the dissolution of the Commissie on April 19, 1850, Schwaner was allowed to serve out his term (which formally would run until 1852), but he died on March 30, 1851.

Appendix 2: Biographies

This biographical section contains short biographies of all members and assistants of the Natuurkundige Commissie in alphabetical order. For the sake of completeness, biographies of Jan Cornelis de Bruijn (who left the Commissie before he departed for the Indies) and Heinrich Bürger (who was associated with and travelled with the Commissie but was never appointed as member or assistant) are included. Each biography is preceded by a list of entries from the set of reference works identified below. In addition, references to previous work on that particular person are included as well. Minor entries in biographical lexica that do not include original research are not included.

Abbreviations used:

BWN	Van der Aa, *Biographisch woordenboek der Nederlanden* (1852–1878)[1]
NNBW	Molhuysen and Blok, *Nieuw Nederlands biografisch woordenboek* (1911–1937)[2]
BPN	*Biografisch Portaal van Nederland* (current)[3]
ENI	Paulus, *Encyclopaedie van Nederlandsch-Indië* (1917–1940)
FHA	Fransen, Holthuis and Adema, "Type-catalogue of the Decapod Crustacea."[4]
MPCC	*Malaysian Plant Collectors and Collections* (1950–1974)[5]
BB	Bastin and Brommer, *Nineteenth Century Prints*
M	Maronier, *Pictures of the Tropics*
NSTO	*Nederlandse schilders en tekenaars in de Oost*
T	Troelstra, *Bibliography of Natural History Travel Narratives*
ADB	Von Liliencron, *Allgemeine Deutsche Biographie* (1875–1912)
NDB	Historische Kommission, *Neue Deutsche Biographie* (1953–2024)
OM	Gebhardt, *Die Ornithologen Mitteleuropas*

1 Available in open access: http://resources.huygens.knaw.nl/retroboeken/vdaa.

2 Available in open access: http://resources.huygens.knaw.nl/retroboeken/nnbw.

3 Available in open access: http://www.biografischportaal.nl/. The biographical entry for each individual can be accessed by adding the BPN number to the URL.

4 Available in open access: https://repository.naturalis.nl/pub/317736.

5 The original publication and the two supplements are available in open access: https://repository.naturalis.nl/pub/532515 (original); https://repository.naturalis.nl/pub/532688 (Supplement I); https://repository.naturalis.nl/pub/532591 (Supplement II).

 | DOI:10.1163/9789004696877_013

1 Jannes Theodorus Bik (1796–1875, Figure 53)

FIGURE 53
Portrait of Johannes Theodorus Bik, lithograph by J. T. Berghaus after a daguerreotype, 1859, KITLV 37B634, Leiden University Libraries, Digital Collections. Public domain
HTTP://HDL.HANDLE.NET/1887.1/ITEM:856397

b. June 12, 1796 (Amsterdam), d. July 3, 1875 (Batavia)
Assistant: March 21, 1823–September 1823

BWN, vol. 2.1, 533 (as J. H. Bik) | NNBW, vol. 8, 108–109 | BPN 98487492 | ENI, vol. 1, 302–303 | BB, 314–315
Scalliet, Marie-Odette. "Beelden van Oost-Indië: De collectie Bik in het Rijksprentenkabinet." *Bulletin van het Rijksmuseum* 49, no. 4 (2001): 342–379.

Jannes Theodorus (Theodoor) Bik was born on June 12, 1796, in Amsterdam to Jan Bik (1757–1827) and Jeannette Butin (1764–1821).[6] Upon his arrival in the Indies in 1816, he was appointed as second draughtsman at Reinwardt's department of agriculture, arts and sciences and worked mostly at 's Lands Plantentuin. Bik accompanied Reinwardt on two of his expeditions, in 1819 on an expedition to Priangan and again in 1821/1822 on Reinwardt's grand expedition that led them all the way to Timor and the Moluccas. When Van Raalten had to return from the Banten expedition due to health issues, governor-general Van der Capellen quickly appointed Bik to take his place.[7] Bik was initially not inclined to travel again, but a personal appeal (and the promise that he would

6 Scalliet, "Beelden van Oost-Indië," 345–346.
7 Res.GG, March 21, 1823, no. 6. NL-HaNA, Koloniën, 1814–1849, 2.10.01, inv.no. 2468. See also Scalliet, "Beelden van Oost-Indië," 376n45.

not forget it) by Van der Capellen after a dinner party on March 16, 1823, convinced Bik to consent.[8]

From March 22 to September 5, 1823, Bik accompanied Van Hasselt on his expedition in Banten, which would end rather suddenly when Van Hasselt fell ill. This also meant the end of Bik's association with the Natuurkundige Commissie. On January 21, 1824, he was appointed as auctioneer ('vendumeester') and stayed in this position until his pension in 1836. Bik would make his fortune in the Indies and died there on July 3, 1875. Theodoor Bik was the younger brother of Adrianus Johannes Bik (1790–1872), who also spent his early career in the Indies as (first) draughtsman with Reinwardt but was never formally attached to the Natuurkundige Commissie.

2 Heinrich Boie (1794–1827, Figure 11)

b. May 4, 1794 (Meldorff, Germany), d. September 4, 1827 (Buitenzorg)
Member: December 5, 1823–September 4, 1827

BWN, vol. 2.1, 776–779 | NNBW, vol. 1, 395–396 | BPN 17929302 | ENI, vol. 1, 342 | FHA, 214–215 | OM, 43–44

Anonymus. "Hendrik Boie, natuuronderzoeker op Java." *Handelingen en geschriften van het Indisch Genootschap* 6 (1858): 81–87.

Susanna, J. A. *Lebens-Abriss und Würdigung der Verdienste Heinrich Boies*. Schleswig: K. Taubstummen-institut, 1830.

Susanna, J. A. *Levensschets van Hendrik Boie, en hulde aan zijne deugden en verdiensten, benevens eenige door hem geschrevene brieven, gedurende zijne reis en verblijf in Oost-Indië*. Amsterdam: P. Meijer Warnars, 1834.

Heinrich Boie was born on May 4, 1794, in Meldorff, Germany, to Heinrich Christian Boie (1744–1806) and Sara(h) Helena von Hugo. Boie studied law in Kiel, Göttingen, and Heidelberg from 1812 to 1817. During his studies he occupied himself with natural history and found a patron in Friedrich Tiedemann (1781–1861) in Heidelberg, who offered him a post as curator at the zoological cabinet at the university in Heidelberg.[9] Boie started working there in 1817 and followed many lectures in natural history in Heidelberg. Both his friendship with Kuhl, whom he first met in 1814, and his correspondence with Temminck, whom he first met in 1818, were decisive factors that led to his appointment in the Natuurkundige Commissie. Boie's experience at the zoological cabinet in Heidelberg and his expertise as an academic zoologist, a testimony to which

8 Bik, "Aanteekeningen op eene reis," 260–261.
9 See Von Leonhard, *Fremdenbuch für Heidelberg*, 97–98, for more information on the cabinet.

was his critical review of Temminck's *Manuel d'ornithologie*,[10] led Temminck to invite Boie to take up the post of curator at 's Rijks Museum with the promise of a future appointment to an expedition to the Indies. Boie accepted and arrived in Leiden in the summer of 1821.

When news of the death of Kuhl (and Keultjes) reached the Netherlands, it was decided that a successor needed to be appointed. Recommendations by both Kuhl himself (made on his deathbed and communicated and supported by governor-general Van der Capellen) and by Temminck, led to Boie's appointment on December 5, 1823,[11] together with Macklot and Müller. Boie's actual departure was postponed when news of Van Hasselt's death reached the Netherlands, if only because Boie's expertise was needed in processing the vast collections of Kuhl and Van Hasselt that were sent to Leiden in the next two years. This resulted in Boie's manuscript of *Erpétologie de Java*.[12]

On December 21, 1825, Boie, Macklot, Müller and Van Oort left the Netherlands on the ship *Dijkzigt*,[13] and arrived in Batavia on June 6, 1826,[14] in a time of turmoil. Both the recent outbreak of the Java War and the chafing relationship between the commissioner-general (Du Bus) and the lieutenant-governor-general (De Kock) caused the Natuurkundige Commissie to drop substantially on the list of priorities. As a result, it was only on February 27, 1827, that the party was formally appointed in the Indies: Boie and Macklot at a salary of f500, Müller and Van Oort at f200. Van Raalten was appointed as administrator at f300, permission was given to hire six non-European employees at f12 each, and the Natuurkundige Commissie was given an annual budget for collecting costs of f6,500.[15]

This does not mean that the Commissie had been idle until that time. In August 1826, they had taken residence in the house that had previously been given to Van Hasselt and departed on their first expedition in December 1826.[16] They stayed mostly within one or two days travel from Buitenzorg. Boie was heavily involved in the administrative side, including a discussion on whether Surabaya was an option as headquarters for the Commissie, an idea which had been raised by the commissioner-general on June 20, 1826, in contradiction to the advice of the governor-general,[17] and was finally

10 Boie, "Manuel d'ornithologie."

11 KB, December 5, 1823, no. 84. NL-HaNA, Staatssecretarie, 2.02.01, inv.no. 1831.

12 The manuscript is available in the archives of Naturalis. See Hoogmoed, "Nomenclatural Problems," 132. Savage, Sparreboom, and Pregill, *Dutch East Indies Commission*, will contain texts and images from Boie's original manuscript with an analysis of its significance.

13 Veth, *Overzicht van hetgeen gedaan is*, 45.

14 Weber and Van Zanen, *Kunstenaar op Java*, 55.

15 Res.CG, February 23, 1827, no. 18, confirmed by Res.GG, February 27, 1827, no. 35. NL-HaNA, Koloniën, 1814–1849, 2.10.01, inv.no. 2501.

16 Weber and Van Zanen, *Kunstenaar op Java*, 58–59.

17 Res.GG, June 20, 1826, no. 2. NL-HaNA, Koloniën, 1814–1849, 2.10.01, inv.no. 2493.

dismissed by Boie and Macklot on May 1, 1827.[18] Boie fell ill on August 26, 1827, while in Buitenzorg, and died there on September 4, 1827.[19]

3 A. A. Brendonck

b. unknown, d. unknown
Assistant: October 30, 1848–December 14, 1848

On October 30, 1848, A. A. Brendonck, a 'flankeur' in the army,[20] was attached to Schwaner as topographical artist. He received an additional salary of f25 and had to move in with Schwaner. Brendonck arrived at Schwaner's house in Bandar Peté(e) on November 7, 1848.[21] He started his work with scaling down a map of Pontianak. Unfortunately, due to problems with his eyes, he had to leave this post already on December 14, 1848, but not without a written positive recommendation and a reward of f25, paid for from Schwaner's own pocket.[22]

4 Jan Cornelis de Bruijn (1795–1862)

b. August 29, 1795 (Amsterdam), d. November 11, 1862 (Amersfoort)
Assistant: February 17, 1824–February 13, 1825

Mac Lean, J. "Gegevens over tekenaars van biologische voorwerpen in het voormalige Nederlandsch Indië." *Scientiarum historia* 15 (1973): 101–114.

Jan Cornelis de Bruijn was born on August 29, 1795, in Amsterdam to Cornelis de Bruijn and Wilhelmina Lusing.[23] De Bruijn was trained as an artist and had made natural history his speciality. He wrote a request to the king that was received on January 10, 1824, accompanied by four drawings of his hand, asking to be appointed as a natural history artist anywhere in the East or West Indies.[24] Van Ewijck, on behalf of minister Falck, proposed to the king to appoint De Bruijn to join Boie, Macklot, and Müller 'to

18 Res.GG, May 1, 1827, no. 8. NL-HaNA, Koloniën, 1814–1849, 2.10.01, inv.no. 2504.
19 Weber and Van Zanen, *Kunstenaar op Java*, 71.
20 'Flankeur' is a kind of infantry role.
21 Res.GG, October 30, 1848, no. 2. NL-HaNA, Koloniën, 1814–1849, 2.10.01, inv.no. 2711.
22 Brendonck, *Mijne reize*, 36–42.
23 Registry card Jan Cornelis de Bruijn. NL-HaNA, Koloniën, 1814–1849, 2.10.01, inv.no. 3094, f. 638.
24 Both the original request and the four drawings are kept with his appointment decree.

draw flowers, fruits, birds and other objects of natural history.'[25] As a test of his skills, De Bruijn had made a few drawings in the presence of Boie and Macklot at 's Rijks Museum, much to their approval. On February 17, 1824, De Bruijn was appointed to accompany Boie and Macklot.[26] It is said that he contributed drawings to Boie's unpublished *Erpétologie de Java*. Within a year, on February 13, 1825, De Bruijn was relieved of his duties upon his own request.[27] It is not clear why De Bruijn made this request, but it was quickly granted, 'because the news about the behaviour and actions of De Bruijn is far from favourable,'[28] and the expectation existed that a suitable replacement could be found, which materialised on April 16, 1825, with the appointment of Pieter van Oort.[29] In the end, De Bruijn was only ever an assistant of the Commissie on paper and never even seems to have received any payment. He died in Amersfoort on November 11, 1862.[30]

5 Heinrich Bürger (1804/1806–1858, Figure 18)

b. February 29 or November 7, 1804, or November 7, 1806, d. March 25, 1858
Associate: May 18, 1833–January 7, 1834

FHA, 218–219 | MPCC, suppl. 2, 18

Kernkamp, P. M. "Heinrich Bürger (Hamelen 1804/1806–Indramajoe 1858) en zijn Duitse en Nederlands-Indische familie." *De Nederlandsche leeuw* 132, no. 3 (2015): 108–132.

Steenis-Kruseman, M. J. van. "Contributions to the History of Botany and Exploration in Malaysia 8: Heinrich Bürger (? 1806–1858), Explorer in Japan and Sumatra." *Blumea* 11, no. 2 (1962): 495–505.

Uéno, Masuzo. "A Japanese Portrait of Heinrich Bürger." *Zoologische mededelingen* 49, no. 10 (1975): 91–94.

Heinrich Bürger was born in Hameln, Germany, to Samuel Salomon and Eva Meyer. The family was Jewish and adopted the name Bürger around 1808. Bürger's date of

---

25 'Tot het afteekenen van bloemen, vruchten, vogels en andere voorwerpen van natuurlijke historie.' Van Ewijck's advice is kept with the appointment decree.

26 KB, February 17, 1824, no. 173. NL-HaNA, Staatssecretarie, 2.02.01, inv.no. 1891.

27 KB, February 13, 1825, no. 86. NL-HaNA, Staatssecretarie, 2.02.01, inv.no. 2161.

28 'Daar de berigten omtrent het gedrag en de handelingen van De Bruijn verre van gunstig zijn.' From the advice from the minister that is kept with the appointment decree.

29 KB, April 16, 1825, no. 120. NL-HaNA, Staatssecretarie, 2.02.01, inv.no. 2209.

30 Death certificate, November 11, 1862, no. 263. NL-UtrUA, Burgerlijke Stand van de gemeenten in de provincie Utrecht 1811–1902, Amersfoort, 481, inv.no. 1035-02.

birth is shrouded in mystery, as three different dates are possible (two in 1804 and one in 1806).[31] Bürger studied mathematics and astronomy in Göttingen. Bürger never earned a doctorate, even though he styled himself throughout his life as dr. Bürger. Bürger left Europe on September 7, 1823, and arrived in Batavia on February 7, 1824, on the ship *Cornelia*, motivated by the drive to earn money in the Indies in order to support his family in Germany.

In Batavia, Bürger took the apothecary exam and was appointed as apothecary third class at the hospital in Weltevreden on January 14, 1825.[32] Already on June 14, 1825, Bürger was appointed as assistant to Philipp Franz von Siebold (1796–1866), who worked as physician in the Dutch trade post Dejima in Japan from 1823 to 1829. Von Siebold was given the special task to collect Japanese naturalia. Bürger assisted him in this task from 1825 to 1827 but fell victim to the budget cuts that commissioner-general Du Bus was forced to make in these years. However, when Von Siebold was expelled from Japan, Bürger was appointed to take over from him on June 30, 1828.[33] Bürger travelled back to Java in 1832 with a consignment of tea seeds and plants, intended for the agricultural establishment in Krawang.

On May 18, 1833, Bürger was ordered to join the members of the Natuurkundige Commissie on their expedition to Sumatra.[34] He was awarded a salary of f500, equal to a member of the Natuurkundige Commissie.[35] Bürger was never appointed as member of the Commissie, but his close association with the Commissie made it easy for people to assume that he was a member. The expectation was that Bürger was to return to his post in Japan as soon as his expedition to Sumatra had finished. From June to September 1833, Bürger seems to have mostly accompanied the members of the Natuurkundige Commissie on Sumatra.[36] The last mention of Bürger by Van Oort is on September 23, 1833.[37] Bürger married Anna Cornelia van Daalen on December 11, 1833, in Padang (Sumatra).[38] The newlyweds left Padang on January 7, 1834, arriving in Batavia on January 12, 1834.[39]

Governor-general Van den Bosch was so happy with Bürger's work on planning a new road on Sumatra that Bürger was appointed as Knight in the Order of the Dutch

31 Kernkamp, "Heinrich Bürger," 112. Even Kernkamp is unsure of the actual date.

32 Res.GG, January 14, 1825, no. 1. NL-HaNA, Koloniën, 1814–1849, 2.10.01, inv.no. 2480.

33 Res.CG, June 30, 1828, no. 33, confirmed by Res.GG, July 1, 1828, no. 1. NL-HaNA, Koloniën, 1814–1849, 2.10.01, inv.no. 2815.

34 Res.GG, May 18, 1833, no. 2. NL-HaNA, Koloniën, 1814–1849, 2.10.01, inv.no. 2550.

35 Res.GG, May 18, 1833, no. 3. NL-HaNA, Koloniën, 1814–1849, 2.10.01, inv.no. 2550.

36 The report of his expedition (Bürger, "Aanmerkingen gehouden op eene reize") only mentions 1833.

37 Weber and Van Zanen, *Kunstenaar op Sumatra*, 141.

38 *Javasche Courant*, no. 4 (January 11, 1834): 5.

39 *Javasche Courant*, no. 5 (January 15, 1834): 6.

Lion on Van den Bosch' recommendation on September 16, 1834.[40] Bürger returned to Japan for a short while and was honourably discharged at his own request on July 22, 1835, and put on redundancy payment,[41] which was prolonged several times and seems to have lasted until the end of June 1838.[42] Bürger spent the period from May 1840 to October 1842 in Europe and returned to Batavia on February 10, 1843.[43] Shortly after, on June 30, 1843, he was awarded a pension according to the existing pension arrangement for members of the Natuurkundige Commissie, because his merits had earned him the right 'to be considered equal to the members of the Natuurkundige Commissie.'[44] He stayed in the Indies as a businessman and died in Indramayu (Java) on March 25, 1858.[45]

6 Lambert Pierre Comblen (1808–1856)

b. November 30, 1808 (Luik), d. August 3, 1856 (Surabaya)
Assistant: August 2, 1843–February 2, 1846

Lambert Pierre Comblen was born on November 30, 1808, in Liège/Luik, Belgium, to Adam Henri Comblen (1776–1825) and Catherine Josèphe Rodberg (1783–1837). He first appears in the Indies in 1843 as a corporal in the army. After writing to the governor-general, introducing himself as an expert on coal mining with a recommendation from the administrator of the coal mines in Manceau Fontaine (Belgium), he was appointed on August 2, 1843, to accompany Schwaner to Borneo and assist him in all practical matters regarding coal exploitation at a daily allowance of f5.[46]

Schwaner, Von Gaffron, and Comblen departed from Batavia on July 14, 1843, on the ship *Tatal Karim*.[47] In the fall of 1844, Schwaner and Comblen travelled to Surabaya in order to test coal samples (they did not have the right equipment on Borneo).[48] On September 19, 1845, Comblen received a gratification of f500 (to be taken from the

40 Kernkamp, "Heinrich Bürger," 116.
41 Res.GG, July 22, 1835, no. 9. NL-HaNA, Koloniën, 1814–1849, 2.10.01, inv.no. 2856.
42 Res.GG, March 17, 1838, no. 10. NL-HaNA, Koloniën, 1814–1849, 2.10.01, inv.no. 2580.
43 Kernkamp, "Heinrich Bürger," 117.
44 'Zou kunnen worden gelijkgesteld met de leden der natuurkundige commissie.' Res.GG, June 30, 1843, no. 9. NL-HaNA, Koloniën, 1814–1849, 2.10.01, inv.no. 2646. For the pension arrangement, see KB, August 14, 1842, no. 69. NL-HaNA, KdK 1841–1897, 2.02.04, inv.no. 177.
45 *Javasche Courant*, April 17, 1858, quoted by Kernkamp, "Heinrich Bürger," 120.
46 Res.GG, August 2, 1843, no. 1. NL-HaNA, Koloniën, 1814–1849, 2.10.01, inv.no. 2648.
47 Res.GG, August 13, 1843, no. 3a. NL-HaNA, Koloniën, 1814–1849, 2.10.01, inv.no. 2648.
48 Res.GG, August 16, 1844, no. 15 and August 29, 1844, no. 5. NL-HaNA, Koloniën, 1814–1849, 2.10.01, inv.no. 2660.

budget of the Natuurkundige Commissie) because of his excellent work and was recommended by the resident for a permanent post in coal exploitation on Borneo.[49] On February 2, 1846, this recommendation led to Comblen's appointment as supervisor of all coal mining activities on Borneo at a salary of f300 and a bonus of f1 for every 1,000 kg of coal.[50] This appointment marks the end of Comblen's formal attachment to the Natuurkundige Commissie.

On July 9, 1848, Comblen was relieved of his duties (he refused to follow orders resulting in a dangerous situation for the workers) and put on a social security benefit of f100 until the end of December 1848.[51] On May 2, 1849, he received a new appointment, now as supervisor of mining activities in Surabaya at a salary of f300.[52] His social security benefits were retroactively prolonged until the beginning of his new appointment.[53] Since the drillings in Surabaya did result in the discovery of usable coal deposits, Comblen was reassigned on April 3, 1850, to undertake similar experiments on Bawean (an island north of Surabaya).[54] On August 26, 1851, after the experimental drillings on Bawean ended, Comblen was thanked for his service.[55] He died on August 3, 1856, in Surabaya.

7 Pierre-Médard Diard (1794–1863, Figures 16 and 33)

b. March 19, 1794 (Chenusson), d. February 16, 1863 (Batavia)
Member: January 31, 1828–August 26, 1841

NNBW, vol. 8, 388–389 | BPN 63879505 | ENI, vol 1, 598 | FHA, 227–229 | MPCC, 136

Anonymus. "De verliezen der wetenschap in Nederlandsch Indië: Dood van P. Diard en J. G. F. Brumund." *Tijdschrift voor Nederlandsch Indië. Nieuwe serie* 1, 1 (1863): 388–392.

Brébion, Antoine. "Diard, naturaliste français dans l'Extrême-Orient." *T'oung Pao. Second Series* 15, no. 2 (1914): 203–213.

49 Res.GG, September 19, 1845, no. 16. NL-HaNA, Koloniën, 1814–1849, 2.10.01, inv.no. 2673.
50 Res.GG, February 22, 1846, no. 2. NL-HaNA, Koloniën, 1814–1849, 2.10.01, inv.no. 2679.
51 Res.GG, July 9, 1848, no. 20. NL-HaNA, Koloniën, 1814–1849, 2.10.01, inv.no. 2708.
52 Res.GG, May 2, 1849, no. 1. NL-HaNA, Koloniën, 1814–1849, 2.10.01, inv.no. 2718.
53 Res.GG, May 23, 1849, no. 11. NL-HaNA, Koloniën, 1814–1849, 2.10.01, inv.no. 2718.
54 Res.GG, April 3, 1850, no. 21. NL-HaNA, Koloniën, 1850–1900, 2.10.30, inv.no. 7108.
55 Res.GG, August 26, 1851, no. 6. NL-HaNA, Koloniën, 1850–1900, 2.10.30, inv.no. 7124. See Comblen's registry card. NL-HaNA, Koloniën / Stamboeken Burgerlijke Ambtenaren, 2.10.36.22, inv.no. 903, f. 61.

Champoiseau, Noel. “Voyages aux Indes Orientales: Notice sur les voyages de M. Diard, naturaliste français.” *Journal des voyages, découvertes et navigations modernes ou archives géographiques du XIXe siècle* 43 (1829): 222–239.

Claeys, Jean-Yves. “Deux Français à la conquête de Singapour: Les naturalistes Alfred Duvaucel et Pierre-Médard Diard.” *France-Asie* 9, no. 10 (1954): 1001–1007.

Cuvier, Georges. “Notice sur les voyages de MM. Diard et Duvaucel, naturalistes français, dans les Indes orientales et dans les îles de la Sonde; extraite de leur correspondence, et lue à l académie des sciences.” *Revue encyclopédique, ou analyses et annonces raisonnées des productions les plus remarquables dans la Littérature, les Sciences et les Arts* 10, no. 30 (1821): 473–482.

Diard, Hippolyte. “Pierre-Médard Diard, naturaliste.” *Annales de la Société d’Agriculture, Sciences, Arts, et Belles-Lettres du Département d’Indre-et-Loire. Deuxième série* 56 (1877): 198–210.

Dorai, Francis, and Martyn E. Y. Low, eds. *Diard & Duvaucel: French Natural History Drawings of Singapore and Southeast Asia (1818–1820) = Dessins d’histoire naturelle français de Singapour et d’Asie du Sud-Est (1818–1820)*. Epigram Books, 2021.

Oury, Clément. “Diard, Duvaucel and the Muséum National d’Histoire Naturelle = Diard, Duvaucel et le Muséum national d’histoire naturelle.” In *Diard & Duvaucel: French Natural History Drawings of Singapore and Southeast Asia (1818–1820) = Dessins d’histoire naturelle français de Singapour et d’Asie du Sud-Est (1818–1820)*, edited by Francis Dorai and Martyn E. Y. Low. Epigram Books, 2021.

Peyssonnaux, J. H. “Vie, voyages et travaux de Pierre Médard Diard: Naturaliste français aux Indes-Orientales, 1794–1863: Voyage dans l’Indochine, 1821–1824.” *Bulletin des amis du vieux Hué* 22, no. 1 (1935): 3–115.

Weiler, Danièle. “Pierre Médard Diard and Alfred Duvaucel: Two French Naturalists in the Service of Sir Stamford Raffles (December 7, 1818–March 27, 1820).” In *Voyageurs, explorateurs et scientifiques: The French and Natural History in Singapore*, edited by Martyn E. Y. Low, Kate Pocklington and Wan F. A. Jusoh. Lee Kong Chian Natural History Museum, 2019.

Pierre-Médard Diard was born on March 19, 1794, in Château de la Brosse (or de Labrosse) in Chenusson, canton Châteaurenault in the Indre-et-Loire Department in France. His medical studies were interrupted when he enlisted in the honour guard where he served in 1813 and 1814. After resuming his studies, Diard was taken under his wing by Georges Cuvier (1769–1832). Several plans to accompany an expedition came to nothing until Diard finally departed for Calcutta from Bordeaux on August 20, 1817. He arrived there on January 5, 1818, and was joined by Alfred Duvaucel (1793–1824) in May 1818. Together, they established themselves in Chandernagor, where their house became a menagerie and museum in one.

On one of their rare visits to Calcutta in November 1818, they met Thomas Stamford Raffles (1781–1826), the newly appointed governor of Bengkulu (Sumatra), who invited them to join him on his expedition. The expedition left Calcutta on December 7, 1818. Via the Malaysian Peninsula, Singapore and Aceh (Sumatra), they arrived in Bengkulu on July 31, 1819. After eight months of peaceful collecting, a conflict with Raffles regarding the ownership of their collections spiralled out of control and led to a breach. Diard and Duvaucel left Bengkulu on April 1, 1820, and decided to each go their own way temporarily. However, their planned rendez-vous never happened, if only because of Duvaucel's death in 1824.

Diard spent a year on Java, from April 7, 1820,[56] to April 29, 1821,[57] when he left Java as part of a French diplomatic mission to Vietnam. During this year he met Reinwardt, Van der Capellen, and Kuhl and Van Hasselt. During his mission to Vietnam Diard undertook several expeditions to areas rarely visited by Europeans. His plan to join a Vietnamese diplomatic mission to Pegu (Burma) in March 1824 was foiled by the British. After a brief stay in Malacca, Diard arrived on Java again on December 7, 1824.[58]

Diard's excellent personal relationship with Van der Capellen led to his appointment to a mission to Bangka in order to inspect the tin mines and recommend improvements,[59] and finally to his appointment as cultivation inspector on December 31, 1825, at a salary of f500, with Maurevert as his assistant.[60] The job of cultivation inspector never really took off and was dissolved by Du Bus on June 6, 1827. Both Diard and Maurevert were honourably discharged.[61] On July 21, 1826, when he was still cultivation inspector, Diard was appointed to undertake an expedition to Borneo,[62] assisted by F. J. Perret.[63] The goal of the expedition was to inspect the current situation on Borneo, suggest improvements, and give advice on trade and cultivation opportunities. Diard left for Borneo from Surabaya (Java) on August 21, 1826,[64] and returned to Semarang (Java) on January 7, 1827.[65]

During these years, Diard worked tirelessly towards creating an experimental agricultural establishment. A first request for a plot of land in Krawang for this purpose

56 *Bataviasche Courant*, no. 16 (April 20, 1820): 1.

57 *Bataviasche Courant*, no. 18 (May 5, 1821): 1.

58 *Bataviasche Courant*, no 50 (December 11, 1824): 1.

59 Res.GG, April 5, 1825, no. 9. NL-HaNA, Koloniën, 1814–1849, 2.10.01, inv.no. 2788. See Anonymus, "Korte schets."

60 Res.GG, December 31, 1825, no. 21. NL-HaNA, Koloniën, 1814–1849, 2.10.01, inv.no. 2791.

61 Res.GG, June 6, 1827, no. 8. NL-HaNA, Koloniën, 1814–1849, 2.10.01, inv.no. 2505.

62 Res.GG, July 21, 1826, no. 11. NL-HaNA, Koloniën, 1814–1849, 2.10.01, inv.no. 2494. Specific instructions can be found in Res.GG, August 1, 1826, no. 35. NL-HaNA, Koloniën, 1814–1849, 2.10.01, inv.no. 2495.

63 Res.GG, August 4, 1826, no. 23. NL-HaNA, Koloniën, 1814–1849, 2.10.01, inv.no. 2495.

64 *Bataviasche Courant*, no. 34 (August 23, 1826): 12.

65 *Bataviasche Courant*, no. 3 (January 17, 1827): 7.

was provisionally denied om June 20, 1826,[66] but a day later he was given the use of part of 's Lands Plantentuin for this purpose.[67] In order to honour his commitment, on August 1, 1826, Du Bus gave Diard a plot of land in Wanayassa in perpetual use as long as he would finance the experiments himself.[68] The issue of the plot of land in Krawang was picked up again in 1827 and based on Diard's proposals, he was finally given permission on March 18, 1828, to found an agricultural establishment in Krawang and appointed as its first director.[69] When Du Bus founded his own botanical garden on August 29, 1828, the director of the establishment in Krawang was made responsible for its development.[70]

Meanwhile, on January 31, 1828, at the request of Macklot, Diard had been appointed as member of the Natuurkundige Commissie (at a salary of f500) in order to replace Boie, who had died on September 4, 1827.[71] Because his responsibilities with the Natuurkundige Commissie started taking up more of his time, on July 9, 1829, Diard gave up his position as director of the agricultural establishment in Krawang.[72]

Despite not being involved with the establishment in Krawang anymore, Diard continued to work on agricultural experiments in various official and unofficial capacities. The three main cultivations that Diard was involved in during the 1830s were sericulture, tea cultivation, and sugar manufacturing. From May 8, 1833,[73] to July 15, 1841,[74] he officially served as sericulture inspector, initially at a salary of f200, but from June 1, 1836, he gave up his salary and continued to serve as director of the Natuurkundige Commissie and sericulture inspector in an honorary capacity.[75] Already from the mid-1820s Diard had been involved in tea experiments and he continued to experiment. On October 18, 1834, he was made responsible for tea plantations in Cheribon,[76] until, on June 26, 1838, he asked to be relieved of this task.[77] Diard's involvement in the sugar industry took place between 1835 and 1840. His first two contracts, in Cheribon, were approved on September 17, 1835, and December 17, 1836, and he sold both contracts

66 Res.GG, June 20, 1826, no. 11. NL-HaNA, Koloniën, 1814–1849, 2.10.01, inv.no. 2794, see also Res.GG, March 28, 1826, no. 5. NL-HaNA, Koloniën, 1814–1849, 2.10.01, inv.no. 2792.

67 Res.GG, June 21, 1826, no. 12. NL-HaNA, Koloniën, 1814–1849, 2.10.01, inv.no. 2493.

68 Res.GG, August 1, 1826, no. 6. NL-HaNA, Koloniën, 1814–1849, 2.10.01, inv.no. 2495.

69 Res.GG, March 18, 1828, no. 2. NL-HaNA, Koloniën, 1814–1849, 2.10.01, inv.no. 2811.

70 Res.GG, August 29, 1828, no. 9. NL-HaNA, Koloniën, 1814–1849, 2.10.01, inv.no. 2816.

71 Res.GG, February 4, 1828, no. 43. NL-HaNA, Koloniën, 1814–1849, 2.10.01, inv.no. 2513, quoting Res.CG, January 31, 1828, no. 22.

72 Res.GG, July 9, 1829, no. 15. NL-HaNA, Koloniën, 1814–1849, 2.10.01, inv.no. 2530.

73 Res.GG, May 8, 1833, no. 2. NL-HaNA, Koloniën, 1814–1849, 2.10.01, inv.no. 2847.

74 Res.GG, July 18, 1841, no. 1. NL-HaNA, Koloniën, 1814–1849, 2.10.01, inv.no. 2620.

75 Res.GG, April 19, 1836, no. 17. NL-HaNA, Koloniën, 1814–1849, 2.10.01, inv.no. 2860.

76 Res.GG, October 18, 1834, no. 6. NL-HaNA, Koloniën, 1814–1849, 2.10.01, inv.no. 2853.

77 Res.GG, June 26, 1838, no. 6. NL-HaNA, Koloniën, 1814–1849, 2.10.01, inv.no. 2583.

again on July 10, 1838.[78] His third contract, in Besuki, was approved on August 2, 1839.[79] On November 27, 1839, Diard bought two additional sugar contracts in Besuki at a public auction, but the sale was cancelled on June 25, 1840, when his financial backers withdrew from the project.[80] Due to illness, Diard had to give up his only remaining sugar contract on September 8, 1840.[81]

In addition to his activities in sericulture, tea cultivation, and sugar manufacturing, Diard occupied himself with various other agricultural and industrial experiments. He received an honourable mention for his indigo at the public agricultural exhibition in 1829, organised by Du Bus.[82] He developed a way to improve the drying of coffee.[83] He developed a new method of manufacturing tiles, which he freely shared with the Dutch government.[84] He was credited with having first introduced merino sheep farming to Java.[85] In short, there was hardly any utilitarian facet of Netherlands Indies society that he did not dabble in. When Diard was given the honour of holding a lecture in the presence of young prince Hendrik on August 27, 1837, at the recently founded Colonial Museum in Batavia, he explained how he considered the natural history museum to be the stimulus towards agricultural development, which benefitted the prosperity of all.[86] The importance of and the appreciation for Diard's contributions is illustrated by the fact that he was appointed Knight in the Order of the Dutch Lion on June 9, 1839.[87]

Diard resigned from all his posts in the Indies (including that of directing member of the Natuurkundige Commissie) on August 26, 1841. In recognition of his work, he was given the title of honorary cultivation inspector, in addition to a gratification of f2,000 to pay for his way back to Europe via Mauritius and Bourbon (La Réunion),[88] where he studied agricultural methods from September 25, 1841, to February 24, 1842.[89] He seems to have spent several years in Europe and returned to the Indies in early 1848. His request for a new sugar contract (made in 1847 in the Netherlands) was denied

78 Res.GG, September 17, 1835, no. 8. NL-HaNA, Koloniën, 1814–1849, 2.10.01, inv.no. 2856; Res.GG, December 17, 1836, no. 20. NL-HaNA, Koloniën, 1814–1849, 2.10.01, inv.no. 2565; Res.GG, July 10, 1838, no. 28. NL-HaNA, Koloniën, 1814–1849, 2.10.01, inv.no. 2584.

79 Res.GG, August 2, 1839, no. 2. NL-HaNA, Koloniën, 1814–1849, 2.10.01, inv.no. 2597.

80 Res.GG, November 27, 1839, no. 2. NL-HaNA, Koloniën, 1814–1849, 2.10.01, inv.no. 2600; Res.GG, June 25, 1840, no. 1. NL-HaNA, Koloniën, 1814–1849, 2.10.01, inv.no. 2607.

81 Res.GG, September 8, 1840, no. 2. NL-HaNA, Koloniën, 1814–1849, 2.10.01, inv.no. 2610.

82 *Javasche Courant*, no. 112 (September 24, 1829): 2.

83 *Nieuwe Rotterdamsche Courant*, no 236 (August 27, 1853): 1.

84 *Nederlandsche Staatscourant*, no. 48 (February 25, 1854): 2–3.

85 *Java-Bode*, no. 72 (September 10, 1862): 4.

86 The speech was published in full in two parts in the *Javasche Courant*, no. 74 (September 16, 1837); 1–2, and no. 75 (September 20, 1837): 1–2.

87 KB, June 9, 1839, no. 61. NL-HaNA, Staatssecretarie, 2.02.01, inv.no. 4488.

88 Res.GG, August 26, 1841, no. 1aa. NL-HaNA, Koloniën, 1814–1849, 2.10.01, inv.no. 2621.

89 Scalliet, *Antoine Payen*, 669.

on January 6, 1849.[90] Another government-sponsored trip, from November 10–12, 1857,[91] to April 21–26, 1859,[92] allowed him to once more inspect the sugar industry on Mauritius and Bourbon, and cinnamon and coffee cultivation on Ceylon. At the end of his life, when he was once again collecting specimens, now for 's Rijks Museum, Diard developed a method of preparing specimens that perfectly preserved suppleness and lustre. Unfortunately, the method involved the lavish use of arsenic and Diard died in Batavia as a result of arsenic poisoning on February 16, 1863.

8 Franciscus Adrianus Fievez dit de Malines (1792–Unknown)

b. December 17, 1792 (Utrecht), d. after 1848
Assistant: February 9, 1824–? and January 25, 1832–?

Mac Lean, J. "Gegevens over tekenaars van biologische voorwerpen in het voormalige Nederlandsch Indië." *Scientiarum historia* 15 (1973): 101–114.

Franciscus Adrianus Fievez dit de Malines[93] was born on December 17, 1792, in Utrecht, to Adrianus Fievez dit de Malines (1750–1819) and Marie Françoise Canon (1752/3–1823).[94] Upon Blume's request, Fievez de Malines was provisionally appointed as second draughtsman at 's Lands Plantentuin on December 10, 1822, at a salary of f80 per month.[95] This appointment was turned permanent on August 30, 1823, at a salary of f120.[96] When he was unable to accompany Blume on an expedition due to health reasons, he was charged on February 9, 1824, with working on copying and finishing the drawings from the Kuhl/Van Hasselt papers, seconded to Van Raalten.[97]

90 Res.GG, January 6, 1849, no. 10. NL-HaNA, Koloniën, 1814–1849, 2.10.01, inv.no. 2714.
91 *Java-Bode*, no. 91 (November 14, 1857): 8. The precise date of Diard's departure is not given.
92 *Bataviaasch Handelsblad*, no. 61 (April 27, 1859): 8. The precise date of Diard's arrival is not given.
93 The spelling of this name differs widely in the official documentation ranging from Fievez Maline to Fieves dit de Malines. Even his baptismal record gives Feviez de Malines as his father's last name. See entry December 17, 1792. NL-UtrUA, Burgerlijke stand gemeente Utrecht, 711, inv.no. 64, f. 272.
94 Date of birth taken from De Malines' entry in the militia register (NL-UtrUA, Gemeentebestuur van Utrecht 1813–1869, 1007–2, inv.no. 11820, f. 103).
95 Res.GG, December 10, 1822, no. 2. NL-HaNA, Koloniën, 1814–1849, 2.10.01, inv.no. 2467.
96 Res.GG, August 30, 1823, no. 14. NL-HaNA, Koloniën, 1814–1849, 2.10.01, inv.no. 2471.
97 Res.GG, February 9, 1824, no. 1. NL-HaNA, Koloniën, 1814–1849, 2.10.01, inv.no. 2474; Res.GG, February 17, 1824, no. 30. NL-HaNA, Koloniën, 1814–1849, 2.10.01, inv.no. 2474.

After the completion of the project on November 25, 1825,[98] Fievez de Malines disappears from view until the Natuurkundige Commissie was requested on January 25, 1832, to hire him as draughtsman.[99] Both Van Oort and Korthals mention Fievez de Malines travelling with the Natuurkundige Commissie. According to Van Oort, Fievez de Malines had become mentally ill ('zinneloos') in 1824/1825. He had spent a couple of years in a hospital in Batavia and finally recovered.[100] After his appointment (at a salary f60 per month, the same as the Chinese artist, Tsing Wang Ho, who had to make way for him), Fievez de Malines reported for duty on March 21, 1832.[101] Van Oort's last mention of Fievez de Malines is on December 3, 1832.[102]

There is no record as to how, why and when Fievez de Malines' association with the Commissie ended. However, he seems to have disappeared from view before the 1833 expedition to Sumatra. He spent another couple of years in the Indies, and, on December 5, 1836, Fievez de Malines received free transport to the Netherlands,[103] where he seems to have made a life for himself in the Betuwe area (he moved from Eck to Maurik in 1845, and from Maurik to Beusichem in 1848).[104]

9 Eltio Alegondus Forsten (1811–1843, Figures 15 and 19)

b. July 2, 1811 (Middelburg), d. January 2, 1843 (Ceram)
Member: February 4, 1838–January 3, 1843

BWN, vol. 6, 175 | NNBW, vol. 4, 613 | BPN 65260210 | FHA, 233–234 | MPCC, 178–179.

Mulyasari, Prima Nurahmi. "A Short Journey to Explore the East: Eltio Alegondas Forsten." *Paramita* 25, no. 1 (2015): 1–12.

Wingerden, Pieter van. "Goudkoorts in de gordel van smaragd." *Wonderkamer*, no. 1 (2020): 46–47.

Wingerden, Pieter van. "Science on the Edge of Empire: E. A. Forsten (1811–1843) and the Natural History Committee (1820–1850) in the Netherlands Indies." *Centaurus* 62, no. 4 (2020): 797–821.

98 Res.GG, November 25, 1825, no. 13. NL-HaNA, Koloniën, 1814–1849, 2.10.01, inv.no. 2489.

99 Res.GG, January 25, 1832, no. 6. NL-HaNA, Koloniën, 1814–1849, 2.10.01, inv.no. 2842.

100 Weber and Van Zanen, *Kunstenaar op Java*, 209.

101 Weber and Van Zanen, *Kunstenaar op Java*, 209.

102 Weber and Van Zanen, *Kunstenaar op Java*, 358.

103 Res.GG, November 10, 1836, no. 6. NL-HaNA, Koloniën, 1814–1849, 2.10.01, inv.no. 2565; Res.GG, December 5, 1836, no. 12. NL-HaNA, Koloniën, 1814–1849, 2.10.01, inv.no. 2565.

104 Registers van ingezetenen. NL-TieRAR, Gemeentebestuur van Maurik, 1811–1945 (1989), 0182, inv.no. 1223, f. 195.

Eltio Alegondus Forsten was born on July 2, 1811, in Middelburg, to Jacob Antoon Forsten (1778/9–1829) and Johanna Gijsberta Knibbe (1786/7–1832).[105] Jacob Antoon was a medical doctor by education and had studied with Reinwardt in Harderwijk. The family moved to Middelburg in the early 1800s where Jacob Antoon was appointed as postmaster. Eltio Alegondus spent his youth in Middelburg and went to the Latin school there. He studied medicine in Leiden from 1829 to 1836. One prize-winning essay and his dissertation were published.[106] Forsten sent a copy of his dissertation to the king along with an open application for membership of the Natuurkundige Commissie on February 27, 1837. Since there was no opening at that particular moment, he was advised to gain experience at 's Rijks Museum until such a time as an opening would occur.[107] This happened shortly after and Forsten was appointed as member on February 4, 1838.[108]

He left the Netherlands in August 1838 and arrived in Batavia on December 14, 1838. Forsten spent a full year in Batavia and Buitenzorg, waiting for administrative permission to embark on the expedition to his designated destination of Celebes. During this year, he undertook some small expeditions on West Java, including a volcano climbing foray with Junghuhn in April 1839.[109] On December 22, 1839, he finally received permission to depart for Celebes. Forsten's salary of f300 was raised to f500 from the day of his departure and received f350 per month collecting costs (from which the salaries of his assistants needed to be paid).[110] Upon Forsten's request, his higher salary was retroactively awarded from September 1, 1839.[111] Long-time Commissie assistant Overdijk was appointed to accompany Forsten.[112] Both Forsten (because he did not like Overdijk) and Overdijk (who wanted a higher salary) objected and, as a result, Overdijk was replaced by Von Gaffron on January 19, 1840.[113] Ten days later, on January 29, 1840, the expedition, consisting of Forsten, Von Gaffron, and seven Javanese assistants,[114] departed on the ship *Charles*.[115] During the preparations for his expedition, on January

105 Birth certificate, no. 229, NLMidZA, Burgerlijke Stand Zeeland (1796) 1811–1980, 1.1.078, inv.no. MDB-G-1811-1.

106 Forsten, *De aethere et naphthis*; Forsten, *De cedrela febrifuga*.

107 KB, February 27, 1837, no. 51. NL-HaNA, Staatssecretarie, 2.02.01, inv.no. 4226.

108 KB, February 4, 1838, no. 98. NL-HaNA, Staatssecretarie, 2.02.01, inv.no. 4334.

109 Junghuhn, "Bijdragen tot de geschiedenis der vulkanen," 107.

110 Res.GG, December 22, 1839, no. 4. NL-HaNA, Koloniën, 1814–1849, 2.10.01, inv.no. 2601.

111 Res.GG, February 16, 1840, no. 2. NL-HaNA, Koloniën, 1814–1849, 2.10.01, inv.no. 2603.

112 Res.GG, December 22, 1839, no. 4. NL-HaNA, Koloniën, 1814–1849, 2.10.01, inv.no. 2601.

113 Res.GG, January 19, 1840, no. 1. NL-HaNA, Koloniën, 1814–1849, 2.10.01, inv.no. 2602.

114 In his diary Forsten only mentions six hunters. However, his financial overview shows that his personal servant Hendrik was not included in this number, making him the seventh Javanese member of the expedition party. See NL-HaNA, Van Beresteyn, 2.21.018, inv.no. 1523.

115 Reisverhaal van Dr. Forsten naar Menado. NCAO, 1075_003.

8, 1840, Forsten asked for and received red cloth, blue cloth, goldthread and glasswork at a total value of f600 to be used as gifts to local leaders.[116] On board the ship, Forsten's baggage and equipment took the space normally occupied by 25 koyangs of cargo.[117]

In his diary, Forsten kept a meticulous record of his itinerary and his activities.[118] Forsten travelled to the residency of Manado on Northeast Celebes, where he arrived on March 22, 1840.[119] He made his way to his headquarters in the village of Tondano from where he traversed the area. He stayed there for over a year, from April 8, 1840,[120] to April 24, 1841.[121] He then travelled to Manado to prepare for an expedition to Gorontalo, on the south side of Northeast Celebes. While still in Manado, Forsten sent his first shipment to Rotterdam on the ship *Bonjol* on May 12, 1841.[122] Forsten first travelled to Kema, then to Belang. He stayed in Belang from June 5 to June 14, 1841.[123] During the preparation of this expedition, on May 18, 1841, Forsten's son, also named Eltio Alegondus, was born. His mother's name was Francina Rottij, and she is identified as a (probably indigenous) Christian woman.[124] It is likely that Forsten met her in Manado, where many indigenous Christians lived.

When sailing to Gorontalo from Belang, unfavourable winds forced the party to divert to Ternate, where they arrived on June 17, 1841.[125] Due to some health issues Forsten had to delay his departure. On September 9, 1841,[126] they finally departed again for Gorontalo where they arrived on September 18, 1841.[127] On November 22, 1841, Forsten left Gorontalo again, to arrive in Kema on November 28, 1841.[128] Forsten alternatively stayed in Manado, Kema, and Tondano, all the while awaiting an opportunity

116 Res.GG, January 8, 1840, no. 2. NL-HaNA, Koloniën, 1814–1849, 2.10.01, inv.no. 2602.

117 Res.GG, March 13, 1840, no. 10. NL-HaNA, Koloniën, 1814–1849, 2.10.01, inv.no. 2604. A koyang is a unit of measurement that denotes the space taken up by approx. 2,000 kg of dry freight like rice, salt or sugar. A precise number is hard to give, but it certainly gives an impression of the staggering amount of baggage and equipment that was required for this expedition.

118 A diary of Forsten's travels that runs from January 29, 1840, to July 31, 1842, is extant in the archives of Naturalis: Reisverhaal van Dr. Forsten naar Menado. NCAO, 1075. An additional document by Von Gaffron (in cursive Fraktur) documents the final months and death of Forsten: Schluss zu der Reise auf Ceram und Todt von den Dr. Forsten auf Amboina 3/1 1843. NCAO, 1472.

119 Reisverhaal van Dr. Forsten naar Menado. NCAO, 1075_012.

120 Reisverhaal van Dr. Forsten naar Menado. NCAO, 1075_014.

121 Reisverhaal van Dr. Forsten naar Menado. NCAO, 1075_076.

122 Reisverhaal van Dr. Forsten naar Menado. NCAO, 1075_076.

123 Reisverhaal van Dr. Forsten naar Menado. NCAO, 1075_077.

124 Inventory of the documents left by Forsten. NL-HaNA, Van Beresteyn, 2.21.018, inv.no. 1524.

125 Reisverhaal van Dr. Forsten naar Menado. NCAO, 1075_077.

126 Reisverhaal van Dr. Forsten naar Menado. NCAO, 1075_088.

127 Reisverhaal van Dr. Forsten naar Menado. NCAO, 1075_094.

128 Reisverhaal van Dr. Forsten naar Menado. NCAO, 1075_107.

to travel to Makassar, which was supposed to be his next destination.[129] During this time, Forsten formally adopted his son on April 5, 1842.[130] When no travel opportunity to Makassar arose for months, he decided to travel to Ambon, departing from Kema on April 15, 1842,[131] and arriving on April 25, 1842.[132] Forsten spent some time on Ambon and the much larger neighbouring island of Ceram, where he died on January 2, 1843.[133]

Von Gaffron took charge of the expedition and organised transport of the collections back to Java, arriving in Surabaya on June 11, in Semarang on June 14,[134] and finally in Batavia on June 25, 1843.[135] Forsten's Javanese assistants were held on by Schwaner for a while to help with preparing the collections for shipping to the Netherlands.[136] Forsten's collections arrived at 's Rijks Museum in four instalments between July 24 and September 24, 1844, comprising a total of fourteen crates and eight barrels.[137] Two lists with shipments are present in the archive of the Natuurkundige Commissie.[138] The first is the list of the *Bonjol* shipment which consisted of six crates and six barrels. The second list mentions three crates and nine barrels. Further research will need to enlighten the relationship between the second shipping list and the four instalments that arrived at 's Rijks Museum in 1844.

10 D[aniel?] Th. François (Unknown–1860)

b. unknown, d. 1860
Assistant: June 14, 1848–December 21, 1848

129 Reisverhaal van Dr. Forsten naar Menado. NCAO, 1075_109.

130 Inventory of the documents left by Forsten. NL-HaNA, Van Beresteyn, 2.21.018, inv.no. 1524. Eltio Alegondus junior served in the army from 1859 to 1866, but disappears from view after that. See military registry card no. 3962. NL-HaNA, Koloniën / Stamboeken Militairen KNIL Oost- en West-Indië, 2.10.50, inv.no. 352.

131 Reisverhaal van Dr. Forsten naar Menado. NCAO, 1075_109.

132 Reisverhaal van Dr. Forsten naar Menado. NCAO, 1075_110.

133 *Javasche Courant*, no. 42 (May 27, 1843): 6.

134 *Javasche Courant*, no. 49 (June 21, 1843): 6.

135 *Javasche Courant*, no. 51 (June 28, 1843): 7.

136 Res.GG, July 27, 1843, no. 15. NL-HaNA, Koloniën, 1814–1849, 2.10.01, inv.no. 2647.

137 Temminck to the minister of internal affairs, January 20, 1848. NL-HaNA, Van Beresteyn, 2.21.018, inv.no. 1525.

138 NCAO, 1461 and 1462. The lists are somewhat mixed up. Inventory number 1461 contains the list from the *Bonjol* (scans 001–002, and 023–034) and the second shipping list (scans 003–022). Inventory number 1462 only contains the list from the *Bonjol*. Details for the second shipment can be found in inventory numbers 1455, 1456, 1458, 1459 and 1460, while 1457 contains some notes by presumably Müller on the *Bonjol* shipment.

D. Th. François' association with the Natuurkundige Commissie was short-lived. On June 14, 1848, he was attached to Schwaner in order to help him in drawing up his reports on his expeditions on Borneo. At that moment, François was on redundancy pay, which he continued to receive during his time with Schwaner. He was awarded an additional daily allowance of f5.[139] According to Schwaner, however, François did not meet the goal ('niet beantwoordt aan het doel') of his appointment. Apparently, he had already ceased his work and was replaced by Brendonck and Strung by the time his appointment was rescinded on December 21, 1848.[140]

His registry card indicates that he held various lower-level positions in the government until his death in 1860.[141] François' registry card also suggests that he was either the son or the husband of Johanna Petronella van Steenbergen. Since Johanna Petronella van Steenbergen (1794–1870) was married to Daniel François (1781–1840), mayor of Loosduinen in het Netherlands, I believe it is likely that he was a son of Van Steenbergen and Daniel François and it is possible that his first name was also Daniel.

11 Heinrich von Gaffron (1813–1880, Figure 30)

b. April 28, 1813 (Grebel), d. November 2, 1880 (Sindanglaya)
Assistant: January 19, 1840–May 19, 1846

NNBW, vol. 8, 578–579 | BPN 08106670 | ENI, vol 1, 728 | M, 24–32 | BB, 317

Heinrich von Gaffron was born on April 28, 1813, in Grebel, Germany, to Heinrich von Gaffron and Christiana von Knobloch. He received his military training in Berlin, where he was appointed as second lieutenant on October 2, 1834. He was released from the army at his own request on April 17, 1838. He arrived at the military recruitment office in Harderwijk on July 25, 1838, where he signed up as a private destined for the Indies. Von Gaffron left for the Indies on November 22, 1838, on the ship *Jacoba Maurina*.[142] What moved Von Gaffron to leave an officer rank in the Prussian army and sign up as a private in Harderwijk is unknown. It is likely that some sort of scandal or episode of misbehaviour was the reason for this, especially since Harderwijk was known as the

139 Res.GG, June 14, 1848, no. 26. NL-HaNA, Koloniën, 1814–1849, 2.10.01, inv.no. 2707.
140 Res.GG, December 21, 1848, no. 9. NL-HaNA, Koloniën, 1814–1849, 2.10.01, inv.no. 2713.
141 Registry card François. NL-HaNA, Koloniën / Stamboeken Burgerlijke Ambtenaren, 2.10.36.22, inv.no. 900, f. 230–231.
142 All dates from Von Gaffron's military registry card. NL-HaNA, Koloniën / Stamboeken Militairen KNIL Oost- en West-Indië, 2.10.50, inv.no. 321, f. 5641.

sewer of Europe where all sorts of desperate characters signed up for military service in the Indies.[143]

In the Indies, Von Gaffron made Forsten's acquaintance some time in 1839. Forsten was impressed by his drawing skills and nominated him to replace Overdijk as expedition companion. On January 19, 1840, Von Gaffron was appointed as assistant-draughtsman to accompany Forsten on his Celebes expedition at a salary of f150.[144] After Forsten's death in Ambon on January 2, 1843, Von Gaffron made sure that their collections and documents arrived safely back in Batavia. On Schwaner's request, Von Gaffron was appointed on July 28, 1843, as assistant-draughtsman to Schwaner's Borneo expedition at his usual salary of f150, which was raised to f200 on March 19, 1846.[145] Von Gaffron would stay attached to Schwaner, undertaking various coal survey expeditions on Borneo, until May 19, 1846, when he was made available to the governor of Borneo for further exploration trips at a salary of f200.[146] This appointment formally ended his association with the Natuurkundige Commissie.

On July 9, 1848, Von Gaffron was appointed as supervisor of coal exploration in Banjarmasin (as replacement for Comblen) at a salary of f300 with a bonus of f0.25 for every ton of coal produced.[147] His exemplary work led to his appointment as director of the new coal mine at Pengaran on May 3, 1849.[148] When the main gallery of this new mine, which was expected to cover the coal requirements in the entire Netherlands Indies, was opened by governor-general Rochussen, the mine was renamed Oranje Nassau.[149] Further highlights from his career include a four-year prison sentence for abuse of power in 1852 for which he received a pardon in 1853, an appointment as assistant-resident of Sintang (Borneo) in 1858, the appointment as Knight fourth class in the Military Order of Willem in 1859,[150] an appointment as assistant-resident of Belitung (an island between Sumatra and Borneo) in 1860, a two-year leave to the Netherlands from 1866 to 1868, and a further appointment as 'algemeen pakhuismeester' [general warehouse manager] in Batavia at a salary of f1,000. Von Gaffron died on November 2, 1880, in Sindanglaya (West Java).[151]

143 Bossenbroek, *Van Holland naar Indië*, 28, 30.
144 Res.GG, January 19, 1840, no. 1. NL-HaNA, Koloniën, 1814–1849, 2.10.01, inv.no. 2602.
145 Res.GG, July 28, 1843, no. 15. NL-HaNA, Koloniën, 1814–1849, 2.10.01, inv.no. 2647; Res.GG, March 19, 1846, no. 7. NL-HaNA, Koloniën, 1814–1849, 2.10.01, inv.no. 2680.
146 Res.GG, May 19, 1846, no. 6. NL-HaNA, Koloniën, 1814–1849, 2.10.01, inv.no. 2682.
147 Res.GG, July 9, 1848, no. 20. NL-HaNA, Koloniën, 1814–1849, 2.10.01, inv.no. 2708.
148 Res.GG, May 3, 1849, no. 4. NL-HaNA, Koloniën, 1814–1849, 2.10.01, inv.no. 2718.
149 Res.GG, September 27, 1849, no. 4a. NL-HaNA, Koloniën, 1814–1849, 2.10.01, inv.no. 2722.
150 Köffler, *De Militaire Willemsorde*, 180.
151 All information on Von Gaffron's career after 1849 from his registry card. NL-HaNA, Koloniën / Stamboeken Burgerlijke Ambtenaren, 2.10.36.22, inv.no. 901, f. 16, and 902, f. 640.

12 Dominicus Hermanus Righard van Gelder (1809–1879)

b. November 7, 1809 (Leiden), d. August 13, 1879 (Rijswijk)
Assistant: July 25, 1830–March 5, 1838

MPCC, 188
Gols, Paul. "Rijke oud-Indië gangers kopen Wateren (versie 4, januari 2018)." (2018). https://www.deruyter.org/uploads/media/606ddeed9f4b6.docx.

Dominicus Hermanus Righard van Gelder was born on November 7, 1809, in Leiden, to Hermanus Righard van Gelder and Geertruida Sala.[152] His baptism registration names him Dominicus Hermanus, but his death certificate identifies him as Dominicus Hermanus Righard van Gelder, which is consistent with the fact that he is persistently identified as D. H. R. van Gelder in all official documents relating to the Natuurkundige Commissie.[153] Not much is known about Van Gelder's early years. He is recorded to have worked as a preparator at 's Rijks Museum between 1825 and 1830.[154] His presence at the nerve centre of the Natuurkundige Commissie led to his appointment as assistant to Korthals on July 25, 1830.[155]

During his first couple of years in the Indies he joined various members on a series of short expeditions on Java. He followed the rest of the Commissie to Sumatra in early 1834. After Van Oort's death, Van Gelder was tasked with the administration of the Commissie.[156] Van Gelder is mentioned regularly in the diaries of Van Oort and Korthals. Upon his return to Java, Van Gelder seems to have mostly stayed there and was heavily involved in preparing and caring for the specimens in the museum of the Bataviaasch Genootschap. On March 5, 1838, after more than seven years of loyal service, he asked to be relieved of his duties to the Natuurkundige Commissie in order to become administrator of an estate in Surakarta, owned by E. A. Fritze and Smulders.[157] On October 15, 1840, he married Johanna Frederika Wilhelmina Quentin.[158] Van Gelder made his fortune in the Indies and returned to the Netherlands with his family in June

152 Baptismal record Van Gelder in Doopboek van de kerk De zon (RK) te Leiden, 3 oktober 1683–1811, Erfgoed Leiden en Omstreken, Leiden, 2004, inv.no. 305.
153 The Latin baptism certificate spells his third name as Richard, while his Dutch death certificate spells it as Righard.
154 Holthuis, *Rijksmuseum van Natuurlijke Historie*, 38.
155 KB, July 25, 1830, no. 40. NL-HaNA, Staatssecretarie, 2.02.01, inv.no. 3437.
156 Res.GG, December 3, 1834, no. 8. NL-HaNA, Koloniën, 1814–1849, 2.10.01, inv.no. 2853.
157 Res.GG, March 5, 1838, no. 3. NL-HaNA, Koloniën, 1814–1849, 2.10.01, inv.no. 2580.
158 *Javasche Courant*, no. 88 (October 31, 1840): 7.

1858, where he invested in property and land in the province of Drenthe.[159] Van Gelder died on August 13, 1879, in Rijswijk.

13 Johan Conrad van Hasselt (1797–1823, Figure 10)

b. June 26, 1797 (Doesburg), d. September 8, 1823 (Buitenzorg)
Member: May 2, 1820–September 8, 1823

BWN, vol. 8.1, 273–275 | NNBW, vol. 3, 548–550 | BPN 90728786 | ENI, vol 2, 65–66 | FHA, 239–240 | MPCC, 219–221

Anonymus. "Johan Coenraad van Hasselt en Heinrich Kühl, leden der Kommissie voor het natuurkundig onderzoek in Nederlandsch Indië." *Handelingen en geschriften van het Indisch Genootschap* 5 (1858): 66–80.

Anonymus. "Levensschets van J. C. van Hasselt, in leven lid der Natuurkundige Commissie." *Indisch magazijn* 2, no. 5–6 (1845): 28–43.

Drost, Aschwin. "De jongen uit de Kloosterstraat." *doesburgvertelt.nl* (2022).[160]

Genealogie, Centraal Bureau voor. "Van Hasselt." *Nederland's Patriciaat* 94 (2015): 225–454.

Greshoff, M. "Kuhl en Van Hasselt: Eene episode in het Nederlandsch-Indisch natuuronderzoek." *Album der natuur* 52, no. 1 (1903): 1–22, 60–88.

Klaver, Charles. *Inseparable Friends in Life and Death: The Life and Work of Heinrich Kuhl (1797–1821) and Johan Conrad van Hasselt (1797–1823), Students of Prof. Theodorus van Swinderen.* Barkhuis, 2007.

Striekwold, Robbert. "Naam maken: Wat de dood van twee jonge negentiende-eeuwse wetenschappers ons leert over de natuurhistorie." *Ex tempore* 37, no. 3 (2018): 213–226.

Swinderen, Theodorus van. "Bijdragen tot eene schets van het leven, het karakter en de verdiensten van wijlen Dr. J. C. van Hasselt." *Almanak der Akademie van Groningen* 13 (1825): 1–67.

159 Paul Gols has traced the history of Wateren in Drenthe, an agricultural colony owned and operated by the Maatschappij van Weldadigheid (founded by Johannes van den Bosch), which included the estates of Zorgvlied and Groot- en Klein Wateren. The latter was purchased by Van Gelder in 1860 and would stay in his ownership until 1870. See Paul Gols, "Rijke oud-Indië gangers," (2018), for a detailed history of this event.

160 This article has since disappeared from its original location, but is fortunately still available through the Internet Archive Wayback Machine.

Swinderen, Theodorus van. “Levensschets van Johan Coenraad van Hasselt.” *Algemeene konst- en letter-bode*, no. 13 (April 1, 1825): 198–202.
Swinderen, Theodorus van. “Levensschets van Johan Coenraad van Hasselt.” *Algemeene konst- en letter-bode*, no. 14 (April 8, 1825): 212–218.
Swinderen, Theodorus van. “Levensschets van Johan Coenraad van Hasselt.” *Algemeene konst- en letter-bode*, no. 15 (April 15, 1825): 230–233.

Johan Conrad van Hasselt was born on June 26, 1797, in Doesburg, the Netherlands, to Barthold van Hasselt (1741–1819) and Bernardina Antonia Rasch (1767–1849).[161] Like Kuhl, he met Theodorus van Swinderen (1784–1851), professor of natural history at Groningen University, who convinced him to study medicine in Groningen.[162] When Van Hasselt and Kuhl met in Groningen they struck up a firm friendship and Van Hasselt accompanied Kuhl on some of his travels (but not to London). On May 2, 1820, less than two weeks before he received his medical doctorate,[163] Van Hasselt was appointed, together with Kuhl, with Keultjes and Van Raalten as assistants, as member of what later became known as the Natuurkundige Commissie.[164]

The expedition party of four left the Netherlands on July 11, 1820,[165] and arrived in Batavia on December 22, 1820, on the ship *Nordloh*.[166] On January 23, 1821, Kuhl and Van Hasselt were appointed at a salary of f400, as of January 1, 1821, in order to ‘travel the Dutch possessions in the East Indies, with the specific goal of expanding the scientific knowledge of the products of nature in those lands.’[167] They were placed under the supervision of Reinwardt and received permission to hire six indigenous employees at a salary of f12 each.[168] This number was expanded to twelve on June 11, 1821.[169] They set up camp in Buitenzorg and spent four months collecting in the close vicinity of Buitenzorg. When a cholera outbreak prevented them from undertaking a planned expedition to Banten, ‘they decided to explore the mountains south of Buitenzorg.’[170]

After Kuhl and Keultjes had died, on September 14 and 15, 1821, Van Hasselt needed some time to regain his composure.[171] On November 7, 1821, governor-general Van

161 Klaver, *Inseparable Friends*, 7.
162 Van Berkel, *Universiteit van het Noorden: Deel 1*, 642–646.
163 Klaver, *Inseparable Friends*, 17.
164 KB, May 2, 1820, no. 10. NL-HaNA, Staatssecretarie, 2.02.01, inv.no. 988.
165 Veth, *Overzicht van hetgeen gedaan is*, 25; Klaver, *Inseparable Friends*, 18.
166 *Bataviasche Courant*, no. 53 (December 30, 1820): 1.
167 ‘Om de Nederlandsche bezittingen in Oost Indien te bereizen, met het bepaalde doel om de wetenschappelijke kennis van de voortbrengselen der natuur in die landen uittebreiden.’
168 Res.GG, January 23, 1821, no. 12A. NL-HaNA, Koloniën, 1814–1849, 2.10.01, inv.no. 2774.
169 Res.GG, June 11, 1821, no. 7. NL-HaNA, Koloniën, 1814–1849, 2.10.01, inv.no. 2460.
170 Klaver, *Inseparable Friends*, 20.
171 Klaver, *Inseparable Friends*, 24.

der Capellen had the former home of the resident of Buitenzorg renovated for use by Van Hasselt and his personnel.[172] In early 1822, he started collecting again in the vicinity of Batavia.[173] After Reinwardt's departure, Van Hasselt took on his natural historical responsibilities and received a raise to f600 on June 11, 1822.[174] In August 1822, Van Hasselt finally left on the planned Banten expedition with Van Raalten (later replaced by Bik) and Maurevert,[175] and was allowed to continue to hire the twelve indigenous employees.[176] When Van Hasselt fell ill in August 1823, he was transported to Buitenzorg by Bik, but eventually succumbed on September 8, 1823.[177] A pension of f100 was awarded to Van Hasselt's mother in recognition of his achievements.[178]

14 Ludwig Horner (1811–1838, Figure 14)

b. March 1, 1811 (Zurich), d. December 7, 1838 (Padang)
Member: February 23, 1835–December 7, 1838

BWN, vol. 8.2, 1271 | NNBW, vol. 4, 784 | BPN 54882307 | ENI, vol. 2, 107 | FHA, 244 | MPCC, 242

Anonymus. "Ludwig Horner." *Verhandlungen der schweizerischen naturforschenden Gesellschaft bei ihrer Versammlung zu Bern, den 5., 6. und 7. August 1839* (1839): 211–232.

Anonymus. "Untitled." *Siebzehntes Neujahrsblatt zum Besten des Waisenhauses auf das Neujahr 1854 von einer Gesellschaft herausgegeben*, no. 76 (1854).

Horner, Friedrich. "Briefe und Tagebuchskizzen des Dr. med. Ludwig Horner (1811–1838) aus Niederländisch-Indien." *Zurcher Taschenbuch* (1919): 183–208.

Horner, Friedrich. "Briefe und Tagebuchskizzen des Dr. med. Ludwig Horner (1811–1838) aus Niederländisch-Indien." *Zurcher Taschenbuch* (1926): 1–45.

Lutz, Nicolau. "Ludwig Horner and Scientific Research in the Dutch East Indies, 1800–1850." B.A., University of Cambridge, 2018.

172 Res.GG, November 7, 1821, no 4. NL-HaNA, Koloniën, 1814–1849, 2.10.01, inv.no. 2463.
173 Klaver, *Inseparable Friends*, 24.
174 Res.GG, June 11, 1822, no. 4. NL-HaNA, Koloniën, 1814–1849, 2.10.01, inv.no. 2776.
175 Klaver, *Inseparable Friends*, 24. See Bik, "Aanteekeningen op eene reis."
176 Res.GG, August 4, 1822, no. 6. NL-HaNA, Koloniën, 1814–1849, 2.10.01, inv.no. 2466.
177 Klaver, *Inseparable Friends*, 25.
178 Res.GG, September 9, 1823, no. 2. NL-HaNA, Koloniën, 1814–1849, 2.10.01, inv.no. 2779.

Ludwig (Louis) Horner was born on March 1, 1811, in Zurich, Switzerland, to Melchior Horner-Horner (1779–1853).[179] He studied medicine in Zurich, and mineralogy in Heidelberg, where he received his medical doctorate in 1832. He returned to Zurich, where he taught at the university. When time permitted, he travelled throughout Switzerland and Germany as naturalist. He travelled to Amsterdam in May 1834 and decided to take the exam for the military medical service. He was preparing to leave to the Indies as surgeon second class of the military medical service, when he met Temminck at 's Rijks Museum in Leiden and was recruited by him.

Horner was appointed as member of the Natuurkundige Commissie on February 23, 1835,[180] and left for the Indies, as planned, on February 26, 1835. He arrived in Batavia on July 8, 1835, on the ship *Johanna*.[181] On August 31, 1835, he was officially appointed at a salary of f300 (which did not befit his status of member),[182] and on October 1, 1835, he was ordered to join the rest of the Commissie in Padang (Sumatra) to take up his role as geologist/mineralogist of the Natuurkundige Commissie.[183] Already on November 14, 1835, the Natuurkundige Commissie was recalled in its entirety from Sumatra,[184] so despite the previous order Horner did not travel to Sumatra in 1835.

On February 27, 1836, Horner was ordered to leave for Borneo with Müller and Korthals on May 1, 1836, at the latest, in order to return in November.[185] He received the additional task of undertaking an expedition to Banten and parts of the Buitenzorg residency in order to inspect the known coal layers.[186] The departure of the Borneo expedition was delayed, and the expeditionary party was allowed an additional two to three months on November 16, 1836.[187] Müller, Korthals, and Horner departed from Banjarmasin on December 17, 1836, on the brig *Barket* and arrived in Batavia on January 9, 1837.[188]

When Müller and Korthals were recalled to the Netherlands on March 31, 1837, Horner was appointed to undertake an expedition to Sumatra in order to inspect the coal deposits in Bengkulu.[189] The goal of Horner's Sumatra expedition was further specified on May 3, 1837, when he was given a specific itinerary, personnel support in

179 Horner official first name was Ludwig, but he signed his letters to his parents as Louis. See Friedrich Horner, "Briefe und Tagebuchskizzen," 188, 191.

180 KB, February 23, 1835, no. 71. NL-HaNA, Staatssecretarie, 2.02.01, inv.no. 4003.

181 *Javasche Courant*, no. 55 (July 11, 1835): 7. Since Horner was part of a military contingent, he is not mentioned as a passenger.

182 Res.GG, August 31, 1835, no. 30. NL-HaNA, Koloniën, 1814–1849, 2.10.01, inv.no. 2856.

183 Res.GG, October 1, 1835, no. 1. NL-HaNA, Koloniën, 1814–1849, 2.10.01, inv.no. 2857.

184 Res.GG, November 14, 1835, no. 2. NL-HaNA, Koloniën, 1814–1849, 2.10.01, inv.no. 2857.

185 Res.GG, February 27, 1835, no. 12. NL-HaNA, Koloniën, 1814–1849, 2.10.01, inv.no. 2858.

186 Res.GG, February 28, 1836, no. 34. NL-HaNA, Koloniën, 1814–1849, 2.10.01, inv.no. 2561.

187 Res.GG, November 16, 1836, no. 7. NL-HaNA, Koloniën, 1814–1849, 2.10.01, inv.no. 2565.

188 *Javasche Courant*, no. 5 (January 18, 1837): 6.

189 Res.GG, March 31, 1837, no. 7. NL-HaNA, Koloniën, 1814–1849, 2.10.01, inv.no. 2568.

the form of Overdijk and four indigenous hunters, and a monthly allowance of f200 for expedition costs.[190] They departed to Padang on the ship *Sumatra* on June 13, 1837.[191] On June 23, 1837, Horner was given a lumpsum of f3,500 instead of a monthly allowance in order to pay for all the expedition costs.[192] Finally, on December 1, 1837, his salary was raised retrospectively from June 1, 1837, to f400, and from January 1, 1838, to f500, bringing him level with the standard for members.[193] Horner died during the Sumatra expedition in Padang on December 7, 1838, leaving Overdijk to transport his papers and collections back to Batavia.

15 Franz Wilhelm Junghuhn (1809–1864, Figure 17)

b. October 26, 1809 (Mansfeld), d. April 24, 1864 (Lebak)
Assistant: July 1, 1838–November 1, 1839
Member: April 1, 1845–June 27, 1855 (?)

BWN, vol. 33, 312–314 | NNBW, vol. 4, 820–822 | BPN 99314660 | ENI, vol. 2, 223–226 | MPCC, 267–269 | BB, 321 | T, 227–228 | ADB, vol. 14, 712–718 | NDB, vol. 10, 685–686

Angerler, Johann. "Franz Wilhelm Junghuhn und die Menschen Indonesiens: Über sein kulturanthropologisches Werk, seine Spiritualität und seine Beziehung zum Kolonialismus." *Anthropos* 107, no. 2 (2012): 407–426.

Bakker, Hans J. *Leven en werk van Franz Wilhelm Junghuhn, 1809–1864: De Humboldt van Java: Een bijdrage tot de geschiedenis van Nederlands-Indië*. Amersfoort: Bakker & De Vries, 2009.

Beekman, E. M. "F. W. Junghuhn (1809–1864): Elevating Tropical Nature." In *Troubled Pleasure: Dutch Colonial Literature from the East Indies, 1600–1950*. Clarendon Press, 1996.

Beekman, E. M. "Junghuhn's Perception of Javanese Nature." *Canadian Journal of Netherlandic Studies* 12, no. 1 (1991): 11–20.

Bel, Jacqueline. "De eeuwige natuur op Java: Aanstekelijke reisverhalen van Franz Wilhelm Junghuhn." In *Een tint van het Indische Oosten: Reizen in Insulinde 1800–1950*, edited by Rick Honings and Peter van Zonneveld. Verloren, 2015.

Blume, C. L. "Opheldering van de inlichtingen van den heer Fr. Junghuhn." *Algemeene konst- en letter-bode*, no. 43 (October 25, 1850): 258–261.

190 Res.GG, May 3, 1837, no. 1. NL-HaNA, Koloniën, 1814–1849, 2.10.01, inv.no. 2570.
191 *Javasche Courant*, no. 47 (June 14, 1837): 5.
192 Res.GG, June 23, 1837, no. 11. NL-HaNA, Koloniën, 1814–1849, 2.10.01, inv.no. 2571.
193 Res.GG, December 1, 1837, no. 11. NL-HaNA, Koloniën, 1814–1849, 2.10.01, inv.no. 2577.

Blume, C. L. "Opheldering van de inlichtingen van den heer Fr. Junghuhn." *Algemeene konst- en letter-bode*, no. 44 (November 1, 1850): 274–279.

Bosma, Ulbe. "Franz Junghuhn's Three-Dimensional and Transcendental Java." In *The Role of Religions in the European Perception of Insular and Mainland Southeast Asia: Travel Accounts of the 16th to the 21st Century*, edited by Monika Arnez and Jürgen Sarnowsky. Cambridge Scholars Publishing, 2016.

Goethe-Institut Indonesien. *Forschen – vermessen – streiten: Franz Wilhelm Junghuhn (1809–1864)*. Regiospectra Verlag, 2014.

Goss, Andrew. *The Floracrats: State-Sponsored Science and the Failure of the Enlightenment in Indonesia*. New Perspectives in Southeast Asian Studies. University of Wisconsin Press, 2011.

Günst, F. "Levensschets van Dr. Franz Wilhelm Junghuhn." In *Licht- en schaduwbeelden uit de binnenlanden van Java: Over het karakter, de mate van beschaving, de zeden en gebruiken der Javanen; over de invoering van het Christendom op Java, het bezigen van vrijen arbeid en andere vragen van den dag: Verhalen en gesprekken der gebroeders Dag en Nacht; verzameld op reizen door gebergten en bosschen, in de woningen van armen en rijken: Medegedeeld door Dr. F. W. Junghuhn*. Zesde herziene en vermeerderde druk. Amserdam: F. Günst, 1867.

Honings, Rick. "De Humboldt van Java: De natuurbespiegelingen van Franz Wilhelm Junghuhn." In *De ontdekking van Insulinde: Op reis in Nederlands-Indië in de negentiende eeuw*. Prometheus, 2023.

Hoppe, Felicitas. "Junghuhn: Eis und Schnee." In *Verbrecher und Versager: Fünf Porträts*. Fischer Taschenbuch Verlag, 2006.

Junghuhn, Franz Wilhelm. *De onuitputtelijke natuur: Een keuze uit zijn geschriften*. Stoa. Edited by Rob Nieuwenhuys and Frits Jacquet. Amsterdam: G. A. van Oorschot, 1966.

Junghuhn-commissie, ed. *Gedenkboek Franz Junghuhn, 1809–1909*. Martinus Nijhoff, 1910.

Kamp, Hans J. van der. "Franz Wilhelm Junghuhn, 1809–1864: Een korte biografie." *De Boekerij: Mededelingenblad van de Vereniging Vrienden der Bibliotheek van de Koninklijke Nederlandsche Akademie van Wetenschappen* 2, no. 2 (1997): 14–16.

Kamp, Hans J. van der. "Junghuhns biografen." *De Boekerij: Mededelingenblad van de Vereniging Vrienden der Bibliotheek van de Koninklijke Nederlandsche Akademie van Wetenschappen* 2, no. 2 (1997): 16–18.

Kamp, Hans J. van der. "Een bijzondere druk van Junghuhn in de Bibliotheek KNAW." *De Boekerij: Mededelingenblad van de Vereniging Vrienden der Bibliotheek van de Koninklijke Nederlandsche Akademie van Wetenschappen* 2, no. 2 (1997): 18–21.

Kamp, Hans J. van der. "Schreef Junghuhn in het Nederlands?". *De Boekerij: Mededelingenblad van de Vereniging Vrienden der Bibliotheek van de Koninklijke Nederlandsche Akademie van Wetenschappen* 8, no. 1 (2003): 13–21, 28.

Kamp, Hans J. van der. "Het grafmonument van Junghuhn." *De Boekerij: Mededelingenblad van de Vereniging Vrienden der Bibliotheek van de Koninklijke Nederlandsche Akademie van Wetenschappen* 7, no. 2–3 (2002–2003): 12–22, 24.

Kamp, Hans J. van der. "Junghuhn: Niet in October 1812 geboren maar in October 1809." *De Boekerij: Mededelingenblad van de Vereniging Vrienden der Bibliotheek van de Koninklijke Nederlandsche Akademie van Wetenschappen* 8, no. 2 (2004): 13–19.

Knecht-Van Eekelen, Annemarie de. "The Debate About Acclimatization in the Dutch East Indies, 1840–1860." In *Medical Geography in Historical Perspective*, edited by Nicholaas A. Rupke. Medical History, Supplement 20. Wellcome Trust Centre for the History of Medicine at UCL, 2000.

Kroon, A. W. "Levensschets van Franz Wilhelm Junghuhn." *De dageraad* 18 (1864): 415–462.

Lach de Bère-Van der Vlies, T. "Junghuhn (1809–1864): Een kijkje in zijn eerste zône." *Tropisch Nederland* 6, no. 11 (1933): 163–165; no. 12 (1933): 179–185; no. 13 (1933): 195–199; no. 15 (1933): 216–221.

Müller, Karl. "Franz Junghuhn: Vortrag, gehalten in Gewerbevereine zu Halle am 6. März 1865." *Die Natur: Zeitung zur Verbreitung naturwissenschaftlicher Kenntnis und Naturanschauung für Leser aller Stände* 14, no. 18 (May 3, 1865): 137–139; no. 19 (May 10, 1865): 145–148; no. 20 (May 17, 1865): 153–156; no. 23 (June 7, 1865): 177–180; no. 24 (June 14, 1865): 190–192.

Nieuwenhuys, Rob, Frits Jacquet, and Franz Wilhelm Junghuhn. *Java's onuitputtelijke natuur: Reisverhalen, tekeningen en fotografieën van Franz Wilhelm Junghuhn*. A. W. Sijthoff, 1980.

Rochussen, H. "Levensbericht van F. W. Junghuhn." *Tijdschrift voor Nederlandsch Indië. Nieuwe serie* 4, 1 (1866): 207–216.

Rochussen, H. "Levensbericht van F. W. Junghuhn." *Natuurkundig tijdschrift voor Nederlandsch Indië* 28 (1865): 342–356.

Schmidt, Max C. P., and Franz Wilhelm Junghuhn. *Franz Junghuhn: Biographische Beiträge zur 100. Wiederkehr seines Geburtstages*. Verlag der Dürr'schen Buchhandlung, 1909.

Schütte, Heinz, and Renate Sternagel. *Der Naturforscher Junghuhn (1809–1864): Zur 200sten Wiederkehr seines Geburtstages*. Working Paper Series: Global and European Studies Institute at the University of Leipzig, 2. Leipziger Universitätsverlag, 2009.

Sep, Peter. "De receptie van *Licht- en schaduwbeelden in de binnenlanden van Java* van F. W. Junghuhn." *Indische letteren* 2 (1987): 53–64.

Sternagel, Renate. *Der Humboldt von Java: Leben und Werk des Naturforschers Franz Wilhelm Junghuhn 1809–1864*. 2., durchgesehene Auflage. Mitteldeutscher Verlag, 2018.

Teijsmann, J. E. "Kritische aanmerkingen op de Bijdragen van doctor Junghuhn, tot de geschiedenis der vulkanen in den Indischen Archipel." *Tijdschrift voor Neêrland's Indië* 5, 1 (1843): 487–508.

Troelstra, Anne S. "De navolgers." In *Tijgers op de Ararat: Natuurhistorische reisverhalen, 1700–1950*. Atlas, 2003.

Veer, Paul van 't. "In de schaduw van de kinaboom: Franz Wilhelm Junghuhn, 1809–1864." In *Geen blad voor de mond: Vijf radicalen uit de negentiende eeuw*. Arbeiderspers, 1958.

Wormser, C. W. *Frans Junghuhn*. Bouwers van Indië, 5. W. van Hoeve, 1943.

Since a short biography like this will never do justice to Junghuhn's many achievements, I limit this biography to a set of general biographical facts and the period of his involvement with the Natuurkundige Commissie, which in the secondary literature on Junghuhn is only part of a much larger story.

Franz Wilhelm Junghuhn was born on October 26, 1809, in Mansfeld (Germany), to Wilhelm Friedrich Junghuhn and Christine Marie Schiele. Junghuhn studied medicine in Halle and Berlin, driven by a desire to occupy himself professionally with natural history. A troublesome relationship with his father, a failed suicide attempt, an unfortunate duel, and an ensuing prison sentence contributed to a difficult early life. After his escape from prison in 1833, he enlisted in the Foreign Legion and briefly served in North Africa. When illness forced him to return to France, he found out he had been pardoned. After briefly visiting his home country, he took his exam for the Dutch medical military service in Utrecht and departed for the Indies on June 3, 1835. He arrived in the Indies on October 13, 1835, as surgeon third class.

On November 7, 1835, Junghuhn was provisionally placed at the hospital in Weltevreden.[194] His work brought him to the attention of E. A. Fritze, the head of the military medical service, and Diard. Upon his own request (supported by a recommendation from Diard), Junghuhn was provisionally attached to the Natuurkundige Commissie on July 1, 1838, for the duration of one year. The collections made during his work were to be considered government property, including everything that he had already collected since his arrival on Java.[195] Junghuhn was not given a specific task, nor was he attached to an expedition. He seems to have mostly spent his time on Java climbing volcanoes and botanising, sometimes accompanied by Forsten. Diard was eager to keep Junghuhn on and held him on after his appointment officially ended on June 30, 1839. Only on September 29, 1839, was Junghuhn's appointment officially withdrawn.[196] Still Diard kept him in service, but finally let him return to his military

194 Res.GG, November 7, 1835, no. 3. NL-HaNA, Koloniën, 1814–1849, 2.10.01, inv.no. 2560.

195 Res.GG, July 10, 1838, no. 13. NL-HaNA, Koloniën, 1814–1849, 2.10.01, inv.no. 2584.

196 Res.GG, September 29, 1839, no. 12. NL-HaNA, Koloniën, 1814–1849, 2.10.01, inv.no. 2598.

post from November 1, 1839, when it became apparent that a prolongation was not going to happen. Junghuhn retrospectively received his salary for his work with the Natuurkundige Commissie from July to October 1839 on April 5, 1840.[197]

Junghuhn was promoted to surgeon second class on February 9, 1840.[198] On October 20, 1840, Junghuhn received an appointment that would occupy him until 1844: he was to undertake a topographical and ethnographical survey of the Batak lands in North Sumatra.[199] The expedition itself lasted less than half a year, from its start on Sumatra on November 20, 1841,[200] to his return to Java for health reasons in March 1842,[201] finally approved on September 13, 1842.[202] On Java, Junghuhn worked on his recovery and on his report. Working on his report on the Sumatra expedition slowly morphed into research on Java. His work on the Sumatra report was deemed to have officially ended (for budgetary purposes) on April 30, 1844.[203] On August 1, 1844, his ongoing research on Java was turned into permission to undertake a research journey on East Java, which would formally continue until March 31, 1845.[204]

In the meantime, Junghuhn had been lobbying for a formal appointment as a member of either a commission for topographical research or the Natuurkundige Commissie. His request had been referred to the Netherlands and on May 5, 1845, he was retrospectively appointed as member of the Natuurkundige Commissie as of April 1, 1845, at a salary of f350, but with no specific task.[205] Junghuhn used his appointment to continue his earlier research on Java, until he was reminded by the governor-general that he had not been given a task, and should therefore send in a report of his current work and a proposal for what he planned to do.[206]

It was not until April 5, 1846, that Junghuhn was finally charged with an expedition on Java that would take him to the residencies of Banten, Batavia, Buitenzorg, Priangan, Krawang and Cheribon. He should concentrate on the areas of geography, topography, geognosy, mineralogy, and meteorology, and should produce a map. He received f150 travel allowance and f350 collecting costs. Due to previous clashes regarding Junghuhn's publications, he was ordered to hand in his reports only to the

197 Res.GG, April 5, 1840, no. 8. NL-HaNA, Koloniën, 1814–1849, 2.10.01, inv.no. 2605.
198 Res.GG, February 9, 1840, no. 1. NL-HaNA, Koloniën, 1814–1849, 2.10.01, inv.no. 2603.
199 Res.GG, October 20, 1840, no. 3. NL-HaNA, Koloniën, 1814–1849, 2.10.01, inv.no. 2611.
200 Res.GG, January 24, 1842, no. 7. NL-HaNA, Koloniën, 1814–1849, 2.10.01, inv.no. 2629.
201 Sternagel, *Humboldt von Java*, 166.
202 Res.GG, September 13, 1842, no. 17. NL-HaNA, Koloniën, 1814–1849, 2.10.01, inv.no. 2637.
203 Res.GG, August 22, 1844, no. 12. NL-HaNA, Koloniën, 1814–1849, 2.10.01, inv.no. 2660.
204 Res.GG, January 15, 1845, no. 2. NL-HaNA, Koloniën, 1814–1849, 2.10.01, inv.no. 2665.
205 Res.GG, May 5, 1845, no. 1. NL-HaNA, Koloniën, 1814–1849, 2.10.01, inv.no. 2669.
206 Res.GG, July 18, 1845, no. 25. NL-HaNA, Koloniën, 1814–1849, 2.10.01, inv.no. 2671.

government and was expressly forbidden from having them published elsewhere.[207] During his expedition, Junghuhn sent information to the governor-general regarding coal, gold and iron deposits,[208] and new methods of wax production.[209] On June 2, 1848, Junghuhn finally received his coveted raise to f500 which he was entitled to after three years of service.[210]

On August 12, 1848, Junghuhn's request for leave to Europe in order to write up his report was approved[211] and he left Java on August 27, 1848 on the ship *Etna*.[212] When the Natuurkundige Commissie was formally dissolved on April 17, 1850,[213] Junghuhn was allowed to finish his reporting in the Netherlands. Two of his publications, his famous work on Java and an initially anonymous spiritual work, published during his Dutch period propelled him to national and international fame.[214] On June 27, 1855, Junghuhn was appointed to continue his natural historical research in the Netherlands Indies.[215] This presumably also meant his official resignation as member of the Natuurkundige Commissie. Junghuhn returned to Java in 1855, where he would be heavily involved in the development of cinchona cultivation.[216] Junghuhn died in Lebak (Java) on April 24, 1864.

16 Gerrit Laurens Keultjes (1786–1821)

b. August 10, 1786 (Utrecht), d. September 15, 1821 (Buitenzorg)
Assistant: May 2, 1820–September 15, 1821

BPN 33871170

Mac Lean, J. "Gegevens over tekenaars van biologische voorwerpen in het voormalige Nederlandsch Indië." *Scientiarum historia* 15 (1973): 101–114.

207 Res.GG, April 5, 1846, no. 27. NL-HaNA, Koloniën, 1814–1849, 2.10.01, inv.no. 2681. The resolution contains extensive written instructions for Junghuhn.

208 See Res.GG, April 17, 1846, no. 33. NL-HaNA, Koloniën, 1814–1849, 2.10.01, inv.no. 2681 for coal; Res.GG, June 20, 1847, no. 6. NL-HaNA, Koloniën, 1814–1849, 2.10.01, inv.no. 2695 for gold; Res.GG, March 16, 1848, no. 22. NL-HaNA, Koloniën, 1814–1849, 2.10.01, inv.no. 2704 for gold and iron.

209 Res.GG, July 14, 1846, no. 1. NL-HaNA, Koloniën, 1814–1849, 2.10.01, inv.no. 2684; Res.GG, December 23, 1846, no. 15. NL-HaNA, Koloniën, 1814–1849, 2.10.01, inv.no. 2689.

210 Res.GG, June 2, 1848, no. 9. NL-HaNA, Koloniën, 1814–1849, 2.10.01, inv.no. 2707.

211 Res.GG, August 12, 1848, no. 1. NL-HaNA, Koloniën, 1814–1849, 2.10.01, inv.no. 2709.

212 Sternagel, *Humboldt von Java*, 246.

213 KB, April 17, 1850, no. 14. NL-HaNA, KdK 1841–1897, 2.02.04, inv.no. 609.

214 Junghuhn, *Java*; Gebroeders Dag en Nacht, *Licht- en schaduwbeelden*.

215 KB, June 27, 1855, no. 100. See *Algemeene konst- en letter-bode*, no. 13 (March 27, 1858): 101.

216 See Goss, *Floracrats*, 33–58.

Gerrit Laurens Keultjes was born on August 10, 1786, in Utrecht to Petrus Josephus Keultjes and Maria Geertruijda Colberg.[217] Keultjes' father died on April 11, 1787. His burial record indicates that he was a widower, so at the age of eight months, Keultjes was left an orphan.[218] He was raised in an orphanage, where he was educated as a painter. He married Johanna van der Rijst (born 1791) on June 8, 1814.[219] A two-year journey through the Mediterranean gave him the opportunity to develop as a seascape painter.[220]

On May 2, 1820, Keultjes was appointed as assistant-draughtsman to accompany Kuhl and Van Hasselt along with Van Raalten.[221] Keultjes' selection was based on good recommendations ('goede getuigschriften'), but it does not say who did the recommending. Perhaps he was already acquainted with Van Raalten and his mentor Van Lidth de Jeude. Kuhl, Van Hasselt, Van Raalten, and Keultjes left the Netherlands on July 11, 1820,[222] and arrived in Batavia on December 22, 1820, on the ship *Nordloh*.[223] Keultjes was accompanied by his wife Johanna. On January 20, 1821, Keultjes was awarded an advance and the extra allowance that was customary for married civil servants.[224] His salary was confirmed at f200 on January 23, 1821.[225] Within a year after arrival, Keultjes died on September 15, 1821.[226] However, the news of his death was eclipsed by the death of Kuhl one day earlier and hardly any attention was paid to it. His widow received a pension in the Indies of f60 from October 1, 1823, with permission to continue receiving the pension after her return to the Netherlands.[227]

17 Georg Joseph August Kollmann (1797–1839)

b. December 13, 1796 (Würzburg), d. March 8, 1839 (Padang)
Assistant: 1824

---

217 The baptismal record for Keultjes gives his name in Latin: Gerardus Laurentius. NL-UtrUA, Burgerlijke stand gemeente Utrecht, 711, inv.no. 72, f. 201.

218 Burial record. NL-UtrUA, Burgerlijke stand gemeente Utrecht, 711, inv.no. 138, f. 460.

219 Marriage certificate, no. 120. NL-UtrUA, Burgelijke Stand van de gemeenten in de provincie Utrecht 1811–1902, 481, inv.no. 273–01.

220 Van Eijnden and Van der Willigen, *Geschiedenis der vaderlandsche schilderkunst: Deel 3*, 272–273. See https://www.rijksmuseum.nl/collectie/SK-A-1376 for an example.

221 KB, May 2, 1820, no. 10. NL-HaNA, Staatssecretarie, 2.02.01, inv.no. 988.

222 Veth, *Overzicht van hetgeen gedaan is*, 25; Klaver, *Inseparable Friends*, 18.

223 *Bataviasche Courant*, no. 53 (December 30, 1820): 1.

224 Res.GG, January 20, 1821, no. 18. NL-HaNA, Koloniën, 1814–1849, 2.10.01, inv.no. 2458.

225 Res.GG, January 23, 1821, no. 12A. NL-HaNA, Koloniën, 1814–1849, 2.10.01, inv.no. 2774.

226 Registry card Gerrit Laurens Keultjes. NL-HaNA, Koloniën, 1814–1849, 2.10.01, inv.no. 3091, f. 247.

227 Res.GG, March 30, 1822, no. 15. NL-HaNA, Koloniën, 1814–1849, 2.10.01, inv.no. 2464.

MPCC, 290–291
Mac Lean, J. "Gegevens over tekenaars van biologische voorwerpen in het voormalige Nederlandsch Indië." *Scientiarum historia* 15 (1973): 101–114.
Steenis-Kruseman, M. J. van. "Contributions to the History of Botany and Exploration in Malaysia 1: Kollmann's Collection of Javan Plants." *Bulletin of the Botanic Gardens, Buitenzorg. Series 3* 18, no. 4 (1950): 463–466.

The story of Kollmann's involvement with the Natuurkundige Commissie is an odd one. He was born on December 13, 1796, in Würzburg, Germany, to Nicolaas Kollmann and Elisabeth Goebel.[228] He served as surgeon in the military medical service from 1821 until his death on March 8, 1839 in Padang (Sumatra). On November 10, 1824, he was awarded a gratification of f800 for his excellent work in copying the Kuhl/Van Hasselt papers and promised a further gratification of f400 if he would keep up the good work.[229]

When he was on leave in Europe, he offered a collection of 4,000 botanical specimens for sale to the Dutch government in June 1836. When Blume inspected the collection in early 1837, in the presence of Forsten and two other witnesses, about ninety percent of the collection turned out to consist of specimens collected by Blume himself and Zipelius that had been left at 's Lands Plantentuin in Buitenzorg. Since these were government-owned already, Blume kept these specimens and returned the remaining 400 specimens to Kollmann's German agents.[230] Nothing further was made of Kollmann's appropriation of the collections of Blume and Zipelius. It is not entirely clear whether Kollmann had originally acquired these collections in good faith, but it seems likely.

18 Pieter Willem Korthals (1807–1892, Figure 13)

b. September 1, 1807 (Amsterdam), d. February 7, 1892 (Haarlem)
Member: July 25, 1830–February 11, 1842

NNBW, vol. 8, 984 | BPN 60135535 | ENI, vol. 2, 437 | MPCC, 295–298
Manse, Maarten. "Kennis is macht: De veelzijdige expedities van botanicus Pieter Willem Korthals (1807–1892)." *Studium* 6, no. 1 (2013): 42–50.

228 Military registry card Georg Joseph August Kollmann. NL-HaNA, Oorlog / Stamboeken Officieren KL, 2.13.04, inv.no. 388, f. 34.
229 Res.GG, November 10, 1824, no. 2. NL-HaNA, Koloniën, 1814–1849, 2.10.01, inv.no. 2479.
230 Van Steenis-Kruseman, "Malaysian Plant Collectors," 290–291.

Pieter Willem Korthals was born on September 1, 1807, in Amsterdam, to Willem Korthals Jacobszoon and Woutera Willemina Korthals (d. 1864). Korthals' education is somewhat of a mystery. He seems to have been trained as an apothecary assistant with some connection to Groningen.[231] In 1825, he was appointed as assistant to Reinwardt in Leiden.[232] When his appointment to the Commissie was discussed in 1830, he was introduced as working at the Herbarium of the Hogeschool Leiden and at 's Rijks Museum, where he was training for his future commission to the Indies.[233] The connection with Reinwardt had clearly helped him to get noticed by Temminck. On Blume's request, Korthals was appointed to catalogue the herbarium of Christiaan Hendrik Persoon (1761–1836) that had been purchased in 1828 by the Dutch government.[234] This commission led Korthals to Paris in May 1829. In the end, the proposal by Temminck, supported by positive recommendations from both Reinwardt and Blume, was approved and Korthals was appointed as member of the Natuurkundige Commissie on July 25, 1830, with assistants Van Gelder and Overdijk.[235]

Korthals arrived in Batavia, along with Van Gelder and Overdijk, on April 13, 1831, on the ship *Jonge Adriana*.[236] On May 25, 1831, he was appointed at a salary of f500 and the Natuurkundige Commissie was instructed to charge Korthals with botanical work.[237] One of his tasks from the Netherlands was to analyse the working of the Hoofdkommissie van Landbouw [Main Committee of Agriculture]. In order to fulfil this task, he was appointed as member of the Hoofdkommissie on November 22, 1831.[238] Korthals had arrived just in time to join the rest of Commissie in their new expedition along Java's northern coast to Madura and Bali and then back along the south coast. The Commissie took it slowly and spent most of 1831 and 1832 in the Priangan and Krawang residencies, undertaking small trips with varying combinations of individuals, continuing this after the death of Macklot in May 1832, until the hastily organised expedition to Sumatra cut this short. Korthals was ordered on May 21, 1833, to join Müller and Van Oort, who had been ordered to go to Sumatra three days before.[239] The quartet (including Bürger) departed on June 4, 1833, on the ship *Hercules*, to Padang, where they arrived on June 16, 1833.[240] On November 21, 1834, Korthals' request for

231 Entry 524 in Inschrijvingsregister voor de Nationale Militie, 1826–1827. NL-RotSAR, Gemeentesecretarie Rotterdam, afdeling Militaire Zaken, 356, inv.no. 90.

232 Karstens and Kleibrink, *Leidse Hortus*, 66.

233 Den Ouden, "C. L. Blume," 89.

234 Den Ouden, "C. L. Blume," 66–67.

235 KB, July 25, 1830, no. 40. NL-HaNA, Staatssecretarie, 2.02.01, inv.no. 3437.

236 *Javasche Courant*, no. 46 (April 16, 1831): 3.

237 Res.GG, May 25, 1831, no. 19. NL-HaNA, Koloniën, 1814–1849, 2.10.01, inv.no. 1839.

238 Res.GG, November 22, 1831, no. 1. NL-HaNA, Koloniën, 1814–1849, 2.10.01, inv.no. 2841.

239 Res.GG, May 21, 1833, no. 1. NL-HaNA, Koloniën, 1814–1849, 2.02.01, inv.no. 2550.

240 Weber and Van Zanen, *Kunstenaar op Sumatra*, 37–38.

leave on Java in order to prepare his collections for shipment to the Netherlands was denied.[241] Only a year later, on November 14, 1835, Korthals was summoned back to Java along with Müller.[242]

Shortly after their return to Java, it was decided that Müller and Korthals, along with Horner, should undertake one final expedition before returning to the Netherlands in February 1837. The destination would be Borneo and they were to go there from May to November 1836.[243] On March 12, 1836, an additional assignment was given to them: they needed to verify the expedition reports of army officer Henrici and recommend to the governor-general whether these expeditions should continue.[244] Due to the delayed departure of the Borneo expedition (they only arrived in Banjarmasin on July 28, 1836),[245] they were allowed an additional two to three months on November 16, 1836.[246] Müller, Korthals, and Horner departed from Banjarmasin on December 17, 1836, on the brig *Barket* and arrived in Batavia on January 9, 1837.[247]

Müller and Korthals were awarded free transport for themselves and their collections to the Netherlands on March 31, 1837.[248] Their baggage included 76 crates, two suitcases and seven barrels of specimens, with an additional thirteen crates with living plants.[249] On April 12, 1837, Müller and Korthals departed from Batavia on the ship *Prinses Marianne*.[250] They arrived in Hellevoetsluis on August 21, 1837, and departed for Leiden the next day.[251] Korthals was put on half-pay and given time to process his collections and write up his reports.[252] Korthals was appointed as Knight in the Order of the Dutch Lion on June 9, 1839.[253] He became the main writer of the botanical volume of the *Verhandelingen*. Upon his own request, Korthals was honourably discharged and awarded an annual pension of f1,200 on February 11, 1842.[254]

After his pension, Korthals does not seem to have taken a job, even though in the decade after his pension he still occupied himself with botany. He undertook a

241 Res.GG, November 21, 1834, no. 19. NL-HaNA, Koloniën, 1814–1849, 2.02.01, inv.no. 2853.
242 Res.GG, November 14, 1835, no. 2. NL-HaNA, Koloniën, 1814–1849, 2.10.01, inv.no. 2857.
243 Res.GG, February 27, 1836, no. 12. NL-HaNA, Koloniën, 1814–1849, 2.10.01, inv.no. 2858.
244 Res.GG, March 12, 1836, no. 1. NL-HaNA, Koloniën, 1814–1849, 2.10.01, inv.no. 2859.
245 Dagboek Korthals, July 28, 1836. NL-LdnNAT, Korthals_5_050. Müller reports that they arrived in the first days of August, see Susanna, *Verhandelingen*, vol. 3, 338.
246 Res.GG, November 16, 1836, no. 7. NL-HaNA, Koloniën, 1814–1849, 2.10.01, inv.no. 2565.
247 *Javasche Courant*, no. 5 (January 18, 1837): 6.
248 Res.GG, March 31, 1837, no. 7. NL-HaNA, Koloniën, 1814–1849, 2.10.01, inv.no. 2568.
249 Res.GG, April 19, 1837, no. 1. NL-HaNA, Koloniën, 1814–1849, 2.10.01, inv.no. 2569.
250 *Javasche Courant*, no. 30 (April 15, 1837): 11.
251 Dagboek Korthals, August 21, 1837. NL-LdnNAT, Korthals_5_361 and Korthals_5_362; *Leydsche Courant*, no. 103 (August 28, 1837): 1.
252 Res.GG, March 31, 1837, no. 7. NL-HaNA, Koloniën, 1814–1849, 2.10.01, inv.no. 2568.
253 KB, June 9, 1839, no. 62. NL-HaNA, Staatssecretarie, 2.02.01, inv.no. 4488.
254 KB, February 11, 1843, no. 65. NL-HaNA, KdK 1841–1897, 2.02.04, inv.no. 148.

botanical journey to North and South America from 1847 to 1849. But it seems that already during these years, his interests moved towards psychology.[255] He married Hermina Nieuwveen (1809–1888) in Leiden on July 1, 1857.[256] Nieuwveen was the widow of Julian Hendrik Molkeboer (1816–1854), fellow botanist and protégé of Reinwardt.[257] Korthals died on February 7, 1892, in Haarlem.[258]

19 Heinrich Kuhl (1797–1821, Figure 9)

b. September 17, 1797 (Hanau), d. September 14, 1821 (Buitenzorg)
Member: May 2, 1820–September 14, 1821

NNBW, vol. 3, 721–722 | BPN 05819235 | ENI, vol. 2, 476 | FHA, 256–258 | MPCC, 303 | OM, 202

Anonymus. "Johan Coenraad van Hasselt en Heinrich Kühl, leden der Kommissie voor het natuurkundig onderzoek in Nederlandsch Indië." *Handelingen en geschriften van het Indisch Genootschap* 5 (1858): 66–80.

Greshoff, M. "Kuhl en Van Hasselt: Eene episode in het Nederlandsch-Indisch natuuronderzoek." *Album der natuur* 52, no. 1 (1903): 1–22, 60–88.

Hildenhagen, Thomas. "Das kurze Leben des Dr. h. c. Heinrich Kuhl (1797–1821) und seine herpetologischen Beiträge." *Sekretär* 10, no. 2 (2010): 40–59.

Hildenhagen, Thomas, and Jakob Hallermann. "Die von Kuhl (1797–1821) beschriebenen und die nach ihm benannten Amphibien- und Reptilien-Taxa." *Sekretär* 10, no. 2 (2010): 59–62.

Hildenhagen, Thomas. "Heinrich Kuhl: Das Leben eines fast vergessenen Naturforscher aus Hanau." *Neues Magazin für Hanauische Geschichte* (2013): 110–214.

Klaver, Charles. *Inseparable Friends in Life and Death: The Life and Work of Heinrich Kuhl (1797–1821) and Johan Conrad van Hasselt (1797–1823), Students of Prof. Theodorus van Swinderen.* Barkhuis, 2007.

Striekwold, Robbert. "Naam maken: Wat de dood van twee jonge negentiende-eeuwse wetenschappers ons leert over de natuurhistorie." *Ex tempore* 37, no. 3 (2018): 213–226.

---

255 The Korthals archive at Naturalis contains a folder with notes on psychological topics dated between 1840 and 1873.

256 Marriage certificate, no. 141. NL-LdnSAL, Stadsbestuur (1816–1929), 516, inv.no. 4848. Hermina died on December 23, 1887, see *Opregte Haarlemsche Courant*, no. 304 (December 28, 1827): 4.

257 Touw, "Wat bezielde onze voorgangers," 19.

258 *Opregte Haarlemsche Courant*, no. 34 (February 10, 1892): 2. Korthals' testament is available as no. 7/6810. NL-HaaNHA, Memories van successie in Noord-Holland, 178, inv.no. 1679.

Swinderen, Theodorus van. "Bijdragen tot eene schets van het leven, het karakter en de verdiensten van wijlen Dr. H. Kuhl." *Almanak der Akademie van Groningen* 11 (1823): 1–71.

Wingerden, Pieter van. "Werk in uitvoering = Work in Progress," *Gauronica* (blog), October 13, 2023. https://www.rug.nl/library/gauronica/blogposts/van-hasselt.

Heinrich Kuhl was born on September 17, 1797, in Hanau, Germany, to Johann Heinrich Kuhl (1757–1830) and Maria Judith Walt(h)er (1770–1810). At the age of 19, he published his first monograph *Die deutschen Fledermäuse*.[259] When Theodorus van Swinderen (1784–1851), professor of natural history at Groningen University, met Kuhl on his travels in Germany, he 'offered Kuhl the opportunity to study natural sciences in Groningen as well as a post as assistant at the museum of natural history.'[260] Kuhl accepted and travelled to Groningen in September 1816, where he immediately struck up a friendship with fellow student Van Hasselt.[261] During his studies in Groningen, Kuhl travelled extensively in the Netherlands and Germany, often with Van Hasselt, meeting Temminck for the first time in the winter of 1816/1817. In 1819, Kuhl was 'chosen to undertake an expedition to the Dutch East Indies,'[262] and was given the financial means to visit natural history collections in London and Paris as preparation for his own expedition.[263] On August 6, 1819, he received a doctorate *honoris causa* from Groningen University, doing away with the need to continue his studies and making time available for him to travel to London and Paris. On May 2, 1820, Kuhl was appointed, together with Van Hasselt, with Keultjes and Van Raalten as assistants, a member of what later became known as the Natuurkundige Commissie.[264]

The expedition party of four left the Netherlands on July 11, 1820,[265] and arrived in Batavia on December 22, 1820, on the ship *Nordloh*.[266] On January 23, 1821, Kuhl and Van Hasselt were appointed at a salary of f400 as of January 1, 1821, in order to 'travel the Dutch possessions in the East Indies, with the specific goal of expanding the scientific knowledge of the products of nature in those lands.'[267] They were placed under the supervision of Reinwardt and received permission to hire six indigenous employees

259 Kuhl, *Deutschen Fledermäuse*.
260 Klaver, *Inseparable Friends*, 7.
261 Van Berkel, *Universiteit van het Noorden: Deel 1*, 642–646.
262 Klaver, *Inseparable Friends*, 13.
263 KB, January 23, 1819, no. 123. NL-HaNA, Staatssecretarie, 2.02.01, inv.no. 757.
264 KB, May 2, 1820, no. 10. NL-HaNA, Staatssecretarie, 2.02.01, inv.no. 988.
265 Veth, *Overzicht van hetgeen gedaan is*, 25; Klaver, *Inseparable Friends*, 18.
266 *Bataviasche Courant*, December 30, 1820, no. 53, p. 1.
267 'Om de Nederlandsche bezittingen in Oost Indien te bereizen, met het bepaalde doel om de wetenschappelijke kennis van de voortbrengselen der natuur in die landen uittebreiden.'

at a salary of f12 each.[268] This number was expanded to twelve on June 11, 1821.[269] They set up camp in Buitenzorg and spent four months collecting in the close vicinity of Buitenzorg. When a cholera outbreak prevented them from undertaking a planned expedition to Banten, 'they decided to explore the mountains south of Buitenzorg.'[270] In August 1821, 'Kuhl became violently ill with diarrhoea and later hepatitis.'[271] Despite the care of his medical doctor friend Van Hasselt, Kuhl died on September 14, 1821 in Buitenzorg, to be followed by Keultjes a day later.[272]

20 Auguste Jean Latour

b. unknown, d. unknown
Assistant: 1828

Mac Lean, J. "Gegevens over tekenaars van biologische voorwerpen in het voormalige Nederlandsch Indië." *Scientiarum historia* 15 (1973): 101–114.

Nothing is known about Latour's early life. He appears when, at Blume's request, he was provisionally appointed as artist at 's Lands Plantentuin as replacement for Bik on March 21, 1823.[273] This appointment was made permanent on February 16, 1824.[274] Latour accompanied Blume on an expedition to West Java, together with G. H. Nagel and William Kent.[275] On March 31, 1825, Latour was warned because of his 'indifference and his culpably diminished work ethics' ('onverschilligheid en laakbare verflauwing van dienstijver'). Shortly after he was put on redundancy pay and left government service to end up as 'pionier' in the army (a military engineering role), stationed on the frigate *Melampus*.[276] A painting of the *Melampus* by Keultjes, made during his pre-Commissie Mediterranean adventure, survives in the Scheepvaartmuseum in Amsterdam.[277]

Latour's association with the Natuurkundige Commissie was short-lived and initiated by Macklot, who requested Latour to be added to the military detachment that

268 Res.GG, January 23, 1821, no. 12A. NL-HaNA, Koloniën, 1814–1849, 2.10.01, inv.no. 2774.
269 Res.GG, June 11, 1821, no. 7. NL-HaNA, Koloniën, 1814–1849, 2.10.01, inv.no. 2460.
270 Klaver, *Inseparable Friends*, 20.
271 Klaver, *Inseparable Friends*, 21.
272 Registry card Heinrich Kuhl. NL-HaNA, Koloniën, 1814–1849, 2.10.01, inv.no. 3091, f. 249.
273 Res.GG, March 21, 1823, no. 6. NL-HaNA, Koloniën, 1814–1849, 2.10.01, inv.no. 2468.
274 Res.GG, February 16, 1824, no. 16. NL-HaNA, Koloniën, 1814–1849, 2.10.01, inv.no. 2474.
275 Res.GG, March 3, 1824, no. 5. NL-HaNA, Koloniën, 1814–1849, 2.10.01, inv.no. 2475.
276 Res.GG, March 31, 1825, no. 3. NL-HaNA, Koloniën, 1814–1849, 2.10.01, inv.no. 2481.
277 https://collectie.hetscheepvaartmuseum.nl/Details/collect/503776.

was to take part in the New Guinea expedition.[278] Latour had to quickly travel from Rembang to Surabaya, in order to meet the expedition members who were travelling on the *Minerva* from Surabaya to Makassar.[279] When the expedition left New Guinea, the Natuurkundige Commissie was dropped off on Timor. There is no mention of Latour in any sources for the Natuurkundige Commissie after this, so it can be presumed that he stayed on New Guinea (or died there) or travelled back with the military detachment to fade from view.

21 Heinrich Christian Macklot (1799–1832)

b. October 20, 1799 (Frankfurt am Main), d. May 12, 1832 (Krawang)
Member: December 5, 1823–May 12, 1832

BWN, vol. 22, 366 | BPN 90563629 | ENI, vol. 2, 632 | FHA, 259–260 | MPCC, 342–343 | ADB, vol. 20, 17 | NDB, vol 15, 625 | OM, 227

Anonymus. "Dr. Macklot." *Frankfurter Konversationsblatt*, no. 68 (1833).

Jacobs, Hans J. "Heinrich Christian Macklot (20.10.1799–12.5.1832) und die Natuurkundige Commissie in Niederländisch Ostindien 1825–32." *Sekretär* 23, no. 2 (2023): 18–51.

Heinrich Christian Macklot was born on October 20, 1799, in Frankfurt am Main, Germany. He worked there as an apothecary from 1815 to 1818. He studied medicine in Heidelberg and obtained his doctorate there in 1822. Through his friendship with Boie, he received an appointment at 's Rijks Museum as curator on February 26, 1822. He was appointed as member of the Natuurkundige Commissie on December 5, 1823,[280] together with Boie and Müller. On December 21, 1825, Boie, Macklot, Müller, and Van Oort left the Netherlands on the ship *Dijkzigt*,[281] and arrived in Batavia on June 6, 1826.[282] Only on February 27, 1827, was the party formally appointed in the Indies: Boie and Macklot at a salary of f500, Müller and Van Oort at f200. Van Raalten was appointed as administrator at f300, permission was given to hire six non-European

278 Res.GG, January 29, 1828, no. 1. NL-HaNA, Koloniën, 1814–1849, 2.10.01, inv.no. 2809; Res.GG, January 31, 1828, no. 2. NL-HaNA, Koloniën, 1814–1849, 2.10.01, inv.no. 2512.

279 Res.GG, February 4, 1828, no. 14. NL-HaNA, Koloniën, 1814–1849, 2.10.01, inv.no. 2513; Res.GG, February 7, 1828, no. 20. NL-HaNA, Koloniën, 1814–1849, 2.10.01, inv.no. 2513.

280 KB, December 5, 1823, no. 84. NL-HaNA, Staatssecretarie, 2.02.01, inv.no. 1831.

281 Veth, *Overzicht van hetgeen gedaan is*, 45.

282 Weber and Van Zanen, *Kunstenaar op Java*, 55.

employees at f12 each, and the Natuurkundige Commissie was given an annual budget for collecting costs of f6,500.[283]

This does not mean that the Commissie had been idle until that time. In August 1826, they had taken residence in the house that had previously been given to Van Hasselt and departed on their first expedition in December 1826.[284] They stayed mostly within one or two days travel from Buitenzorg. Macklot received a special commission to investigate the coal deposits in Lebak (Banten), but his report was filed away because exploitation was deemed too difficult and costly.[285] He departed to Lebak on August 2, 1827, and returned on September 4, 1827, the morning of Boie's death.[286] As the sole remaining member, Macklot became responsible for the Natuurkundige Commissie as its directing member.

On December 31, 1827, Macklot was attached to the expedition to New Guinea.[287] Upon Macklot's request, the full Commissie was allowed to join the expedition, Latour was seconded to the Commissie and permission was received to take fourteen indigenous employees along (including eight natural history assistants).[288] Macklot, Müller, Van Oort, and Van Raalten left Batavia on the *Minerva* on February 2, 1828.[289] When the Commissie had already left Batavia, it was confirmed that the Commissie should travel back from New Guinea via Timor in order to perform a geological survey of copper and gold deposits on the island.[290] On February 10, on board the *Minerva*, between Batavia and Samarang an uprising of the exiles that were on board took place, but it could be quelled and not one of the passengers was hurt.[291] From Surabaya (arrival on February 19),[292] where Latour joined, the *Minerva* travelled to Makassar.

On March 15, 1828,[293] the Natuurkundige Commissie (Macklot, Müller, Van Oort, Van Raalten, Zipelius,[294] Latour) boarded the *Triton* and departed for their first big

283 Res.CG February 23, 1827, no. 18, confirmed by Res.GG, February 27, 1827, no. 35. NL-HaNA, Koloniën, 1814–1849, 2.10.01, inv.no. 2501.

284 Weber and Van Zanen, *Kunstenaar op Java*, 58–59.

285 Res.GG, December 20, 1827, no. 2. NL-HaNA, Koloniën, 1814–1849, 2.10.01, inv.no. 2511.

286 Weber and Van Zanen, *Kunstenaar op Java*, 70–71.

287 Res.GG (secret), December 31, 1827, no. 8. NL-HaNA, Koloniën, 1814–1849, 2.10.01, inv.no. 4515.

288 Res.GG, January 29, 1828, no. 1. NL-HaNA, Koloniën, 1814–1849, 2.10.01, inv.no. 2809.

289 *Javasche Courant*, no. 16 (February 5, 1828): 3.

290 Originally, Diard was supposed to have done this, but this arrangement was more convenient. Res.GG, December 24, 1827, no. 29. NL-HaNA, Koloniën, 1814–1849, 2.10.01, inv.no. 2808; Res.GG, January 29, 1828, no. 5. NL-HaNA, Koloniën, 1814–1849, 2.10.01, inv.no. 2809; Res.GG, February 5, 1828, no. 1. NL-HaNA, Koloniën, 1814–1849, 2.10.01, inv.no. 2810.

291 *Javasche Courant*, no. 23 (February 21, 1828): 1.

292 *Javasche Courant*, no. 26 (February 28, 1828): 4.

293 Mees, "Vogelkundig onderzoek," 6.

294 Zipelius had travelled to Makassar on the different ship (*Johanna Maria*), because he had some business to arrange in Batavia that prevented him from joining the rest of the Commissie from the outset.

expedition to the Outer Islands. They travelled to Ambon, from where they departed for New Guinea on April 21, 1828.[295] After a brief stop on Banda, they finally reached New Guinea on May 21, 1828.[296] During the first six weeks they travelled up the coast looking for a suitable location for a settlement, and they spent the next six weeks at this location, where a fortress was built (Fort Du Bus) and a settlement was founded (Merkusoord). The land was officially claimed for the Netherlands on August 24, 1828.

On August 30, 1828, the Triton left for Ambon with the Natuurkundige Commissie on board. They spent almost a month there before the Triton sailed to Timor to drop off the Natuurkundige Commissie. During their stay on Timor, where the Commissie undertook a series of smaller expeditions which included the required copper and gold surveys, Zipelius and Van Raalten died. The remaining trio of Macklot, Müller, and Van Oort arrived in Batavia on December 11, 1829.[297] In the meantime, a false rumour had reached Batavia that Macklot had died, which was an important reason for cancelling a follow-up expedition to New Guinea.[298] Not much is known of the activities of Macklot, Müller, and Van Oort in 1830, but it can be assumed that they worked processing their collections from New Guinea and Timor.[299] Macklot received a raise of f100 on January 8, 1830.[300]

Macklot's proposal for a new expedition that would take the Commissie along the northern side of Java to Madura and Bali and then back along the southside was approved on April 9, 1831.[301] The Commissie (now bolstered by the arrival of Korthals, Van Gelder and Overdijk) took it slowly and spent most of 1831 and 1832 in the Priangan and Krawang residencies, undertaking small trips with varying combinations of individuals. Macklot himself got caught up in an uprising of Chinese workers and was killed on May 12, 1832, in Krawang. A few days before, Macklot's residence was torched by the Chinese workers which resulted in the devastating loss of almost all the documents that Boie and Macklot had created during their work.[302] Macklot's collection of ethnographic objects was saved and sold by his heirs to Von Siebold and is currently part of the collection of Wereldmuseum Leiden.[303]

295 Mörzer Bruyns, *Met de Triton en Iris*, 38.
296 Mörzer Bruyns, *Met de Triton en Iris*, 38.
297 *Javasche Courant*, no. 147 (December 15, 1829): 4.
298 Res.GG, November 27, 1829, no. 2. NL-HaNA, Koloniën, 1814–1849, 2.10.01, inv.no. 2831.
299 Macklot had nine crates and half a barrel ready for transport to the Netherlands in May. See Res.GG, May 24, 1830, no. 1. NL-HaNA, Koloniën, 1814–1849, 2.10.01, inv.no. 2538.
300 Res.GG, January 8, 1830, no. 4. NL-HaNA, Koloniën, 1814–1849, 2.10.01, inv.no. 2833.
301 Res.GG, April 9, 1831, no. 24. NL-HaNA, Koloniën, 1814–1849, 2.10.01, inv.no. 2839.
302 Weber and Van Zanen, *Kunstenaar op Java*, 254–257. Another report can be found in a letter, dated July 15, 1832, from Abraham Botter (1791–1845) to Temminck. NL-LdnNAT, Wetcor_831_0004 and Wetcor_831_0005.
303 Mörzer Bruyns, *Met de Triton en Iris*, 345.

22 Antoine Maurevert

b. unknown (Pondicherry), d. unknown
Assistant: March 16, 1821–1823

MPCC, 351.
Mac Lean, J. "Gegevens over tekenaars van biologische voorwerpen in het voormalige Nederlandsch Indië." *Scientiarum historia* 15 (1973): 101–114.

Antoine Maurevert, sometimes spelled as Meaurevert, or even Morvert, is described by Jannes Theodorus Bik as a 'polite, quiet young man of Pondicherry, Bengal.'[304] Much about his life, including his birth and death dates, is unknown. Boie states that he arrived with Diard and Duvaucel ('venu avec MM Diard et Duvaucel').[305] Diard arrived in the Indies on April 7, 1820, so Maurevert must have been there since that date. Upon the request of Kuhl and Van Hasselt, Maurevert was appointed as draughtsman at the department of agriculture, arts, and sciences on March 16, 1821, at a salary of f120 per month, presumably with the express goal of working with them.[306] He received a raise to f150 on July 30, 1821, with a further raise to f200 on June 11, 1822.[307] He accompanied Van Hasselt on the 1823 Banten expedition.[308]

On February 9, 1824, he was appointed to accompany Blume on an expedition, but was already released on February 26, 1824, since his health would not allow him to travel.[309] In 1825, Maurevert had undertaken a journey to Bengal and Ceylon, together with John Hunt (also originally from India), and had returned with opium seeds and silkworms. As a result of this, on December 31, 1825, he was appointed as 'cultivation overseer', directly under Diard at a salary of f200.[310] This dual appointment of Diard and Maurevert by Van der Capellen on his last day in office never amounted to much,

304 'Een braaf, stil jong mensch, van Pondischrie in Bengalen.' Bik, "Aanteekeningen op eene reis," 271.

305 Letter Boie to Temminck, June 22, 1826, NCAO, 1200_001. Boie mistakenly assumes that Duvaucel had also been in the Indies, but Diard and Duvaucel had already split before the former's arrival in the Indies.

306 Res.GG, March 16, 1821, no. 12. NL-HaNA, Koloniën, 1814–1849, 2.10.01, inv.no. 2459.

307 Res.GG, July 30, 1821, no. 16. NL-HaNA, Koloniën, 1814–1849, 2.10.01, inv.no. 2461; Res.GG, June 11, 1822, no. 4. NL-HaNA, Koloniën, 1814–1849, 2.10.01, inv.no. 2776.

308 See Bik, "Aanteekeningen op eene reis."

309 Res.GG, February 9, 1824, no. 1. NL-HaNA, Koloniën, 1814–1849, 2.10.01, inv.no. 2474; Res.GG, February 17, no. 2. NL-HaNA, Koloniën, 1814–1849, 2.10.01, inv.no. 2474; Res.GG, February 26, 1824, no. 24. NL-HaNA, Koloniën, 1814–1849, 2.10.01, inv.no. 2475.

310 Res.GG, December 31, 1825, no. 21. NL-HaNA, Koloniën, 1814–1849, 2.10.01, inv.no. 2791. In Dutch, 'opziener der cultures.' Diard was appointed as 'inspecteur der cultures' (cultivation inspector).

due to the fact that no clear instructions were present, and the goal of their posts got lost in transition between Van der Capellen and Du Bus. On May 23, 1826, Maurevert even requested to be honourably discharged because the government did not seem to want to give him a practical task.[311] On September 9, 1826, Maurevert was granted a two-year leave on an allowance of f25 per month in order to travel to Bengal as long as he remained available should the government call upon him.[312] On June 6, 1827, the positions of cultivation overseer and cultivation inspector were dissolved and both Maurevert and Diard were honourably discharged.[313] Presumably, Maurevert was in Bengal at the time. It is unknown what further became of him.

23 Salomon Müller (1804–1863)

b. April 7, 1804 (Heidelberg), d. December 29, 1863 (Freiburg im Breisgau)
Assistant: December 5, 1823–June 3, 1833
Member: June 3, 1833–July 19, 1850

ENI, vol. 2, 792. | FHA, 266 | MPCC, 376–377 | OM, 251 | T: 315–316

Glaubrecht, Matthias. 'The 'Experience' of Nature: From Salomon Müller to Ernst Mayr, or the Insights of Travelling Naturalists Toward a Zoological Geography and Evolutionary Biology.' In *Die Entstehung biologischer Disziplinen II: Beiträge zur 10. Jahrestagung der DGGTB in Berlin 2001*, edited by Uwe Hoßfeld and Thomas Junker. Verhandlungen zur Geschichte und Theorie der Biologie, 9. VWB – Verlag für Wissenschaft und Bildung, 2001.

Jacobs, Hans J. 'Salomon Müller – ein unterschätzter deutscher Naturforscher.' *Sekretär* 22 (2022): 21–56.

Keurs, Pieter ter. 'Agency, Prestige and Politics: Dutch Collecting Abroad and Local Responses.' In *Unpacking the Collection: Networks of Material and Social Agency in the Museum*, edited by Sarah Byrne, Anne Clarke, Rodney Harrison and Robin Torrence. One World Archaeology. Springer, 2011.

Wiedtemann, Leonhard. 'Korrespondenz.' *Didaskalia: Blätter für Geist, Gemüth und Publizität*, no. 240 (1837): 4.

Salomon Müller was born on April 7, 1804, in Heidelberg, Germany, to Johann Gottlieb Müller and Anna Elisabetha Helfrich. Müller's father was a saddler and innkeeper at

311 Res.GG, May 23, 1826, no. 5. NL-HaNA, Koloniën, 1814–1849, 2.10.01, inv.no. 2793.

312 Res.GG, August 29, 1826, no. 1. NL-HaNA, Koloniën, 1814–1849, 2.10.01, inv.no. 2795; Res.GG, September 9, 1826, no. 1. NL-HaNA, Koloniën, 1814–1849, 2.10.01, inv.no. 2496.

313 Res.GG, June 6, 1827, no. 8. NL-HaNA, Koloniën, 1814–1849, 2.10.01, inv.no. 2505.

Sattler Müller's Garten (Wirtschaft zum Riesenstein).[314] Boie and Macklot (who both studied in Heidelberg) at one point visited the inn and were served by the young Salomon Müller. When Müller showed interest in their hunting discussions, they started talking and found out that Müller had become a decent preparator without any training.[315] Boie invited Müller to work as preparator for the zoological cabinet. During his time in Leiden, Boie did not forget about Müller and when it was time to choose an assistant for their expedition to the Indies, Müller was appointed as assistant-preparator of the Natuurkundige Commissie on December 5, 1823.[316]

On December 21, 1825, Boie, Macklot, Müller, and Van Oort left the Netherlands on the ship *Dijkzigt*,[317] and arrived in Batavia on June 6, 1826.[318] Only on February 27, 1827, was the party formally appointed in the Indies: Boie and Macklot at a salary of f500, Müller and Van Oort at f200. Müller's salary was raised to f250 on August 3, 1827.[319] After their return from the New Guinea/Timor expedition,[320] Müller received another raise of f50 to f300 and was awarded the title 'curator of natural history collections' ('conservateur der natuurkundige verzamelingen').[321] Müller spent 1830 to mid-1833 on Java, together with the rest of the Commissie, travelling and collecting all the while. After Macklot's death in May 1832, the Commissie continued with the planned expedition along the northern coast of Java. This expedition was cut short (and thus never reached Madura or Bali) by the resolution of May 18, 1833, that ordered Müller and Van Oort to join a planned military expedition to Sumatra.[322] Bürger and Korthals were also to join them.

The quartet departed on June 4, 1833, on the ship *Hercules*, to Padang, where they arrived on June 16, 1833.[323] Just before this expedition, directing member Diard had requested Müller and Van Oort to be appointed as members and given the appropriate salary of f500. This request was referred to the Netherlands with a positive recommendation by the governor-general.[324] Without awaiting a reply, Müller and Van Oort received a raise of f150 to f450 on June 3, 1833.[325] Because of their rushed departure,

314 Von Leonhard, *Fremdenbuch für Heidelberg*, 116. See also http://www.s197410804.online.de/ABC/ABCGasthaeuser.htm.

315 Veth, *Overzicht van hetgeen gedaan is*, 41–42.

316 KB, December 5, 1823, no. 84. NL-HaNA, Staatssecretarie, 2.02.01, inv.no. 1831.

317 Veth, *Overzicht van hetgeen gedaan is*, 45.

318 Weber and Van Zanen, *Kunstenaar op Java*, 55.

319 Res.GG, August 3, 1827, no. 2. NL-HaNA, Koloniën, 1814–1849, 2.10.01, inv.no. 2507.

320 See the biography of Macklot for a more detailed report on the New Guinea/Timor expedition.

321 Res.GG, January 8, 1830, no. 4. NL-HaNA, Koloniën, 1814–1849, 2.10.01, inv.no. 2830.

322 Res.GG, May 18, 1833, no. 2. NL-HaNA, Koloniën, 1814–1849, 2.10.01, inv.no. 2550.

323 Weber and Van Zanen, *Kunstenaar op Sumatra*, 37–38.

324 Res.GG March 30, 1833, no. 7. NL-HaNA, Koloniën, 1814–1849, 2.10.01, inv.no. 2846.

325 Res.GG, June 3, 1833, no. 23. NL-HaNA, Koloniën, 1814–1849, 2.10.01, inv.no. 2847.

Müller was given permission to return to Java in order to pick up the required books and instruments.[326] He left Sumatra on October 17, 1833, again on the *Hercules*,[327] to return on January 9, 1834, on the bark *Boeroeng*, with Van Gelder and Overdijk.[328] Because the king had in the meantime decided that a raise for Müller and Van Oort could only be approved if the funds could be found within the existing budget,[329] Müller also returned with the bad news that their raise of f150 had been rescinded as of January 1, 1834.[330]

During the first part of the Sumatra expedition, the Natuurkundige Commissie was limited in their movements and mainly followed the army without being able to collect much. After Müller's return, it became gradually possible for them to undertake expeditions and collect again. When Van Oort's death (in September 1834) created space in the budget, Müller received a retrospective raise of f150 to f450 from October 1, 1834.[331] A further retrospective raise of f50 as per January 1, 1835, finally gave him the salary that was in accordance with his status.[332] After Van Oort's death, Müller, Korthals, Overdijk, and Van Gelder continued travelling and collecting on Sumatra (Bürger had already left in January 1834), until they were recalled on November 14, 1835.[333] Müller, Korthals, and Van Gelder departed from Padang on January 9, 1836, travelled via Bengkulu[334] and arrived in Batavia on January 28, 1836, on the brig *Harriet*.[335]

Shortly after their return to Java, it was decided that Müller and Korthals, along with Horner, should undertake one final expedition before returning to the Netherlands in February 1837. The destination would be Borneo and they were to go there from May to November 1836.[336] On March 12, 1836, an additional assignment was given to them: they needed to verify the expedition reports of army officer Henrici and recommend to the governor-general whether these expeditions should continue.[337] Due to the delayed departure of the Borneo expedition (they only left Batavia on July 11, 1836,[338]

326 Res.GG, November 15, 1833, no. 14. NL-HaNA, Koloniën, 1814–1849, 2.10.01, inv.no. 2849.

327 Weber and Van Zanen, *Kunstenaar op Sumatra*, 147.

328 Weber and Van Zanen, *Kunstenaar op Sumatra*, 151.

329 KB, December 25, 1833, no. 11. NL-HaNA, Staatssecretarie, 2.02.01, inv.no. 3873.

330 Res.GG, December 12, 1833, no. 4. NL-HaNA, Koloniën, 1814–1849, 2.10.01, inv.no. 2849.

331 Res.GG, March 3, 1835, no. 8. NL-HaNA, Koloniën, 1814–1849, 2.10.01, inv.no. 2853.

332 Res.GG, August 31, 1835, no. 29. NL-HaNA, Koloniën, 1814–1849, 2.10.01, inv.no. 2856.

333 Res.GG, November 14, 1835, no. 2. NL-HaNA, Koloniën, 1814–1849, 2.10.01, inv.no. 2857.

334 Dagboek Korthals, January 9, 1836, NL-LdnNAT, Korthals_5_004.

335 *Javasche Courant*, no. 9 (January 30, 1836): 4.

336 Res.GG, February 27, 1836, no. 12. NL-HaNA, Koloniën, 1814–1849, 2.10.01, inv.no. 2858.

337 Res.GG, March 12, 1836, no. 1. NL-HaNA, Koloniën, 1814–1849, 2.10.01, inv.no. 2859.

338 *Javasche Courant*, no. 56 (July 13, 1836): 4. The passengers of the schooner *Anadyomene* are not mentioned, but Müller mentions the name of their ship in his travel report, even though he mentions July 10 as their departure date: Susanna, *Verhandelingen*, vol. 3, 324.

and arrived in Banjarmasin on July 28, 1836),[339] they were allowed an additional two to three months on November 16, 1836.[340] Müller, Korthals, and Horner departed from Banjarmasin on December 17, 1836, on the brig *Barket* and arrived in Batavia on January 9, 1837.[341]

Müller and Korthals were awarded free transport for themselves and their collections to the Netherlands on March 31, 1837.[342] Their baggage included 76 crates, two suitcases and seven barrels of specimens, with an additional thirteen crates with living plants.[343] On April 12, 1837, Müller (with his daughter Marie Elisabetha Petronella Müller, b. October 5, 1830)[344] and Korthals departed from Batavia on the ship *Prinses Marianne*.[345] They arrived in Hellevoetsluis on August 21, 1837, and departed to Leiden the next day.[346] Müller was put on half-pay and given time to process his collections and write up his reports.[347] His report on the Borneo expedition led to the cancellation of any further expeditions by Henrici.[348] Müller received his doctorate from Heidelberg University on December 21, 1837,[349] and was appointed as Knight in the Order of the Dutch Lion on June 9, 1839.[350]

Müller became one of the main writers (with Schlegel) of the zoological volume of the *Verhandelingen* and the sole writer of the ethnographical volume. After the dissolution of the Natuurkundige Commissie on April 17, 1850,[351] Müller was honourably discharged on July 19, 1850, and given an annual pension of f2,000 from January 1, 1850.[352] He married Amalie Wilhelmine Rummer (b. 1823) in Heidelberg on April 12, 1850, and spent the rest of his days in Freiburg im Breisgau. The couple had two

339 Dagboek Korthals, July 28, 1836. NL-LdnNAT, Korthals_5_050. Even though Müller confirms this arrival date in Susanna, *Verhandelingen*, vol. 3, 324. He later reports that they arrived in the first days of August: Susanna, *Verhandelingen*, vol. 3, 338.

340 Res.GG, November 16, 1836, no. 7. NL-HaNA, Koloniën, 1814–1849, 2.10.01, inv.no. 2565.

341 *Javasche Courant*, no. 5 (January 18, 1837): 6.

342 Res.GG, March 31, 1837, no. 7. NL-HaNA, Koloniën, 1814–1849, 2.10.01, inv.no. 2568.

343 Res.GG, April 19, 1837, no. 1. NL-HaNA, Koloniën, 1814–1849, 2.10.01, inv.no. 2569.

344 Jacobs, "Salomon Müller," 49.

345 *Javasche Courant*, no. 30 (April 15, 1837): 11. The report writes 'S. Muller en dochtertje' ('S. Muller and little daughter'). This is the only contemporary mention of Müller's daughter that I have seen.

346 Dagboek Korthals, August 21, 1837. NL-LdnNAT, Korthals_5_361 and Korthals_5_362; *Leydsche Courant*, no. 103 (August 28, 1837): 1.

347 Res.GG, March 31, 1837, no. 7. NL-HaNA, Koloniën, 1814–1849, 2.10.01, inv.no. 2568.

348 Res.GG, November 3, 1837, no. 14. NL-HaNA, Koloniën, 1814–1849, 2.10.01, inv.no. 2576.

349 Jacobs, "Salomon Müller," 45.

350 KB, June 9, 1839, no. 62. NL-HaNA, Staatssecretarie, 2.02.01, inv.no. 4488.

351 KB, April 17, 1850, no. 14. NL-HaNA, KdK 1841–1897, 2.02.04, inv.no. 609.

352 Veth, *Overzicht van hetgeen gedaan is*, 124; KB, July 19, 1850, no. 10. NL-HaNA, KdK 1841–1897, 2.02.04, inv.no. 624.

children: Carl Friedrich (b. 1852) and Viktoria (b. 1853).[353] In 1860, Müller sold his ethnographic collection to 's Rijks Japansch Museum (now Wereldmuseum Leiden) for f4,000.[354] Müller died in Freiburg im Breisgau on December 29, 1863.

24 Carolus Augustus Oehler (1818–1885)

b. December 27, 1818 (Vienna), d. August 18, 1885
Assistant: April 1, 1848–March 1, 1851

Carolus Augustus Oehler was born on December 27, 1818, in Vienna to Franz Oehler and Rosine Jiedeker. He was a sergeant in the Netherlands Indies army and received an honourable mention for his actions on Bali in June 1846.[355] On October 30, 1848, he was appointed as topographical artist for the Natuurkundige Commissie as of April 1, 1848, at a salary of f100. Oehler had already been working with Schwaner since that date to assist him in drawing maps.[356] When Schwaner reported in March 1850 that his maps of Borneo were completed, Oehler was allowed to continue as topographical artist for the Natuurkundige Commissie for the whole of 1850, but he was temporarily seconded to the mapmaking department of the corps of engineers, where he was to copy the Borneo maps for use by the local resident.[357] When the Natuurkundige Commissie was dissolved, Oehler was appointed as artist at the committee for the improvement of naval maps as of March 1, 1851.[358] Oehler turned mapmaking into quite a career and finished his career as lithographer at the Netherlands Indies navy.[359] In 1863, he published *De schoolkaart van Java en den Indischen Archipel*, with F. Cronenberg.[360] Oehler died on August 18, 1885, in the Indies.[361]

---

353 Jacobs, "Salomon Müller," 48.
354 See Mörzer Bruyns, *Met de Triton en Iris*, 345–353 (Bijlage 3).
355 https://proxy.archieven.nl/2231/CE9360AEDA6D4E1BA0CF666B372E2F0A.
356 Res.GG, October 30, 1848, no. 6. NL-HaNA, Koloniën, 1814–1849, 2.10.01, inv.no. 2711.
357 Res.GG, March 12, 1850, no. 7. NL-HaNA, Koloniën, 1850–1900, 2.10.30, inv.no. 7107.
358 Res.GG, April 10, 1850, no. 2. NL-HaNA, Koloniën, 1850–1900, 2.10.30, inv.no. 7120.
359 *Bataviaasch Handelsblad*, no. 281 (November 27, 1872): 4. See Oehler's registry card for more details on his career. NL-HaNA, Koloniën / Stamboeken Burgerlijke Ambtenaren, 2.10.36.22, inv.no. 902, f. 561.
360 *Java-Bode*, no. 76 (September 13, 1863): 2. WorldCat suggests this was titled *Schoolkaart van Nederlandsch Oost-Indië*, see https://www.worldcat.org/title/1249953290.
361 *Java-Bode*, no. 193 (August 19, 1885): 1.

25 Pieter van Oort (1804–1834, Figure 12)

b. October 10, 1804 (Utrecht), d. September 2, 1834 (Padang)
Assistant: April 16, 1825–June 3, 1833
Member: June 3, 1833–September 2, 1834

BWN, vol. 14, 140–142 | BPN 34502386 | MPCC, 394 | BB, 322–323 | NSTO, 29–30

Mac Lean, J. "Gegevens over tekenaars van biologische voorwerpen in het voormalige Nederlandsch Indië." *Scientiarum historia* 15 (1973): 101–114.

Weber, Andreas, and Sylvia van Zanen. *Kunstenaar op Java: De reisdagboeken en natuurtekeningen van Pieter van Oort (1825–1833)*. Werken uitgegeven door de Linschoten-Vereeniging, 119. Walburg, 2021.

Weber, Andreas, and Sylvia van Zanen. *Kunstenaar op Sumatra: De reisdagboeken en natuurtekeningen van Pieter van Oort (1833–1834)*. Werken uitgegeven door de Linschoten-Vereeniging, 120. Walburg, 2021.

Weber, Andreas, and Sylvia van Zanen. "Matters of Perspective: Local Visual Expertise and Natural History Drawings in Java, 1820–1850." *Wacana* 23, no. 1 (2022): 38–63.

Zanen, Sylvia van, and Andreas Weber. "Schreeuwende apen, zingende vogels en fluitende kogels: Natuuronderzoek op Sumatra in oorlogstijd (1833–1834)." *Indische letteren* 36, no. 4 (2021): 279–300.

Pieter van Oort was born on October 10, 1804, in Utrecht to Hendrik van Oort (1775–1847) and Adriana Stekelenburg (1780–1832). Van Oort was educated as an artist and specialised in natural history drawings. He frequently worked with Theodoor Gerard van Lidth de Jeude (1788–1853), through whom his work was introduced to Temminck. When the Commissie was in need of a new artist, Van Oort's name came up and he was appointed as artist-assistant on April 16, 1825.[362] On December 21, 1825, Boie, Macklot, Müller, and Van Oort left the Netherlands on the ship *Dijkzigt*,[363] and arrived in Batavia on June 6, 1826.[364] Only on February 27, 1827, was the party formally appointed in the Indies: Boie and Macklot at a salary of f500, Müller and Van Oort at f200. At his own request Van Oort's salary was raised to f250 on May 12, 1827.[365] After their return from

362 KB, April 16, 1825, no. 120. NL-HaNA, Staatssecretarie, 2.02.01, inv.no. 2209.

363 Veth, *Overzicht van hetgeen gedaan is*, 45.

364 Weber and Van Zanen, *Kunstenaar op Java*, 55.

365 Res.GG, May 12, 1827, no. 7. NL-HaNA, Koloniën, 1814–1849, 2.10.01, inv.no., 2504; see Res.GG, April 24, 1827, no. 6. NL-HaNA, Koloniën, 1814–1849, 2.10.01, inv.no. 2800, for a discussion regarding Van Oort's salary.

the New Guinea/Timor expedition,[366] Van Oort received another raise of f50 to f300.[367] Van Oort spent 1830 to mid-1833 on Java, together with the rest of the Commissie, travelling and collecting all the while. After Macklot's death in May 1832, Van Oort was appointed as administrator of the Commissie in his place in July 14, 1832.[368] The Commissie continued with the planned expedition along the northern coast of Java, but it was cut short (and thus never reached Madura or Bali) by the resolution of May 18, 1833, that ordered Müller and Van Oort to join a planned military expedition to Sumatra.[369] Bürger and Korthals were also to join them. The quartet departed on June 4, 1833, on the ship *Hercules*, to Padang, where they arrived on June 16, 1833.[370]

Just before this expedition, directing member Diard had requested Müller and Van Oort to be appointed as members and given the appropriate salary of f500. This request was referred to the Netherlands with a positive recommendation by the governor-general.[371] Without awaiting a reply, Müller and Van Oort received a raise of f150 to f450 on June 3, 1833.[372] The king, however, decided that a raise for Müller and Van Oort could only be approved if the funds could be found within the existing budget.[373] As a result, the raise of f150 was rescinded as of January 1, 1834.[374] During the first part of the Sumatra expedition, the Natuurkundige Commissie was limited in their movements and mainly followed the army without being able to collect much. In 1834, it became gradually possible for them to undertake expeditions and collect again. On September 2, 1834, Van Oort died in Padang (Sumatra). After Van Oort's death, Diard requested a pension to be awarded to his father.[375] This request was denied in the Netherlands.[376] Van Oort left behind a son on Java.[377]

---

366 See the biography of Macklot for a more detailed report on the New Guinea/Timor expedition.

367 Res.GG, January 8, 1830, no. 4. NL-HaNA, Koloniën, 1814–1849, 2.10.01, inv.no. 2830.

368 Res.GG, July 14, 1832, no. 3. NL-HaNA, Koloniën, 1814–1849, 2.10.01, inv.no. 2844.

369 Res.GG, May 18, 1833, no. 2. NL-HaNA, Koloniën, 1814–1849, 2.10.01, inv.no. 2550.

370 Weber and Van Zanen, *Kunstenaar op Sumatra*, 37–38.

371 Res.GG March 30, 1833, no. 7. NL-HaNA, Koloniën, 1814–1849, 2.10.01, inv.no. 2846.

372 Res.GG, June 3, 1833, no. 23. NL-HaNA, Koloniën, 1814–1849, 2.10.01, inv.no. 2847.

373 KB, December 25, 1833, no. 11. NL-HaNA, Staatssecretarie, 2.02.01 inv.no. 3873.

374 Res.GG, December 12, 1833, no. 4. NL-HaNA, Koloniën, 1814–1849, 2.10.01, inv.no. 2849.

375 Res.GG, December 3, 1834, no. 8. NL-HaNA, Koloniën, 1814–1849, 2.10.01, inv.no. 2853.

376 Res.GG, September 22, 1835, no. 30. NL-HaNA, Koloniën, 1814–1849, 2.10.01, inv.no. 2856.

377 Born on September 8, 1831, to a local Javanese woman called Sipit. It is unknown what became of both mother and child after Van Oort's death. See Weber and Van Zanen, *Kunstenaar op Java*, 140, 268.

26 Bartholomeus Nicolaas Overdijk (1807–[1851])[378]

b. August 24, 1807 (Leiden), d. [March 20, 1851]
Assistant: July 25, 1830–January 19, 1840
Assistant: December 21, 1848–August 23, 1849

Bartholomeus Nicolaas Overdijk was born on August 24, 1807, in Leiden to Johannes Overdijk en Johanna de Groot.[379] Nothing is known about his youth. He started working as preparator at 's Rijks Museum in 1824, a position that would lead to his appointment as assistant to Korthals on July 25, 1830.[380] He arrived in Batavia, along with Korthals and Van Gelder, on April 13, 1831, on the ship *Jonge Adriana*.[381] He was given a salary of f200 from the day of his arrival.[382] Overdijk spent his first two years on Java. During the uprising in Krawang in May 1832, which cost Macklot his life, Overdijk lost many of his possessions in the fire set by the insurgents on May 9, 1832.[383]

On December 5, 1833, Overdijk was ordered to join the rest of the Commissie on Sumatra,[384] where he arrived on January 9, 1834, together with Müller and Van Gelder.[385] Overdijk returned to Java in the last quarter of 1835 because of ill-health. His return was approved on November 14, 1835, in the same resolution that ordered the return to Java of Müller and Korthals.[386] On May 3, 1837, Overdijk was appointed to accompany Horner on his expedition to Sumatra.[387] They departed to Padang on the ship *Sumatra* on June 13, 1837.[388] Horner died on December 7, 1838, leaving Overdijk on his own on Sumatra. Overdijk was recalled to Java on January 10, 1839, in order to accompany Forsten on his impending expedition to Celebes. Overdijk was awarded

378 Overdijk must not be confused with Cornelis Overdijk who worked at 's Rijks Museum from 1824 to 1838 and then left for the Indies and occasionally sent back specimens to 's Rijks Museum. See Holthuis, *Rijksmuseum van Natuurlijke Historie*, 38.

379 The baptismal record for Overdijk gives his name and the names of his parents in Latin: Bartholomaeus Nicolaus, Joannes, and Joanna. See Doopboek van de kerk De zon (RK) te Leiden, 3 oktober 1683–1811, Erfgoed Leiden en Omstreken, Leiden, 2004, inv.no. 305.

380 KB, July 25, 1830, no. 40. NL-HaNA, Staatssecretarie, 2.02.01, inv.no. 3437.

381 *Javasche Courant*, no. 46 (April 16, 1831): 3.

382 Res.GG, May 21, 1831, no. 19. NL-HaNA, Koloniën, 1814–1849, 2.10.01, inv.no. 2839.

383 Veth, *Overzicht van hetgeen gedaan is*, 72. Res.GG, June 14, 1832, no. 4. NL-HaNA, Koloniën, 1814–1849, 2.10.01, inv.no. 2546.

384 Res.GG, December 5, 1833, no. 4. NL-HaNA, Koloniën, 1814–1849, 2.10.01, inv.no. 2849.

385 Weber and Van Zanen, *Kunstenaar op Sumatra*, 151.

386 Res.GG, November 14, 1835, no. 5. NL-HaNA, Koloniën, 1814–1849, 2.10.01, inv.no. 2857.

387 Res.GG, May 3, 1837, no. 1. NL-HaNA, Koloniën, 1814–1849, 2.10.01, inv.no. 2570.

388 *Javasche Courant*, no. 47 (June 14, 1837): 5.

f600 to arrange for packing up and transporting his and Horner's papers and collections to Java.[389] He arrived on Java on March 5, 1839, on the bark *The Corsair*.[390]

Overdijk met Forsten in Buitenzorg but did not favourably impress him: 'he has a wilful, proud, and lazy character, is unable to prepare better than the indigenous hunters and is entirely without scientific knowledge.'[391] In 1839, Overdijk was appointed by the Bataviaasch Genootschap as preparator of the natural history department of their museum. He received an additional salary of f50 from the Genootschap in addition to his government salary of f200.[392] When he was appointed to accompany Forsten to Celebes at a salary of f200 on December 22, 1839,[393] Overdijk demanded a salary of f300 and a travel compensation of f75 and invited the government to discharge him should they be disinclined to pay this. The governor-general duly complied and Overdijk was honourably discharged from his position with the Natuurkundige Commissie on January 19, 1840.[394]

He continued in his position at the museum of the Genootschap, who took Overdijk on at a salary of at least f200.[395] In 1844, after Overdijk was notified that his museum position would be dissolved after all naturalia had been shipped to Leiden, he rather impolitely indicated his displeasure to Van Hoëvell, chairman of the Genootschap, and starting wreaking havoc with the remaining collections that were to be prepared for auction. As a result, Overdijk was put on half-pay for three months and referred back to the government.[396] On October 15, 1844, he was awarded a redundancy payment of f66.66 until the end of March 1845,[397] which was prolonged on April 4, 1845, until the end of September 1845,[398] and again on October 17, 1845, until the end of March 1846.[399]

Unfortunately, this was not a good period in Overdijk's life. Financial difficulties and alcohol abuse even led to his six children being put in the Roman Catholic orphanage in Semarang.[400] On December 21, 1848, there was finally some good news for Overdijk.

389 Res.GG, January 10, 1839, no. 4. NL-HaNA, Koloniën, 1814–1849, 2.10.01, inv.no. 2590.
390 *Javasche Courant*, no. 20 (March 9, 1839): 6.
391 Kort verslag van het in Indie verrigtte van 14 December 1838 tot April 1840. NCAO, 1076_003.
392 Kort verslag van het in Indie verrigtte van 14 December 1838 tot April 1840. NCAO, 1076_005.
393 Res.GG, December 22, 1839, no. 4. NL-HaNA, Koloniën, 1814–1849, 2.10.01, inv.no. 2601.
394 Res.GG, January 19, 1840, no. 1. NL-HaNA, Koloniën, 1814–1849, 2.10.01, inv.no. 2602.
395 Groot, *Van Batavia naar Weltevreden*, 298, 326.
396 See Groot, *Van Batavia naar Weltevreden*, 326, for the full story.
397 Res.GG, October 15, 1844, no. 7. NL-HaNA, Koloniën, 1814–1849, 2.10.01, inv.no. 2662.
398 Res.GG, April 15, 1845, no. 24. NL-HaNA, Koloniën, 1814–1849, 2.10.01, inv.no. 2668.
399 Res.GG, October 17, 1845, no. 23. NL-HaNA, Koloniën, 1814–1849, 2.10.01, inv.no. 2674.
400 Res.GG, April 18, 1847, no. 23. NL-HaNA, Koloniën, 1814–1849, 2.10.01, inv.no. 2693. The children's names are Caecilia Catharina (age 11), Adriana (age 10), Johannes (age 7), Hendrik (age 5), Petrus (age 4), Albertus (age 2). Their mother was an indigenous woman from

At his own request, he was appointed as assistant to Schwaner for a period of six months at a salary of f200, with the prospect of a possible permanent appointment.[401] Unfortunately, his request for a permanent appointment was denied on August 23, 1849, mostly because he had been too ill in the previous six months to be of any value to Schwaner.[402] On December 18, 1849, Overdijk was awarded a social security benefit of f25 until further notice. Overdijk reportedly died on March 20, 1851.[403]

27 Auguste van Pers (1815–1871)

b. June 15, 1815 (Tournai/Doornik), d. May 27, 1871
Assistant: May 19, 1846–April 27, 1847

BPN 57809199 | M, 56–58 | BB, 326–327 | NSTO, 32

Auguste van Pers was born in Tournai/Doornik (Belgium) on June 15, 1815. He departed for the Indies on November 1, 1836. Even though he spent most of his life in the Indies, his association with the Natuurkundige Commissie was short-lived. On May 19, 1846, he was appointed to assist Schwaner as artist as replacement for Von Gaffron at a salary of f150 and he received free transport to Banjarmasin (Borneo).[404] Van Pers left Batavia on August 2, 1846, on the bark *Balheer*,[405] arrived in Semarang on August 8, 1846,[406] and in Surabaya on August 18, 1846.[407] Van Pers presumably reached Banjarmasin in late August or early September, 1846, where he joined Schwaner.[408] He arrived back in Banjarmasin on February 25, 1847. Due to health problems, Van Pers was unable to accompany Schwaner further on his travels and his honourable discharge and permission to return to Java as granted by the resident was confirmed by the governor-general on April 27, 1847.[409] Not many drawings made by Van Pers during his association

Nias, named Pekiti, as becomes apparent from Albert's registry card: NL-HaNA, Koloniën / Stamboeken Burgerlijke Ambtenaren, 2.10.36.22, inv.no. 901, f. 436.

401 Res.GG, December 21, 1848, no. 3. NL-HaNA, Koloniën, 1814–1849, 2.10.01, inv.no. 2713.

402 Res.GG, August 23, 1849, no. 9. NL-HaNA, Koloniën, 1814–1849, 2.10.01, inv.no. 2721.

403 http://www.nuotrauka.nl/families/F367.html. I have not found any independent confirmation for this date.

404 Res.GG, May 19, 1846, no. 6. NL-HaNA, Koloniën, 1814–1849, 2.10.01, inv.no. 2682.

405 *Javasche Courant*, no. 62 (August 5, 1846): 9.

406 *Javasche Courant*, no. 66 (August 19, 1846): 14.

407 *Javasche Courant*, no. 68 (August 26, 1846): 11.

408 His ship *Balheer* left Surabaya on August 22, 1846, with destination Banjarmasin via Sumenep (Madura). See *Javasche Courant*, no. 70 (September 2, 1846): 7.

409 Res.GG, April 27, 1847, no. 33. NL-HaNA, Koloniën, 1814–1849, 2.10.01, inv.no. 2693.

with Schwaner are known.[410] Van Pers continued living in the Indies and held various positions as artist, contributing to the illustrated publication *Nederlandsch Oost-Indische Typen*.[411] He died on May 27, 1871.

28 Gerrit van Raalten (1797–1829)

b. August 20, 1797 (Harderwijk), d. April 17, 1829 (Timor)
Assistant: May 2, 1820–April 17, 1829

BWN, vol. 16, 6–9 | BPN 01724393 | BB, 328–329
Anonymus. "Gerrit Van Raalten." *Handelingen en geschriften van het Indisch Genootschap* 5 (1858): 88–91.

Gerrit van Raalten was born on August 20, 1797, in Harderwijk to Reijer van Raalten and Cornelia Schutte.[412] Nothing is known about his youth. He was appointed as assistant to Kuhl and Van Hasselt along with Keultjes on May 2, 1820.[413] Van Raalten's credentials consisted of three years' experience as preparation assistant in the zoological cabinet of Theodoor Gerard van Lidth de Jeude (1788–1853). Kuhl, Van Hasselt, Van Raalten, and Keultjes left the Netherlands on July 11, 1820,[414] and arrived in Batavia on December 22, 1820, on the ship *Nordloh*.[415] Van Raalten was appointed, in accordance with the royal decree from the Netherlands, as assistant preparator to Kuhl and Van Hasselt as of January 1, 1821, at a salary of f200.[416] On July 30, 1821, he received a raise of f50,[417] and on June 11, 1822, his salary was set at f300.[418]

After the death of Kuhl, Keultjes, and Van Hasselt, Van Raalten was made responsible for copying the Kuhl/Van Hasselt papers and shipping their collections to the

410 Maronier, *Pictures of the Tropics*, 56–58, lists ten drawings dated 1845–1847, with two of those depicting birds. Scans of these drawings are available in the Southeast Asian & Caribbean Images (KITLV) digital collection of Leiden University (under the filter author/creator 'Pers, Auguste van.') via https://digitalcollections.universiteitleiden.nl/.

411 Van Pers and Hageman, *Nederlandsch Oost-Indische typen*.

412 Harderwijk, Doopboek – Nederduits Gereformeerde gemeenten, 1793–1811. NL-ArnGA, Retroacta Burgerlijke Stand / Doop-, Trouw- en Begraafboeken, Gelderland, 0176, inv.no. 832.

413 KB, May 2, 1820, no. 10. NL-HaNA, Staatssecretarie, 2.02.01, inv.no. 988.

414 Veth, *Overzicht van hetgeen gedaan is*, 25; Klaver, *Inseparable Friends*, 18.

415 *Bataviasche Courant*, no. 53 (December 30, 1820): 1.

416 Res.GG, January 23, 1821, no. 12A. NL-HaNA, Koloniën, 1814–1849, 2.10.01, inv.no. 2774.

417 Res.GG, July 30, 1821, no. 16. NL-HaNA, Koloniën, 1814–1849, 2.10.01, inv.no. 2461.

418 Res.GG, June 11, 1822, no. 4. NL-HaNA, Koloniën, 1814–1849, 2.10.01, inv.no. 1776.

Netherlands. He was assisted by Fievez de Malines.[419] Van Raalten finished preparing the Kuhl/Van Hasselt papers and collections for shipment on November 25, 1825.[420] It is not clear whether Van Raalten's salary continued in 1826, but he had the necessary means to financially support the newly arrived members and assistants.[421] Van Oort testifies in October 1826: 'We should surely have fallen into poverty, if the good Van Raalten had not taken care of us. Thanks be to this philanthropist.'[422] Van Raalten joined them on their first expedition in December 1826.[423] In February 1827, Van Raalten was appointed as administrator of the Natuurkundige Commissie at a salary of f300,[424] to be raised to f400 as of May 1, 1827, because of his exemplary work in shipping the Kuhl/Van Hasselt papers and collections (consisting of 90 crates) to the Netherlands.[425]

Van Raalten is probably most famous for his unfortunate encounter with a Javan rhinoceros on April 1, 1827.[426] Out on a hunting trip with some of their indigenous hunters, he followed a fresh rhinoceros trail. Unexpectedly, he stumbled upon the beast who charged and repeatedly threw him in the air. Only the brave action of one of the hunters chased the rhinoceros away, but Van Raalten was left seriously wounded. It took several months for him to fully recover from this ordeal.[427] Later in the year, he ended up on the bad side of commissioner-general Du Bus, who suspected him (probably wrongly) of complaining to the minister when the Natuurkundige Commissie was not allowed to join Diard on his 1826 expedition to Borneo.[428]

419 Res.GG, February 17, 1824, no. 30. NL-HaNA, Koloniën, 1814–1849, 2.10.01, inv.no. 2474.

420 Res.GG, November 25, 1825, no. 12. NL-HaNA, Koloniën, 1814–1849, 2.10.01, inv.no. 2489.

421 Boie, Macklot, Müller and Van Oort had arrived in June 1826.

422 'En wij zouden zeker in de grootste armoede vervallen zijn, had de goede Van Raalten niet voor ons gezorgd. Dank zij deze menschenvriend!' Weber and Van Zanen, *Kunstenaar op Java*, 58.

423 Weber and Van Zanen, *Kunstenaar op Java*, 59.

424 Res.GG, January 19, 1827, no. 33. NL-HaNA, Koloniën, 1814–1849, 2.10.01, inv.no. 2798; Res.GG, February 27, 1827, no. 35. NL-HaNA, Koloniën, 1814–1849, 2.10.01, inv.no. 2501.

425 Res.GG, April 10, 1827, no. 10. NL-HaNA, Koloniën, 1814–1849, 2.10.01, inv.no. 2800; Res.GG, April 30, 1827, no. 12. NL-HaNA, Koloniën, 1814–1849, 2.10.01, inv.no. 2503. See also Res.GG, May 22, 1827, no. 1. NL-HaNA, Koloniën, 1814–1849, 2.10.01, inv.no. 2504.

426 Van Oort suggests this happened on April 1 (Weber and Van Zanen, *Kunstenaar op Java*, 65), while Müller (Susanna, *Verhandelingen*, vol. 1, 186) mentions April 3. Given the fact that Van Oort actually was in the neighbourhood and the first European Commissie person to meet with Van Raalten, I give credence to the date he mentions in his diary.

427 Several accounts are available: Weber and Van Zanen, *Kunstenaar op Java*, 65–66; Susanna, *Verhandelingen*, vol. 3, 186–187; Anonymus, "Gerrit van Raalten." Probably two thirds of this last source is dedicated to a dramatic description of the attack.

428 Van Oort mentions this in Weber and Van Zanen, *Kunstenaar op Java*, 73.

When Macklot was ordered to join the expedition to New Guinea, he requested to be allowed to take the full Commissie, including Van Raalten, with him.[429] This request was granted and in March 1828[430] Van Raalten left with the Commissie on an expedition that would take them via Makassar (Celebes) and Ambon to New Guinea and Timor.[431] Van Raalten would never return. After the Commissie had been dropped off on Timor, they undertook various smaller expeditions. In 1829, Van Raalten and Macklot had been on an expedition to Atapupu. On the return trip, Van Raalten fell ill and died at sea on April 17, 1829.[432]

29 Carl Anton Ludwig Maria Schwaner (1817–1851)

b. February 16, 1817 (Mannheim), d. March 30, 1851 (Batavia)
Member: January 9, 1842–March 30, 1851

ENI, vol. 3, 729 | FHA, 294 | MPCC, 478–479 | M, 67–69 | ADB, vol. 33, 183–184 | NDB, vol. 23, 782–783

Anonymus. "Dr. C. A. L. M. Schwaner, natuurkundige." *Handelingen en geschriften van het Indisch Genootschap* 1 (1854): 154–155.

Anonymus. "Levensberigt van Dr. C. A. L. M. Schwaner." *Natuurkundig tijdschrift voor Nederlandsch Indië* 2 (1851): 189–191.

Anonymus. "Reis van Dr. Schwaner van Koeti naar Pontianak." *Tijdschrift voor Nederlandsch Indie* 10, no. 1 (1848): 164–166.

Helbig, Karl. "Carl Anton Ludwig Maria Schwaner: Ein Forscher in unbekannter Wildnis." *Übersee-Rundschau*, no. 1 (1951): 84–85.

Maesel, Markus A. "Ein Mannheimer in Indonesien – der Naturforscher Dr. Schwaner (1817–1851)." *Kina*, no. 1/05 (2005): 82–91.

Pfaff-Giesberg, Robert. "Ein Mannheimer erforschte Borneo." *Amtsblatt für den Stadtkreis Mannheim* 7, no. 1 (January 5, 1951): 3.

Carl Anton Ludwig Maria Schwaner was born on February 16, 1817, in Mannheim (Germany), to Philipp Jakob Schwaner and Clara Baader. Schwaner's father was an apothecary, and his maternal grandfather was keeper of the naturalist cabinet at the castle in Mannheim. Schwaner studied in Heidelberg and Bonn (medicine). He

429 Res.GG, January 29, 1828, no. 1. NL-HaNA, Koloniën, 1814–1849, 2.10.01, inv.no. 2809.

430 Mörzer Bruyns, *Met de Triton en Iris*, 33.

431 See the biography of Macklot for a more detailed overview of the expedition.

432 *Javasche Courant*, no. 75 (June 27, 1829, Bijvoegsel): 1; Veth, *Overzicht van hetgeen gedaan is*, 67–68; Susanna, *Verhandelingen*, vol. 3, 246.

followed courses in geology, mineralogy, zoology, chemistry, anatomy, and botany. He earned his doctorate in Heidelberg in 1840 with a thesis on a geological topic. In the autumn of 1841, he ended up in the Netherlands. It is uncertain how he and Temminck got acquainted, but since both Müller (who was back in the Netherlands) and Schwaner studied in Heidelberg it is not inconceivable that Schwaner was introduced to Temminck via Müller.[433]

Schwaner was appointed as member of the Natuurkundige Commissie on January 9, 1842.[434] Schwaner left in the last days of April 1842 on the ship *Rhoon en Pendrecht*[435] and arrived in Batavia on August 6, 1842. He was awarded a salary of f350 and invited to send in a proposal about where he intended to continue Horner's geological research.[436] On May 30, 1843, Schwaner finally received instructions for an expedition to Borneo where he should mainly work on coal exploitation but should also devote attention to platinum and gold mining opportunities.[437] Von Gaffron was appointed as his draughtsman-assistant on July 28, 1843,[438] while Comblen was appointed as assistant for all practical matters regarding coal exploitation on August 2, 1843.[439]

The expedition party left Batavia on August 14, 1843, on the ship *Tatal Karim*.[440] In the last quarter of 1844, Schwaner and Comblen briefly travelled back and forth to Surabaya, in order to conduct some coal experiments for which they did not have the right equipment on Borneo.[441] Schwaner again travelled back and forth to Surabaya in August/September 1845, where he met with general-major Carel van der Wijck (1797–1852) of the engineering corps, who was involved in coal mining efforts on behalf of the army. Schwaner's proposal for the opening of a mine met with Van der Wijck's approval and was officially approved by the governor-general on October 26, 1845.[442] Both Comblen (February 2, 1846) and Von Gaffron (May 19, 1846) would be given official positions in the further development of mining activities on Borneo. Von Gaffron was formally replaced by Van Pers, but his association with the Natuurkundige Commissie was not a lucky one and he was already honourably discharged on April 27, 1847.[443]

433 I must note that I do not have anything other than circumstantial evidence that emboldens me to make this tentative claim.

434 KB, January 9, 1842, no. 66. NL-HaNA, KdK 1841–1897, 2.02.04, inv.no. 79.

435 *Opregte Haarlemsche Courant*, no. 53 (May 3, 1842): 2.

436 Res.GG, September 8, 1842, no. 1a. NL-HaNA, Koloniën, 1814–1849, 2.10.01, inv.no. 2637.

437 Res.GG, May 30, 1843, no. 10. NL-HaNA, Koloniën, 1814–1849, 2.10.01, inv.no. 2645. This resolution contains extensive written instructions for Schwaner.

438 Res.GG, July 28, 1843, no. 15. NL-HaNA, Koloniën, 1814–1849, 2.10.01, inv.no. 2647.

439 Res.GG, August 2, 1843, no. 1. NL-HaNA, Koloniën, 1814–1849, 2.10.01, inv.no. 2648.

440 Res.GG, August 13, 1843, no. 3a. NL-HaNA, Koloniën, 1814–1849, 2.10.01, inv.no. 2648.

441 Res.GG, August 16, 1844, no. 15. NL-HaNA, Koloniën, 1814–1849, 2.10.01, inv.no. 2660; Res.GG, August 29, 1844, no. 5. NL-HaNA, Koloniën, 1814–1849, 2.10.01, inv.no. 2660.

442 Res.GG, October 26, 1845, no. 3. NL-HaNA, Koloniën, 1814–1849, 2.10.01, inv.no. 2674.

443 Res.GG, April 27, 1847, no. 33. NL-HaNA, Koloniën, 1814–1849, 2.10.01, inv.no. 2693.

Schwaner's most famous expedition, which took him from Banjarmasin to Pontianak through the unknown interior of Borneo, took place from November 2, 1847, to February 2, 1848.[444] After this, Schwaner returned to Batavia where he worked tirelessly on his reports, assisted by François, Strung, Brendonck, and Oehler. In order to process his zoological collections, Schwaner temporarily received Overdijk as assistant for a six-month period between December 12, 1848, and August 23, 1849, but this arrangement did not work out.

When the Natuurkundige Commissie was formally dissolved on April 17, 1850, Schwaner was allowed to complete his term, but without his planned stay in the Netherlands.[445] The members were supposed to spend six years in the Indies, three years in the Netherlands, four years in the Indies and two final years in the Netherlands.[446] This means that Schwaner would have been allowed to continue working as a member of the Natuurkundige Commissie until 1852. Schwaner was instrumental in the founding of the Natuurkundige Vereeniging in Nederlandsch Indië on July 19, 1850, and became its first president.[447] On February 4, 1851, Schwaner was appointed by the governor-general to undertake a new expedition, together with Hermann von de Wall (1807–1873),[448] first to Kusan in Southeast Borneo, then to Tanah Laut in order to investigate mineral deposits and give advice on the options for exploitation.[449] Schwaner died on March 30, 1851, in the middle of preparing for this new expedition.[450]

30 Johan Baptist Spanoghe (1798–1838)

b. 1796 (India), d. April 22, 1838 (Java)
Assistant: 1830

MPCC, 495–496
Anonymus. "Johan Baptist Spanoghe, resident van Timor." *Handelingen en geschriften van het Indisch Genootschap* 2 (1855): 393–400.

Spanoghe's contribution to the history of the Natuurkundige Commissie is a limited one. He was born in 1796 in British India, arrived in the Indies in 1816 and spent most

444 Schwaner, *Borneo*, vol. 2, ii.
445 KB, April 17, 1850, no. 14. NL-HaNA, KdK 1841–1897, 2.02.04, inv.no. 609.
446 KB, August 14, 1842, no. 69. NL-HaNA, KdK 1841–1897, 2.02.04, inv.no. 117.
447 Bleeker, "Algemeen verslag der werkzaamheden," 20.
448 Juynboll, "Wall."
449 Res.GG, February 4, 1851, no. 12–14. NL-HaNA, Koloniën, 1850–1900, 2.10.30, inv.no. 7118.
450 Anonymus, "Levensberigt," 190.

of his administrative career in Banten (West Java) and on Timor.[451] An avid botanist, Spanoghe corresponded regularly with Blume and contributed plants to Blume's collections.[452] While he was in Buitenzorg for health reasons, he was appointed on April 3, 1830, at Macklot's request, to copy the manuscripts of Zipelius, who had died on Timor the year before.[453] It was through Spanoghe's efforts that Zipelius' collections reached 's Rijks Herbarium in an organised condition.[454] Spanoghe was appointed to a government post on Timor in 1831, where he would continue to botanise. He died on April 22, 1838, on Java.

31 G. Strung

b. unknown, d. unknown
Assistant: 1848

On October 30, 1848, G. Strung, a sergeant at the corps of engineers in the army ('sergeant bij het korps sappeurs'), was detached to Schwaner as topographical draughtsman together with Brendonck. He received an additional salary of f25 and had to move in with Schwaner.[455] Nothing is known about how long this commission lasted.

32 Alexander Zipelius (1797–1828)[456]

b. 1797 (Wontfurt), d. December 31, 1828 (Timor)
Assistant: May 25, 1827–December 31, 1828

ENI, vol. 4, 861 | MPCC, 592–593
Steenis, C. G. G. J. van, and M. J. van Steenis-Kruseman. "Contributions to the History of Botany and Exploration in Malaysia 5: Biographical Note on Alexanders Zipelius." *Bulletin of the Botanic Gardens, Buitenzorg. Series 3* 18, no. 4 (1950): 470–471.
Weitzel, A. W. P. "Vergaderingen der Natuurkundige Vereeniging in Nederlandsch Indië." *Natuurkundig tijdschrift voor Nederlandsch Indië* 16 (1858–1859): 362–364.

451 Van Steenis-Kruseman, "Malaysian Plant Collectors," 495.
452 Anonymus, "Johan Baptist Spanoghe," 393.
453 Res.GG, April 3, 1830, no. 9. NL-HaNA, Koloniën, 1814–1849, 2.10.01, inv.no. 2835.
454 Anonymus, "Johan Baptist Spanoghe," 393–394.
455 Res.GG, October 30, 1848, no. 2. NL-HaNA, Koloniën, 1814–1849, 2.10.01, inv.no. 2711.
456 The spelling Zippelius seems to be much more common in secondary sources. However, Van Steenis and Van Steenis-Kruseman have found enough archival evidence to make it very plausible that Zipelius is the correct spelling as used by Zipelius himself.

Alexander Zipelius was born on in 1797 in Wontfurt, Germany, to Johan Georg Zipelius (d. 1847) and Maria Eva Betz (d. 1848). He was appointed as third *hortulanus* at 's Lands Plantentuin on December 6, 1823, at a salary of f150.[457] On March 31, 1825, some of Latour's withheld salary was used to pay Zipelius some extra.[458] On June 28, 1825, Zipelius' provisional appointment as *hortulanus* was turned into a permanent one at a salary of f150.[459] On June 26, 1826, as a provision for Blume's absence, Zipelius was appointed as co-director of 's Lands Plantentuin together with the other *hortulanus* James Hooper.[460]

On Boie's request, Zipelius was added to the Natuurkundige Commissie as 'assistant for botanical work' on May 25, 1827, as long as he would keep 's Land Plantentuin in its current state.[461] Because of some business with a shipment of plants from Japan, Zipelius could not immediately join the New Guinea expedition along with the other members and assistants of the Natuurkundige Commissie but travelled on the ship *Johanna Maria* directly to the Moluccas to join up with them there.[462] On December 31, 1828, shortly after the Natuurkundige Commissie had been dropped off on Timor, Zipelius died.

457 Res.GG, December 6, 1823, no. 6. NL-HaNA, Koloniën, 1814–1849, 2.10.01, inv.no. 2473.

458 Res.GG, March 31, 1825, no. 3. NL-HaNA, Koloniën, 1814–1849, 2.10.01, inv.no. 2481.

459 Res.GG, June 28, 1825, no. 7. NL-HaNA, Koloniën, 1814–1849, 2.10.01, inv.no. 2789.

460 Res.GG, June 26, 1826, no. 2. NL-HaNA, Koloniën, 1814–1849, 2.10.01, inv.no. 2493.

461 Res.GG, May 25, 1827, no. 1. NL-HaNA, Koloniën, 1814–1849, 2.10.01, inv.no. 2504.

462 Res.GG, February 1, 1828, no. 18. NL-HaNA, Koloniën, 1814–1849, 2.10.01, inv.no. 2810; Res.GG, February 5, 1828, no. 1. NL-HaNA, Koloniën, 1814–1849, 2.10.01, inv.no. 2810.

Appendix 3: Publications by Members and Assistants of the Natuurkundige Commissie

This bibliography contains all publications currently known to me by members and assistants of the Natuurkundige Commissie, except for the papers published as part of the Verhandelingen. *These can be found in Appendix 4. Joint publications have been listed with both authors. If a member or assistant is not listed below, no publications by their hand are known to me.*[1]

Bik, Johannes Theodorus

Bik, Johannes Theodorus. "Aanteekeningen op eene reis door het westelijk gedeelte van de residentie Bantam, met de Heeren Van Hasselt en Meaurevert in het jaar 1823." *Tijdschrift voor Indische taal-, land- en volkenkunde* 16 (1867): 260–276.

Boie, Heinrich

*The entries marked with * have been taken from a bibliography that was kindly shared with me by Charles Klaver. Several French (in* Bulletin des sciences naturelles et de géologie) *and German (in* Isis von Oken) *translations of Dutch publications have been omitted.*

Boie, Heinrich. "Manuel d'ornithologie, ou tableau systèmatique des oiseaux qui se trouvent en Europe par C. J. Temminck." *Heidelbergische Jahrbücher der Literatur* 25–26 (1816): 387–416.

Boie, Heinrich. "Ueber den höckerschnäbeligten Eidervogel (*Anas spectabilis* Linn.) von Otton Fabricius." *Neue Annalen der Wetterauischen Gesellschaft für die Gesammte Naturkunde zu Hanau* 4 (1818–1819): 329–341.*

Boie, Heinrich. "Vorrede." In *Tagebuch gehalten auf einer Reise durch Norwegen im Jahre 1817 von F. Boie*, edited by Heinrich Boie, v–viii. Schleswig: Königl. Taubstummen-Institut, 1822.

1 In compiling this bibliography, I have gratefully made use of the following volumes from the Klein Repertorium series: Van Delden, *Indisch Genootschap*; Van Delden, *Tijdschrift voor Nederlandsch Indië*; Van Delden, *Acht algemene tijdschriften*. I have also made use of The and Van der Veur, *Verhandelingen van het Bataviaasch Genootschap*.

 | DOI:10.1163/9789004696877_014

Boie, Heinrich. “Zur Caups Aufsatz über Lurche.” *Isis von Oken* 1825: 1089–1090.*

Boie, Heinrich. “Bemerkungen über die von Hr. von Spix abgebildeten brasilianischen Saurier.” *Isis von Oken* 1826: 117–120.*

Schlegel, Hermann. “Notice sur l’Erpétologie de l’ile de Java; par M. Boie (Ouvrage manuscrit).” *Bulletin des sciences naturelles et de géologie* 9 (1826): 233–240. This is a summary of Boie’s unpublished manuscript.

Boie, Heinrich. “Merkmale einiger japanischen Lurche.” *Isis von Oken* 1826: 203–216.*

Boie, Heinrich. “Kenteekenen van eenige Japansche amphibiën.” *Bijdragen tot de natuurkundige wetenschappen* 2 (1827): 243–273.

Boie, Heinrich, and Heinrich Christian Macklot. “Aanteekeningen en uittreksels uit brieven, door de Heeren Boie en Macklot van Java, aan den Heer Temminck, directeur van ‘s Rijks Museum, gezonden.” *Bijdragen tot de natuurkundige wetenschappen* 2 (1827): 315–328.

Boie, Heinrich. “H. Boie an Wagler. Buitenzorg auf Java den 25. Aug. 1825.” *Isis von Oken* 20 (1827): 724–726.

Boie, Heinrich, and Hermann Schlegel. “Uittreksels uit brieven van Heinrich Boie van Java aan H. Schlegel, conservator animalium vertebratorum aan ‘s Rijks Museum te Leyden.” *Bijdragen tot de natuurkundige wetenschappen* 3 (1828): 231–252.

Boie, Heinrich. “Briefe von Heinrich Boie geschrieben aus Ostindien und auf der Reise dahin.” *Neues staatsbürgerliches Magazin, mit besonderer Rücksicht auf die Herzorgthümer Schleswig, Holstein und Lauenburg* 1 (1832): 126–218, 440–500.

Brendonck, A. A.

Brendonck, A. A. *Mijne reize naar Nederlandsch-Indië: Militaire loopbaan op Java en Bali en mijn terugtogt naar het vaderland.* Dordrecht: Lagerweij, 1859.

Bürger, Heinrich

Bürger, Heinrich. “Aanmerkingen gehouden op eene reize door eenige districten der Padangsche bovenlanden.” *Verhandelingen van het Bataviaasch Genootschap van Kunsten en Wetenschappen* 16 (1836): 157–226.

Bürger, Heinrich. “Beschrijving der Japansche kopermijnen en der bereiding van het koper.” *Verhandelingen van het Bataviaasch Genootschap van Kunsten en Wetenschappen* 16 (1836): 1–28.

Diard, Pierre-Médard

Diard, Pierre-Médard, and Alfred Duvaucel. "Notices on Zoological Subjects." *Malayan Miscellanies* 1 (1820): 1–21.

Diard, Pierre-Médard and Alfred Duvacuel. "Sur le dugong." *Journal de physique, de chimie, d'histoire naturelle et des arts* 91, no. August (1820): 159–160.

Diard, Pierre-Médard, Alfred Duvaucel, and Thomas Hardwicke. "On the Sorex Glis." *Asiatick Researches* 14 (1822): 471–475.

Farquhar, William, A. Seton, Pierre-Médard Diard, and G. J. Siddons. "Account of a New Species of Tapir Found in the Peninsula of Malacca." *Asiatick Researches* 13 (1820): 417–427.

Diard, Pierre-Médard. "Beschrijving van eene manier om verschillende aardsoorten te bereiden, waardoor alle, naar gelang van ieders bijzonderen aard, bij uitstek geschikt worden voor de vervaardiging van metselsteenen, dakpannen, estrikken, van vuurbestendige steenen en van alle soorten van aardewerk en porselein." *Tijdschrift ter bevordering van nijverheid* 16, no. 6 (1853): 678–684.

Diard, Pierre-Médard. "Beschrijving van zekere verbeteringen aangebragt aan eene soort van ovens, waarvan de Sinezen zich in het algemeen bedienen tot het bakken van metselsteenen en tegels." *Tijdschrift ter bevordering van nijverheid* 16, no. 6 (1853): 685–694.

Diard, Pierre-Médard. "Beschrijving van eenen nieuwen toestel en van eene wijziging van de gewone vormen, met behulp waarvan men zeer regelmatige dakpannen kan maken." *Tijdschrift ter bevordering van nijverheid* 16, no. 6 (1853): 695–702.

Diard, Pierre-Médard. "Beschrijving van eene soort van Spoorwegen, gemaakt van eenen zeer-harden steen en op welke de wielen moeten loopen, voorzien van opstaande houten klossen over den ganschen omtrek hunner velgen." *Tijdschrift ter bevordering van nijverheid* 16, no. 6 (1853): 703–704.

Diard, Pierre-Médard. "Mededeeling over eene nieuwe suikerhoudende plant in Frankrijk." *Tijdschrift voor nijverheid in Nederlandsch-Indië* 2 (1855): 282–284.

Diard, Pierre-Médard. "Chine et Indo-Chine: Faits commerciaux, No. 34: Cochinchine." *Annales du commerce extérieur* troisième série, no. 1466 (1863).

Forsten, Eltio Alegondus

Forsten, Eltio Alegondus. *Dissertatio botanico-pharmaceutico-medica inauguralis de cedrela febrifuga.* Leiden: C. C. van der Hoek, 1836.

Forsten, Eltio Alegondus. *Responsio de aethere et naphthis.* Leiden: S. & J. Luchtmans, 1832.

Horner, Ludwig, and E. A. F. "Geologische en mineralogische aanteekeningen over het eiland Borneo." *Tijdschrift voor natuurlijke geschiedenis en physiologie* 4 (1837): 359–369.

Gaffron, Heinrich von

Gaffron, Heinrich von. "Verslag over de goudmijnen in het westelijke gedeelte van Tanah Lawut (eiland Borneo), opgemaakt naar een rapport van H. von Gaffron." *Natuurkundig tijdschrift voor Nederlandsch Indië* 2 (1851): 30–40.

Gaffron, Heinrich von. "Mededeeling aangaande den ijzererts, gevonden ten noorden van kampong Tambaga in Tanah Laut (Z. O. kust van Borneo)." *Natuurkundig tijdschrift voor Nederlandsch Indië* 5 (1853): 225–232.

Gaffron, Heinrich von. "Geognostische tabel der rotssoorten van den berg Pangaron." *Natuurkundig tijdschrift voor Nederlandsch Indië* 6 (1854): 145–150.

Gaffron, Heinrich von. "Beschrijving van den grooten diamant te Mattan op Borneo." *Natuurkundig tijdschrift voor Nederlandsch Indië* 6 (1854): 165–167.

Gaffron, Heinrich von. "Beknopt verslag over het voorkomen van minerale bronnen in de adsistent residentie Sintang." *Natuurkundig tijdschrift voor Nederlandsch Indië* 20 (1859–1860): 130–134.

Gaffron, Heinrich von. "Over menschen met staarten op Borneo." *Natuurkundig tijdschrift voor Nederlandsch Indië* 20 (1859–1860): 227–232.

Gaffron, Heinrich von, and J. Pijnappel Gz. "Beschrijving van het westelijke gedeelte van de Zuid- en Oosterafdeling van Borneo (de afdeeling Sampit en de Zuidkust)." *Bijdragen tot de taal-, land- en volkenkunde* 7 (1860): 243–346.

Martin, K. "Neue Fundpunkte von Tertiaer-gesteinen im indischen Archipel: Nach Sammlungen von Horner, Korthals, Macklot, Müller und Reinwardt." *Sammlungen des Geologischen Reichs-Museums in Leiden: Serie 1, Beiträge zur Geologie Ost-Asiens und Australiens* 1, no. 1 (1881): 131–193.[2]

Hasselt, Johan Conrad van

Kuhl and Van Hasselt corresponded widely. Several (summaries of their) letters were published in German in Isis von Oken *and in French in* Bulletin des sciences naturelles et de géologie *and* Annales générales des sciences physiques. *I have not included these letters in this bibliography. Please consult the bibliography in Klaver,* Inseparable Friends, *for these references.*

2 This publication includes a geological map by Heinrich von Gaffron.

Hasselt, Johan Conrad van. “Responsum ad quaestionem, ab ordine disciplinarum mathematicarum et physicarum propositam: Cum, licet naturae corpora vario modo inter se differant, ex huc usque cognitis observationibus tamen constare videatur, ita comparatam esse rerum naturam, ut lento quasi passu ab una specie ad alteram progrediatur, atque sic continuam quasi catenam efficiat ex variis quidem annulis, intime tamen iunctis, compositam, haec catena in mammalium classe demonstranda quaeritur.” *Annales Academiae Groninganae* 1816–1817 (1818): 127–166.

Hasselt, Johan Conrad van. *Dissertatio medico-anatomica inauguralis exhibens observationes de metamorphosi quarundam partium ranae temporariae*. Groningen: [s.n.], 1820.

Hasselt, Johan Conrad van, and Heinrich Kuhl. *Beiträge zur Zoologie und vergleichenden Anatomie: Zweite Abtheilung: Beiträge zur vergleichenden Anatomie*. Frankfurt am Main: Verlag der Hermannschen Buchhandlung, 1820.

Kuhl, Heinrich, and Johan Conrad van Hasselt. “Uittreksel uit eenen brief van de heeren Kuhl en Van Hasselt, aan den heer Temminck, directeur van 's Rijks Museum van Nat. Hist.” *Algemeene konst- en letter-bode*, no. 12 (March 16, 1821): 180–181.

Kuhl, Heinrich, and Johan Conrad van Hasselt. “Uittreksel van eenen brief van de heeren Kuhl en Van Hasselt aan de heeren F. Nees Phil. Dr. te Bonn en W. de Haan Phil Stud. te Leijden.” *Algemeene konst- en letter-bode*, no. 12 (March 16, 1821): 181–184.

Kuhl, Heinrich, and Johan Conrad van Hasselt. “Uittreksels uit brieven van de heeren Kuhl en Van Hasselt, aan de heeren C. J. Temminck, Th. van Swinderen en W. de Haan, geschreven in de straat Sunda, den 17den December 1820.” *Algemeene konst- en letter-bode*, no. 6 (February 8, 1822): 82–88.

Kuhl, Heinrich, and Johan Conrad van Hasselt. “Uittreksels uit brieven van de heeren Kuhl en Van Hasselt, aan de heeren C. J. Temminck, Th. van Swinderen en W. de Haan.” *Algemeene konst- en letter-bode*, no. 7 (February 15, 1822): 99–104.

Kuhl, Heinrich, and Johan Conrad van Hasselt. “Uittreksels uit brieven van de heeren Kuhl en Van Hasselt, aan de heeren C. J. Temminck, Th. van Swinderen en W. de Haan.” *Algemeene konst- en letter-bode*, no. 8 (February 22, 1822): 115–119.

Kuhl, Heinrich, and Johan Conrad van Hasselt. “Uittreksels uit een brief van de heeren Kuhl en Van Hasselt, aan den heer D. J. van Ewyck, secretaris bij het Ministerie van Onderwijs, de Nationale Nijverheid en de Kolonien.” *Algemeene konst- en letter-bode*, no. 9 (March 1, 1822): 132–135.

Kuhl, Heinrich, and Johan Conrad van Hasselt. “Uittreksels uit een' brief van de heeren Kuhl en Van Hasselt, aan Zijne Excellentie den Ministerie voor het openbaar Onderwijs, de Nationale Nijverheid en de Kolonien.” *Algemeene konst- en letter-bode*, no. 10 (March 8, 1822): 149–153.

Hasselt, Johan Conrad van. “Brief van dr. J. C. van Hasselt aan professor Van Swinderen, te Groningen.” *Algemeene konst- en letter-bode*, no. 44 (October 18, 1822): 275–279.

Hasselt, Johan Conrad van. "Brief van dr. J. C. van Hasselt aan professor Van Swinderen te Groningen." *Algemeene konst- en letter-bode*, no. 45 (October 25, 1822): 290–294.

Hasselt, Johan Conrad van. "Uittreksel uit een' brief van dr. J. C. van Hasselt aan den heer C. J. Temminck." *Algemeene konst- en letter-bode*, no. 20 (May 16, 1823): 315–317.

Hasselt, Johan Conrad van. "Uittreksel uit een' brief van dr. Van Hasselt aan prof. Van Swinderen." *Algemeene konst- en letter-bode*, no. 23 (June 6, 1823): 360–362.

Hasselt, Johan Conrad van. "Verslag eener reis naar den westhoek van Java." *Bataviasche courant*, no. 33 (August 16, 1823): 3–4.

Hasselt, Johan Conrad van. "Uittreksel uit een' brief van den heer J. C. van Hasselt, aan den heer C. J. Temminck, geschreven uit Tjecande, Residentie Bantam, den 29sten December 1822." *Algemeene konst- en letter-bode*, no. 35 (August 29, 1823): 130–133.

Hasselt, Johan Conrad van. "Uittreksel uit een' brief van dr. J. C. van Hasselt aan professor Van Swinderen." *Algemeene konst- en letter-bode*, no. 41 (October 10, 1823): 230–233.

Hasselt, Johan Conrad van. "Uittreksel uit een' brief van dr. J. C. van Hasselt aan professor Van Swinderen." *Algemeene konst- en letter-bode*, no. 42 (October 17, 1823): 242–246.

Hasselt, Johan Conrad van. "Uittreksel uit een' brief van dr. J. C. van Hasselt aan professor Van Swinderen." *Algemeene konst- en letter-bode*, no. 43 (October 24, 1823): 264–266.

Hasselt, Johan Conrad van. "Uittreksel uit een' brief van dr. J. C. van Hasselt." *Algemeene konst- en letter-bode*, no. 48 (November 28, 1823): 341–342.

Hasselt, Johan Conrad van. "Uittreksel uit eenen brief van dr. J. C. van Hasselt, aan prof. Van Swinderen." *Algemeene konst- en letter-bode*, no. 2 (January 9, 1824): 20–24.

Hasselt, Johan Conrad van. "Uittreksel uit eenen brief van dr. J. C. van Hasselt, aan prof. Van Swinderen." *Algemeene konst- en letter-bode*, no. 3 (January 16, 1824): 34–39.

Hasselt, Johan Conrad van. "Uittreksel uit eenen brief van dr. J. C. van Hasselt, aan prof. Van Swinderen." *Algemeene konst- en letter-bode*, no. 4 (January 23, 1824): 54–55.

Hasselt, Johan Conrad van. "Brief van wijlen dr. C. J. van Hasselt aan W. de Haan." *Algemeene konst- en letter-bode*, no. 15 (April 9, 1824): 227–232.

Hasselt, Johan Conrad van. "Verslag eener reize naar den westhoek van Java." *Indisch magazijn* 2, no. 5–6 (1845): 85–91.

Horner, Ludwig

Horner, Ludwig, and E. A. F. "Geologische en mineralogische aanteekeningen over het eiland Borneo." *Tijdschrift voor natuurlijke geschiedenis en physiologie* 4 (1837): 359–369.

Horner, Ludwig. "Beiträge zur Geologie des Indischen Archipels." *Neues Jahrbuch für Mineralogie, Geognosie, Geologie und Petrefaktenkunde* (1838): 1–12.

Horner, Ludwig. "Geologische gesteldheid van de vulkaan Gedé op Java." *Verhandelingen van het Bataviaasch Genootschap van Kunsten en Wetenschappen* 17 (1839): 1–28.

Horner, Ludwig. "Verslag van een geologisch onderzoek van het Zuid-Oostelijke gedeelte van Borneo." *Verhandelingen van het Bataviaasch Genootschap van Kunsten en Wetenschappen* 17 (1839): 87–119.

Horner, Ludwig. "Verslag van eene mineralogische reis in de residentie Bantam." *Verhandelingen van het Bataviaasch Genootschap van Kunsten en Wetenschappen* 17 (1839): 29–59.

Horner, Ludwig. "De beklimming van den berg Ophir." *Tijdschrift voor Neêrland's Indië* 2, 2 (1839): 605–622; 3, 1 (1840): 71–72.

Horner, Ludwig. "Batoe-eilanden, ten westen van Sumatra gelegen." *Tijdschrift voor Neêrland's Indië* 3, 1 (1840): 313–371.

Horner, Ludwig. "Verslag van eene mineralogische reis in de residentie Bantam." *Indisch magazijn* 1, 2 (1844): 334–355.

Horner, Ludwig, Salomon Müller, and H. L. Osthoff. "Barometer-waarnemingen in de jaren 1834 tot 1839, verrigt op Sumatra door Dr. Horner, S. Müller en H. L. Osthoff, verzameld en berekend door H. L. Osthoff." *Natuurkundig tijdschrift voor Nederlandsch Indië* 2 (1851): 153–156.

Müller, Salomon, and Ludwig Horner. "Fragmenten uit de reizen en onderzoekingen in Sumatra, gedaan op last der Nederlandsch-Indische regering, in de jaren 1833 tot 1838: I." *Bijdragen tot de taal-, land- en volkenkunde* 2 (1854): 212–242.

Müller, Salomon, and Ludwig Horner. "Fragmenten uit de reizen en onderzoekingen in Sumatra, gedaan op last der Nederlandsch-Indische regering, in de jaren 1833 tot 1838: II." *Bijdragen tot de taal-, land- en volkenkunde* 2 (1854): 243–271.

Müller, Salomon, and Ludwig Horner. "Fragmenten uit de reizen en onderzoekingen in Sumatra, gedaan op last der Nederlandsch-Indische regering, in de jaren 1833 tot 1838: III." *Bijdragen tot de taal-, land- en volkenkunde* 3 (1855): 65–72.

Müller, Salomon, and Ludwig Horner. "Fragmenten uit de reizen en onderzoekingen in Sumatra, gedaan op last der Nederlandsch-Indische regering, in de jaren 1833 tot 1838: IV." *Bijdragen tot de taal-, land- en volkenkunde* 3 (1855): 193–228.

Müller, Salomon, and Ludwig Horner. "Fragmenten uit de reizen en onderzoekingen in Sumatra, gedaan op last der Nederlandsch-Indische regering, in de jaren 1833 tot 1838: V." *Bijdragen tot de taal-, land- en volkenkunde* 3 (1855): 229–249.

Müller, Salomon, and Ludwig Horner. "Fragmenten uit de reizen en onderzoekingen in Sumatra, gedaan op last der Nederlandsch-Indische regering, in de jaren 1833 tot 1838: VI." *Bijdragen tot de taal-, land- en volkenkunde* 3 (1855): 313–346.

Müller, Salomon, and Ludwig Horner. "Fragmenten uit de reizen en onderzoekingen in Sumatra, gedaan op last der Nederlandsch-Indische regering, in de jaren 1833 tot 1838: VII." *Bijdragen tot de taal-, land- en volkenkunde* 3 (1855): 347–373.
Müller, Salomon, and Ludwig Horner. "Fragmenten uit de reizen en onderzoekingen in Sumatra, gedaan op last der Nederlandsch-Indische regering, in de jaren 1833 tot 1838: VIII." *Bijdragen tot de taal-, land- en volkenkunde* 3 (1855): 374–388.
Müller, Salomon, and Ludwig Horner. *Reizen en onderzoekingen in Sumatra, gedaan op last der Nederlandsch Indische regering, tusschen de jaren 1833 en 1838*. 's-Gravenhage: K. Fuhri, 1855.
Horner, Friedrich. "Briefe und Tagebuchskizzen des Dr. med. Ludwig Horner (1811–1838) aus Niederländisch-Indien." *Zurcher Taschenbuch* (1919): 183–208.
Horner, Friedrich. "Briefe und Tagebuchskizzen des Dr. med. Ludwig Horner (1811–1838) aus Niederländisch-Indien." *Zurcher Taschenbuch* (1926): 1–45.

Junghuhn, Franz Wilhelm

In compiling this bibliography, I have made grateful use of three existing bibliographies: Schmidt, "Biographische Skizze"; Muller, "Junghuhn-bibliographie," and the Wikipedia-entry on Junghuhn: https://de.wikipedia.org/wiki/Franz_Wilhelm_Junghuhn#Literatur. *I have omitted several translations, later editions, and pre-publications.*

Junghuhn, Franz Wilhelm. "Observationes mycologicae in species fungorum tam novas tam male cognitas." *Linnaea* 5 (1830): 388–410.
Junghuhn, Franz Wilhelm. "Communications sur Java faites à MM. Wirtgen et Nees d'Esenbeck par M. Junghuhn." *Annales des sciences naturelles* 7 (1837): 167–171.
Junghuhn, Franz Wilhelm. "Uittreksel uit een' brief van den heer Frans Junghuhn, gedagteekend Djocjokarta den 2 februarij 1837, aan den hoogleraar Blume te Leiden." *Algemeene konst- en letter-bode*, no. 49 (October 27, 1837): 275–277.
J., F. "Goenong Salak." *Tijdschrift voor Neêrland's Indie* 1, 2 (1838): 486–507.
Junghuhn, Franz Wilhelm. "Praemissa in floram cryptogamicam Javae insulae." *Verhandelingen van het Bataviaasch Genootschap van Kunsten en Wetenschappen* 17 (1839): 1–86.
Junghuhn, Franz Wilhelm. "Gunong Salak." *Flora oder allgemeine botanische Zeitung* 23, no. 29 (August 7, 1840): 449–460.
Junghuhn, Franz Wilhelm. "Gunong Salak." *Flora oder allgemeine botanische Zeitung* 23, no. 30 (August 14, 1840): 465–477.
Junghuhn, Franz Wilhelm. "Nova genera et species plantarum florae Javanicae." *Tijdschrift voor natuurlijke geschiedenis en physiologie* 7 (1840): 285–317.

Junghuhn, Franz Wilhelm. "Herr Poggendorff las Hrn. Dr. Fr. Junghuhn's geognostische Bemerkungen über das Gebirge Di-eng auf Java (mitgetheilt von Herrn Prof. v. Schlechtendal)." *Monatsberichte über die Verhandlungen der Gesellschaft für Erdkunde zu Berlin* 2 (1841): 167–174.

Junghuhn, Franz Wilhelm. "Herr Poggendorff legte Herrn Dr. Fr. Junghuhn's barometrische Höhenmessungen auf der Insel Java vor." *Monatsberichte über die Verhandlungen der Gesellschaft für Erdkunde zu Berlin* 2 (1841): 174–180.

Junghuhn, Franz Wilhelm. "Über Javan'sche Balanophoreen." *Nova Acta Academiae Caesareae Leopoldino-Carolinae Naturae Curiosorum = Verhandlungen der Kaiserlichen Leopoldinisch-Carolinischen Akademie der Naturforscher* 18, Suppl. 1 (1841): 201–225.

Junghuhn, Franz Wilhelm, and W. H. de Vriese. "Uitstapje naar de bosschen van de gebergten Malabar, Wayang en Tilu, op Java." *Tijdschrift voor natuurlijke geschiedenis en physiologie* 8 (1841): 349–412.

Junghuhn, Franz Wilhelm. "De gematigde en koude streken van Java, met de aldaar voorkomende warme bronnen: uit een natuur-, aardrijks- en geneeskundig oogpunt beschouwd, als stellende een middel daar ter voorkoming en genezing van die ziekten, waaraan Europeanen, ten gevolge van hun lang verblijf in heete luchtstreken, gewoonlijk lijden." *Tijdschrift voor Neêrland's Indië* 4, 2 (1842): 81–121.

Junghuhn, Franz Wilhelm. "Hoogten boven de oppervlakte der zee van onderscheidene plaatsen in den Indischen Archipel: Eerste afdeeling: Java." *De kopiïst* 1, 2 (1842): 68–78.

Junghuhn, Franz Wilhelm. "Über die gegenwärtigen Bewohner des Gebirges Di-eng und die einheimischen Thiere." *Monatsberichte über die Verhandlungen der Gesellschaft für Erdkunde zu Berlin* 3 (1842): 83–87.

Junghuhn, Franz Wilhelm. "Pflanzenphysiognomie des Gebirges." *Monatsberichte über die Verhandlungen der Gesellschaft für Erdkunde zu Berlin* 3 (1842): 88–100.

Junghuhn, Franz Wilhelm. "Mineralogische Notizen über das Gebirge Di-eng." *Monatsberichte über die Verhandlungen der Gesellschaft für Erdkunde zu Berlin* 3 (1842): 100–102.

Junghuhn, Franz Wilhelm. "Reise nach Di-eng 1840." *Monatsberichte über die Verhandlungen der Gesellschaft für Erdkunde zu Berlin* 3 (1842): 141–151.

Junghuhn, Franz Wilhelm. "Topographische (physikalisch-geogr.) Beschreibung des Gebirges Di-eng." *Monatsberichte über die Verhandlungen der Gesellschaft für Erdkunde zu Berlin* 3 (1842): 151–180.

Junghuhn, Franz Wilhelm. "Abreise von Di-eng." *Monatsberichte über die Verhandlungen der Gesellschaft für Erdkunde zu Berlin* 3 (1842): 180–186.

Junghuhn, Franz Wilhelm. "Bericht über Hrn. Dr. Junghuhn's (vorgelegtes) meteorologisches Beobachtungs-Journal auf einer Reise in's Innere von Java und dessen klimatologische Bemerkungen über das Gebirge Di-eng." *Monatsberichte über die Verhandlungen der Gesellschaft für Erdkunde zu Berlin* 3 (1842): 187–190.

Junghuhn, Franz Wilhelm. "Bijdragen tot de geschiedenis der vulkanen in den Indischen archipel, tot 1842." *Tijdschrift voor Neêrland's Indië* 5 (1843): 97–133, 185–227, 257–280, 614–626, 745–763.

Junghuhn, Franz Wilhelm. "Kritiek over kritische aanmerkingen." *Tijdschrift voor Neêrland's Indië* 5, 1 (1843): 1–15.

Junghuhn, Franz Wilhelm. "Hindoe-oudheden in de Preanger-regentschappen." *Indisch magazijn* 1, 1 (1844): 228–231.

Junghuhn, Franz Wilhelm. "Voortgezette kronijk der vulkanen." *Indisch magazijn* 1, 1 (1844): 332–347.

Junghuhn, Franz Wilhelm. "Oproeping en beleefd verzoek aan Nederlandsch-Indië's ingezetenen." *Indisch magazijn* 1, 1 (1844): 358–362.

Junghuhn, Franz Wilhelm. "Bijdragen tot de geschiedenis der vulkanen in den Indischen archipel, tot en met het jaar 1842: Eerste afdeling: Java." *Indisch magazijn* 1, 2 (1844): 41–83, 163–176, 287–315.

Junghuhn, Franz Wilhelm. "Voortgezette kronijk der vulkanen." *Indisch magazijn* 1, 2 (1844): 84–100.

Junghuhn, Franz Wilhelm. "Nog een woord, gevoegd bij het Iets over acclimatisatie van dr. Swaving." *Indisch magazijn* 1, 2 (1844): 316–333.

Junghuhn, Franz Wilhelm. "Bijdragen tot de geschiedenis der vulkanen in den Indischen archipel, tot en met het jaar 1842: Eerste afdeling: Java." *Indisch magazijn* 1, 3 (1844): 64–94.

Junghuhn, Franz Wilhelm. *Prospectus van Eene beschrijving der Batta-landen op Sumatra.* Batavia, 1844.

Junghuhn, Franz Wilhelm. "Ruïnen van Java: Korte ontleding der steenen gedenkteekenen, welke de hindoes op Java hebben achtergelaten." *Tijdschrift voor Neêrland's Indië* 6, 2 (1844): 341–381.

Junghuhn, Franz Wilhelm. "Ein Wort zur Beurtheilung von Prof. Dr. Walter's Aufsatz: Über die Wasserergüsse der Vulkane, in dieser Zeitschrift f. vergl. Erdkunde, Bd. 1, S. 503." *Zeitschrift für vergleichende Erdkunde* 3 (1844): 481–491.

Junghuhn, Franz Wilhelm. "Chronologisch overzigt der aardbevingen en uitbarstingen van vulkanen in Neêrland's Indië, in vergelijkende zamenstelling met elkander." *Tijdschrift voor Neêrland's Indië* 7, 1 (1845): 27–68.

Junghuhn, Franz Wilhelm. "Kaart van het gebergte Di-eng." *Indisch magazijn* 2, 2 (1845).

Junghuhn, Franz Wilhelm. "Phijsiognomie van de flora der toppen van Javasche bergen, benevens plantenbeschrijvingen." *Natuur- en geneeskundig archief voor Neêrland's-Indië* 2 (1845): 20–33.

Junghuhn, Franz Wilhelm. "Diagnoses et adumbrationes stirpium nonnularum novarum vel non satis cognitarum florae javanicae alpinae indigenarum." *Natuur- en geneeskundig archief voor Neêrland's-Indië* 2 (1845): 34–54.

Junghuhn, Franz Wilhelm. "Schetsen, ontworpen op eene nieuwe reis over Java, voor topographische en natuurkundige navorschingen, aan het einde van het jaar 1844." *Tijdschrift voor Neêrland's Indië* 7, 1 (1845): 69–92, 121–203, 315–350.

Junghuhn, Franz Wilhelm. "Schetsen, ontworpen op eene nieuwe reis over Java, voor topographische en natuurkundige navorschingen, aan het einde van het jaar 1844." *Tijdschrift voor Neêrland's Indië* 8, 2 (1845): 1–38, 203–250, 369–380.

Junghuhn, Franz Wilhelm. *Topographische und naturwissenschaftliche Reisen durch Java.* Magdeburg: Verlag von Emil Baensch, 1845.

Junghuhn, Franz Wilhelm. *Die Battaländer auf Sumatra: Im Auftrage Sr. Excellenz des General-Gouverneurs von Niederländisch-Indien Hrn P. Merkus in den Jahren 1840 und 1841 untersucht und beschrieben.* 2 vols. Berlin: G. Reimer, 1847.

Junghuhn, Franz Wilhelm. "Ruines de Java: Analyse succincte des monuments en pierre que les Hindus ont laissés dans l'île de Java." In *Le moniteur des indes orientales et occidentales, recueil de mémoires et de notices scientifiques et industriels, de nouvelles et de faits importants concernant les possessions néerlandais d'Asie et d'Amérique*, edited by Philipp Franz von Siebold and Pieter Melvill van Carnbee, vol. 1, 18–21, 137–146. La Haye: Belinfante Frères, 1847–1849.

Junghuhn, Franz Wilhelm. "Inlichtingen aangeboden aan het publiek over zeker geschrift van den heer C. L. Blume, en antwoord aan dien heer." *Algemeene konst- en letter-bode*, no. 41 (October 11, 1850): 232–240.

Junghuhn, Franz Wilhelm. *Java, deszelfs gedaante, bekleeding en inwendige structuur.* 3 vols. Amsterdam: P. N. van Kampen, 1850–1854.

Junghuhn, Franz Wilhelm. "Terugreis van Java naar Europa, met de zoogenaamde Engelsche overlandpost, in de maanden september en october, 1848." *Tijdschrift voor Nederlandsch Indië* 13, 1 (1851): 193–238, 345–387, 425–463.

Junghuhn, Franz Wilhelm. *Rückreise von Java nach Europa mit der sogenannten englischen Überlandpost im September und Oktober 1848.* Leipzig: Arnoldische Buchhandlung, 1852.

Junghuhn, Franz Wilhelm. *Java, seine Gestalt, Pflanzendecke und innere Bauart: Nach der zweiten, verbesserten Auflage des holländische Original in's Deutsche übertragen von J. K. Hasskarl.* 3 vols. Leipzig: Arnoldische Buchhandlung, 1852–1854.

Gebroeders Dag en Nacht. *Licht- en schaduwbeelden uit de binnenlanden van Java: Over het karakter, de mate van beschaving, de zeden en gebruiken der Javanen; over de invoering van het Christendom op Java, het bezigen van vrijen arbeid en andere vragen van den dag: Verhalen en gesprekken, verzameld op reizen door gebergten en bosschen, in de woningen van armen en rijken.* Leiden: Jacs. Hazenberg Corns. zoon, 1854.

Junghuhn, Franz Wilhelm. *Catalog der geologischen Sammlung von Java: Oder Verzeichnis der Felsarten gesammelt zur Erläuterung des geologischen Baues dieser Insel, niedergelegt und geordnet im Reichs-museum für Naturgeschichte zu Leiden.* 's Gravenhage: C. W. Mieling, 1854.

[Junghuhn, Franz Wilhelm]. "Die Javanesen." *Zeitschrift für allgemeine Erdkunde* 2, no. 2 (1854): 81–125.

[Junghuhn, Franz Wilhelm]. "De wolf onder de schapen." In *Jaarboekje van het Koninklijk Zoologisch Genootschap Natura Artis Magistra*, 97–128: M. Westerman, 1854.

Junghuhn, Franz Wilhelm. "Kaart van het eiland Java, uitgegeven op last van en opgedragen aan Z. E. den minister van Koloniën Ch. F. Pahud, door Dr. F. Junghuhn, te zamengesteld uit de waarnemingen en opmetingen door hem gedaan, gedurende zijne onderzoekingsreizen op dat eiland in de jaren 1835–1848." 's Gravenhage: C. W. Mieling, 1855.

Junghuhn, Franz Wilhelm. "Fr. Junghuhn's neueste Untersuchungen in Java: Aus seinem Schreiben Fr. Junghuhn's an Herrn A. v. Humboldt d.d. Tjiandjur 8 December 1856." *Zeitschrift für allgemeine Erdkunde* 2 (1857): 506–517.

Junghuhn, Franz Wilhelm. "Over de fossiele zoogdierbeenderen te Pati-hajam, in de residentie Japara." *Natuurkundig tijdschrift voor Nederlandsch Indië* 14 (1857): 215–219.

Junghuhn, Franz Wilhelm. "Toestand der aangekweekte Kinaboomen op het eiland Java, tijdens het bezoek van den gouverneur generaal Chs. F. Pahud in het laatst der maand Juny en het begin July 1857." *Natuurkundig tijdschrift voor Nederlandsch Indië* 15 (1858): 23–138.

Junghuhn, Franz Wilhelm. "Voorloopige handleiding voor de proefkinakultuur, zamengesteld naar een oorspronkelijk geschrift van F. W. Junghuhn." *Natuurkundig tijdschrift voor Nederlandsch Indië* 18 (1859): 97–110.

Junghuhn, Franz Wilhelm. "Aanteekeningen in verband staande met de voorloopige handleiding voor de proef-kinakultuur, zamengesteld naar een oorspronkelijk geschrift van F. W. Junghuhn." *Natuurkundig tijdschrift voor Nederlandsch Indië* 18 (1859): 111–141.

Junghuhn, Franz Wilhelm. "Over fossiele zoogdiertanden van Bonthain." *Natuurkundig tijdschrift voor Nederlandsch Indië* 20 (1859–1860): 385–386.

Junghuhn, Franz Wilhelm. "Berigt over den toestand der aangekweekte Kinaboomen uit een kruidkundig en kultuurgeschiedkundig oogpunt beschouwd." *Natuurkundig tijdschrift voor Nederlandsch Indië* 21 (1860): 179–249.

Junghuhn, Franz Wilhelm. "Brief over het voorkomen van mergelaarde geschikt voor hydraulischen kalk." *Natuurkundig tijdschrift voor Nederlandsch Indië* 24 (1862): 225–226.

Junghuhn, Franz Wilhelm. "Open brief van Dr. F. Junghuhn, aan de Heeren Directeuren der Natuurkundige Vereeniging in Nederlandsch-Indië, te Batavia." *Java-bode*, no. 71 (September 6, 1862): 3–4.

Junghuhn, Franz Wilhelm. "Open brief van Dr. F. Junghuhn, aan de Heeren Directeuren der Natuurkundige Vereeniging in Nederlandsch-Indië, te Batavia." *Java-bode*, no. 72 (September 10, 1862): 3–4.

Junghuhn, Franz Wilhelm. "Jaarlijksch berigt over 1860 aangaande den toestand der kinakultuur op Java." *Natuurkundig tijdschrift voor Nederlandsch Indië* 25 (1863): 1–7.

Junghuhn, Franz Wilhelm. "Jaarlijksch berigt over 1861 aangaande den toestand der kinakultuur op Java." *Natuurkundig tijdschrift voor Nederlandsch Indië* 25 (1863): 12–28.

Junghuhn, Franz Wilhelm. "Berigt over de waarde en het alkaloid gehalte van Cinchona Pahudiana Howard." *Natuurkundig tijdschrift voor Nederlandsch Indië* 25 (1863): 34–40.

Junghuhn, Franz Wilhelm. "Jaarlijksch berigt over 1861, aangaande de op Java gekweekte groene indigo-planten uit China, (Rhamnus utilis en chloroforus), aan Zijne Excellentie den Gouverneur-Generaal van Nederlandsch-Indie, door den inspekteur voor de natuurkundige onderzoekingen in Nederlandsch-Indië, teven belast met de leiding der kina-kultuur op Java." *Natuurkundig tijdschrift voor Nederlandsch Indië* 25 (1863): 44–46.

Junghuhn, Franz Wilhelm. "Jaarlijksch berigt over 1862, aangaande den toestand der kinakultuur op Java." *Natuurkundig tijdschrift voor Nederlandsch Indië* 26 (1863): 125–133.

Junghuhn, Franz Wilhelm. "Jaarlijksch berigt over 1862, aangaande de op Java aangekweekte, zoogenoemde groene-indigo-planten uit China, Rhamnus utilis en Rhamnus chlorophorus." *Natuurkundig tijdschrift voor Nederlandsch Indië* 26 (1863): 145–146.

Junghuhn, Franz Wilhelm. "Jaarlijksch verslag over 1863 aangaande den toestand der kinakultuur." *Natuurkundig tijdschrift voor Nederlandsch Indië* 27 (1864): 341–346.

Junghuhn, Franz Wilhelm. "Berigt omtrent het alkaloid-gehalte van den bast en de bladeren der op Java gekweekte kinaboomen." *Natuurkundig tijdschrift voor Nederlandsch Indië* 27 (1864): 347–369.

Junghuhn, Franz Wilhelm. "Jaarlijksch berigt over 1863, aangaande de op Java aangekweekte, z.g. groene indigo-planten." *Natuurkundig tijdschrift voor Nederlandsch Indië* 27 (1864): 370–371.

Junghuhn, Franz Wilhelm. *Licht- und Schattenbilder aus dem Innern von Java: Ueber den Charakter, den Bildungsgrad, die Sitten und Gebräuche der Javanen; über die Einführung des Christenthums auf Java, die Freigebung der Arbeit und andere Fragen der Zeit: Erzaehlungen und Gespraeche, gesammelt auf Reisen durch Berge und*

Wälder, durch die Wohnungen der Armen und Reichen, zwischen den Gebrüdern TAG *und* NACHT, *mitgetheilt von Dr. F. Junghuhn, nach der Vierten Holländische Auflage übersetzt von* ***. Amsterdam: F. Gunst, 1866.

Junghuhn, Franz Wilhelm. "Meine Flucht nach Afrika (1834)." In *Franz Junghuhn: Biographische Beiträge zur 100. Wiederkehr seines Geburtstages*, edited by Max C. P. Schmidt. Verlag der Dürr'schen Buchhandlung, 1909.

Junghuhn, Franz Wilhelm. "Die menschenfressenden Batakken: Etwas aus Junghuhns Reise durch Java und Sumatra, mitgeteilt von Nees v. Esenbeck (1841)." In *Franz Junghuhn: Biographische Beiträge zur 100. Wiederkehr seines Geburtstages*, edited by Max C. P. Schmidt. Leipzig: Verlag der Dürr'schen Buchhandlung, 1909.

Junghuhn, Franz Wilhelm. *Die Insel der Vulkane: Reisen und Wanderungen durch Java*. Leipzig: Brockhaus, 1950.[3]

Junghuhn, Franz Wilhelm. *De onuitputtelijke natuur: Een keuze uit zijn geschriften*. Stoa. Edited by Rob Nieuwenhuys and Frits Jacquet. G. A. van Oorschot, 1966.

Nieuwenhuys, Rob, Frits Jacquet, and Franz Wilhelm Junghuhn. *Java's onuitputtelijke natuur: Reisverhalen, tekeningen en fotografieën van Franz Wilhelm Junghuhn*. A. W. Sijthoff, 1980.

Korthals, Pieter Willem

Korthals, Pieter Willem. "Over de ontdekking van eenige vervalschingen van de Sulphas quininae." *Bijdragen tot de natuurkundige wetenschappen* 3 (1828): 35–36.

Korthals, Pieter Willem. "Iets aangaande de afscheiding van het iödium en de bereiding der Hydriodas-potassae." *Bijdragen tot de natuurkundige wetenschappen* 3 (1828): 41.

Korthals, Pieter Willem. "Flora Brasiliae meridionalis, auctoribus Aug. de Saint-Hilaire. … Adr. de Jussieau. … Jac. Cambessedes, etc. Paris 1824–1829, etc.". *Bijdragen tot de natuurkundige wetenschappen* 5 (1830): 11–15.

Korthals, Pieter Willem. "Opmerkingen over het geslacht Nepenthes." *Bijdragen tot de natuurkundige wetenschappen* 5 (1830): 371–376.

Korthals, Pieter Willem. "Geognostische opmerkingen op eene reis in julij 1833." *Tijdschrift voor natuurlijke geschiedenis en physiologie* 1 (1834): 190–192.

Korthals, Pieter Willem. "Uittreksels uit eenen brief van den Heer P. W. Korthals aan den hoogleraar C. L. Blume." *Tijdschrift voor natuurlijke geschiedenis en physiologie* 1 (1834): 290–294.

3 Extracts from *Topographische und naturwissenschaftliche Reisen* (1845) with drawings by Erich Gruner.

Korthals, Pieter Willem. "Aanteekeningen over de vijftig Kotta's in de Padangsche bovenlanden op Sumatra." *Tijdschrift voor natuurlijke geschiedenis en physiologie* 2 (1835): 6–26.

Korthals, Pieter Willem. "Uittreksels uit brieven van den Heer P. W. Korthals, uit Sumatra, aan H. C. van Hall." *Tijdschrift voor natuurlijke geschiedenis en physiologie* 3 (1836): 7–15.

Korthals, Pieter Willem. "Tridia en Pellacalyx, twee nieuwe planten-geslachten." *Tijdschrift voor natuurlijke geschiedenis en physiologie* 3 (1836): 16–22.

Korthals, Pieter Willem. "Aanteekeningen over eenige soorten van Loranthus." *Tijdschrift voor natuurlijke geschiedenis en physiologie* 3 (1836): 187–202.

Korthals, Pieter Willem. "Over de oorzaak van het branden der Physalia." *Tijdschrift voor natuurlijke geschiedenis en physiologie* 4 (1837): 209–210.

Korthals, Pieter Willem. "Over het omhulsel van het stigma der Scaevolaceae en Goodeniaceae." *Tijdschrift voor natuurlijke geschiedenis en physiologie* 4 (1837): 370–374.

Korthals, Pieter Willem. "Uittreksel uit eenen brief van den heer P. W. Korthals, lid der Natuurkundige Commissie in de Oost-Indiën; aan H. C. van Hall, hoogleeraar te Groningen." *Algemeene konst- en letter-bode*, no. 25 (June 2, 1837): 387–393.

Korthals, Pieter Willem. "Aanteekeningen over Borneo's Zuidoostkust." *Algemeene konst- en letter-bode*, no. 47 (October 13, 1837): 244–249.

Korthals, Pieter Willem. "Opmerking over Gnetum en eenige Coniferae." *Algemeene konst- en letter-bode*, no. 50 (November 3, 1837): 290–293.

Korthals, Pieter Willem. *Observationes de Naucleis indicis*. Bonn: Carolus Georgius, 1839.

Korthals, Pieter Willem. "Aanteekeningen over Paederia." *Tijdschrift voor natuurlijke geschiedenis en physiologie* 6 (1839): 55–61.

Korthals, Pieter Willem. "Over het geslacht Tupeia." *Verhandelingen van het Bataviaasch Genootschap van Kunsten en Wetenschappen* 17 (1839): 185–196.

Korthals, Pieter Willem. "Verhandeling over de op Java, Sumatra en Borneo verzamelde Loranthaceae." *Verhandelingen van het Bataviaasch Genootschap van Kunsten en Wetenschappen* 17 (1839): 197–288.

Oort, Pieter van, and Pieter Willem Korthals. "Verslag van eenen togt van Padang naar de Boengas- en Setans- of houtbaaijen." *Tijdschrift voor Neêrland's Indië* 4, 1 (1842): 451–461.

Korthals, Pieter Willem. "Coup d'oeil sur la constitution physique et la végétation d'une partie de l'île de Sumatra." In *Le moniteur des indes orientales et occidentales, recueil de mémoires et de notices scientifiques et industriels, de nouvelles et de faits importants concernant les possessions néerlandais d'Asie et d'Amérique*, edited by Philipp Franz von Siebold and Pieter Melvill van Carnbee, vol. 1, 205–212. La Haye: Belinfante Frères, 1847–1849.

Korthals, Pieter Willem. *Topographische schets van een gedeelte van Sumatra*. Leiden: H.W. Hazenberg, 1847.

Korthals, Pieter Willem. "Aanteekeningen over een gedeelte van Borneo's Zuid-Oostkust." *Nederlandsch kruidkundig archief* 1 (1848): 20–45.

Korthals, Pieter Willem. "Blik op een gedeelte van Sumatra." *Nederlandsch kruidkundig archief* 1 (1848): 58–83.

Korthals, Pieter Willem. "Waarnemingen aangaande den berg Gédé op Java." *Nederlandsch kruidkundig archief* 1 (1848): 117–133.

Korthals, Pieter Willem. "De groeiplaats van Dipterocarpus Baudii." *Nederlandsch kruidkundig archief* 1 (1848): 134–139.

Korthals, Pieter Willem. "Bijdrage tot de kennis der Myrtaceae." *Nederlandsch kruidkundig archief* 1 (1848): 185–206.

Korthals, Pieter Willem. "Bijdrage tot de kennis der Ranunculaceae van Nederlandsch Oost-Indie." *Nederlandsch kruidkundig archief* 1 (1848): 207–211.

Korthals, Pieter Willem. "Overzigt der Sterculiaceae en Buttneriaceae van de Nederlandsche Oost-Indische bezittingen." *Nederlandsch kruidkundig archief* 1 (1848): 301–313.

Korthals, Pieter Willem. "Aanteekeningen op de familie der Violarieae van den Indischen Archipel." *Nederlandsch kruidkundig archief* 1 (1848): 356–363.

Korthals, Pieter Willem. "Bijdrage tot de kennis der Indisch Magnoliaceae." *Nederlandsch kruidkundig archief* 2 (1851): 93–98.

Korthals, Pieter Willem. "Overzigt der Rubiaceën van de Nederlandsch-Oostindische Kolonien." *Nederlandsch kruidkundig archief* 2 (1851): 99–136, 145–269.

Korthals, Pieter Willem. "Bijdrage tot de kennis der Chrysobalaneae van Nederlandsch Oostindië." *Natuurkundig tijdschrift voor Nederlandsch Indië* 7 (1854): 206–212.

Korthals, Pieter Willem. "Bijdrage tot de kennis der Chrysobalaneae van Nederl. Oostindie." *Nederlandsch kruidkundig archief* 3 (1855): 379–389.

Kuhl, Heinrich

Kuhl and Van Hasselt corresponded widely. Several (summaries of their) letters were published in German in Isis von Oken *and in French in* Bulletin des sciences naturelles et de géologie *and* Annales générales des sciences physiques. *I have not included these letters in this bibliography. Please consult the bibliography in Klaver,* Inseparable Friends, *for these references.*

Kuhl, Heinrich. *Die deutschen Fledermäuse*. Hanau, 1817.

Kuhl, Heinrich. "Responsum ad quaestionem, ab ordine disciplinarum mathematicarum et physicarum propositam: 'Cum, licet naturae corpora vario modo inter se differant, ex huc usque cognitis observationibus tamen constare videatur, ita comparatam esse rerum naturam, ut lento quasi passu ab una specie ad alteram progrediatur, atque sic continuam quasi catenam efficiat ex variis quidem annulis, intime tamen iunctis, compositam, haec catena in mammalium classe demonstranda quaeritur'" *Annales Academiae Groninganae* 1816–1817 (1818): 85–125.

Kuhl, Heinrich. *Beiträge zur Zoologie und vergleichenden Anatomie: Erste Abtheilung: Beiträge zur Zoologie.* Frankfurt am Main: Verlag der Hermannschen Buchhandlung, 1820.

Hasselt, Johan Conrad van, and Heinrich Kuhl. *Beiträge zur Zoologie und vergleichenden Anatomie: Zweite Abtheilung: Beiträge zur vergleichenden Anatomie.* Frankfurt am Main: Verlag der Hermannschen Buchhandlung, 1820.

Kuhl, Heinrich. *Buffoni et Daubentoni figurarum avium coloratarum nomina systematica.* Edited by Theodorus van Swinderen. Groningen: J. Oomkens, 1820.

Kuhl, Heinrich. "Conspectus psittacorum: cum specierum definitionibus, novarum descriptionibus, synonymis et circa patriam singularum naturalem adversariis, adjecto indice museorum, ubi earum artificiosae exuviae servantur." *Verhandlungen der Kaiserlichen Leopoldinisch-Carolinischen Academie der Naturforscher* 2 = *Nova acta physico-medica Academiae Caesareae Leopoldino-Carolinae naturae curiosorum* 10 (1820): 1–104.

Kuhl, Heinrich. "Uittreksel uit eenen brief van dr. H. Kuhl, van de Kaap de Goede Hoop, den 22sten October 1820, verzonden aan prof. Van Swinderen te Groningen." *Algemeene konst- en letter-bode*, no. 7 (February 9, 1821): 102–105.

Kuhl, Heinrich, and Johan Conrad van Hasselt. "Uittreksels uit brieven van de heeren Kuhl en Van Hasselt, aan de heeren C. J. Temminck, Th. van Swinderen en W. de Haan, geschreven in de straat Sunda, den 17den December 1820." *Algemeene konst- en letter-bode*, no. 6 (February 8, 1822): 82–88.

Kuhl, Heinrich, and Johan Conrad van Hasselt. "Uittreksels uit brieven van de heeren Kuhl en Van Hasselt, aan de heeren C. J. Temminck, Th. van Swinderen en W. de Haan." *Algemeene konst- en letter-bode*, no. 7 (February 15, 1822): 99–104.

Kuhl, Heinrich, and Johan Conrad van Hasselt. "Uittreksels uit brieven van de heeren Kuhl en Van Hasselt, aan de heeren C. J. Temminck, Th. van Swinderen en W. de Haan." *Algemeene konst- en letter-bode*, no. 8 (February 22, 1822): 115–119.

Kuhl, Heinrich, and Johan Conrad van Hasselt. "Uittreksels uit een brief van de heeren Kuhl en Van Hasselt, aan den heer D. J. van Ewyck, secretaris bij het Ministerie van Onderwijs, de Nationale Nijverheid en de Kolonien." *Algemeene konst- en letter-bode*, no. 9 (March 1, 1822): 132–135.

Kuhl, Heinrich. "Uittreksel, uit een brief van den heer Kuhl, geschreven aan boord der Noordlah, onder 24 °. Z. B. en 12 °. lengte v. Greenwich, aan Dr. Boie, te Leyden, den 8sten Oct. 1820." *Algemeene konst- en letter-bode*, no. 10 (March 8, 1822): 147–149.

Kuhl, Heinrich, and Johan Conrad van Hasselt. "Uittreksels uit een' brief van de heeren Kuhl en Van Hasselt, aan Zijne Excellentie den Ministerie voor het openbaar Onderwijs, de Nationale Nijverheid en de Kolonien." *Algemeene konst- en letter-bode*, no. 10 (March 8, 1822): 149–153.

Kuhl, Heinrich, and Johan Conrad van Hasselt. "Uittreksel uit eenen brief van de heeren Kuhl en Van Hasselt, aan den heer Temminck, directeur van 's Rijks Museum van Nat. Hist.". *Algemeene konst- en letter-bode*, no. 12 (March 16, 1821): 180–181.

Kuhl, Heinrich, and Johan Conrad van Hasselt. "Uittreksel van eenen brief van de heeren Kuhl en Van Hasselt aan de heeren F. Nees Phil. Dr. te Bonn en W. de Haan Phil Stud. te Leijden." *Algemeene konst- en letter-bode*, no. 12 (March 16, 1821): 181–184.

Macklot, Heinrich Christian

Boie, Heinrich, and Heinrich Christian Macklot. "Aanteekeningen en uittreksels uit brieven, door de Heeren Boie en Macklot van Java, aan den Heer Temminck, directeur van 's Rijks Museum, gezonden." *Bijdragen tot de natuurkundige wetenschappen* 2 (1827): 315–328.

Macklot, Heinrich Christian. "Uittreksels uit brieven van Dr. H. C. Macklot, natuuronderzoeker in Oost-Indië, aan Dr. C. J. Temminck, directeur, en J. A. Susanna, administrateur van 's Rijks Museum te Leiden." *Bijdragen tot de natuurkundige wetenschappen* 4 (1829): 298–309.

Macklot, Heinrich Christian. "Verslag van het Land, de Bewoners en Voortbrengselen van eenige plaatsen op de kust van Nieuw Guinea, welke in den loop van het jaar 1828, door de Natuurkundige kommissie in Oost-Indië, aan boord van Z. M. korvet Triton zijn bezocht, alsmede van de voorwerpen van Natuurlijke Historie, welke gedurende de reis, van den 20 Mei tot den 30 Augustus, op en langs die kust, door gemelde kommissie zijn verzameld." *Bijdragen tot de natuurkundige wetenschappen* 5 (1830): 142–182.

Müller, Salomon

Müller, Salomon. "Aanteekeningen over de natuurlijke gesteldheid van een gedeelte der westkust en binnenlanden van Sumatra; met bijvoeging van eenige waarnemingen en beschrijvingen van verscheidene, op dit, en andere Sunda-eilanden voorkomende dieren." *Tijdschrift voor natuurlijke geschiedenis en physiologie* 2 (1835): 315–355.

Oort, Pieter van, and Salomon Müller. "Aanteekeningen gehouden op eene reize over een gedeelte van het eiland Java, in den aanvang van het jaar 1833." *Verhandelingen van het Bataviaasch Genootschap van Kunsten en Wetenschappen* 16 (1836): 81–156.

Berigten over Sumatra, met eene kaart van een gedeelte van hetzelve, voornamelijk aantoonende de wegen en rivieren, welke uit de Padangsche binnenlanden naar de oostkust afloopen, getrokken uit het dagboek van den Heer S. Muller, Lid van de Nederlandsche Natuurkundige Commissie in Oost-Indie. Amsterdam: G. J. A. Beijerinck, 1837.

Müller, Salomon. "Waarnemingen over de Indische krokodillen en beschrijving van eene nieuwe soort." *Tijdschrift voor natuurlijke geschiedenis en physiologie* 5 (1838): 61–87.

Müller, Salomon. "Over eenige nieuwe zoogdieren van Borneo." *Tijdschrift voor natuurlijke geschiedenis en physiologie* 5 (1838): 134–150.

Müller, Salomon. "Über die auf den Sunda-Inseln lebenden ungeschwänzten Affen-Arten." *Archiv für Naturgeschichte* 11, 1 (1845): 72–90.

Müller, Salomon. *Bijdragen tot de kennis van Sumatra, bijzonder in geschiedkundig en ethnographisch opzigt.* Leiden: S. & J. Luchtmans, 1846.

Müller, Salomon. "Ueber den Charakter der Thierwelt auf den Inseln des indischen Archipels, ein Beitrag zu zoologischen Geographie." *Archiv für Naturgeschichte* 12, 1 (1846): 109–128.

Müller, Salomon. "Nog iets over Timor." *Algemeen Handelsblad*, no. 4526 (May 25, 1846): 4.

Müller, Salomon. "Rapport concernant une partie de la côte méridionale et des districts de l'intérieur de Bornéo." In *Coup-d'oeil général sur les possessions néerlandaises dans l'Inde archipélagique: Tome second*, edited by Coenraad Jacob Temminck, 344–380. Leiden: A. Arnz, 1847.

Müller, Salomon. "Notices sur quelques points de la Nouvelle-Guinée." In *Le moniteur des indes orientales et occidentales, recueil de mémoires et de notices scientifiques et industriels, de nouvelles et de faits importants concernant les possessions néerlandais d'Asie et d'Amérique*, edited by Philipp Franz von Siebold and Pieter Melvill van Carnbee, vol. 1, 71–77, 117–122, 213–218; vol. 2, 45–47, 376–383. La Haye: Belinfante Frères, 1847–1849.

Müller, Salomon. "Tableau du caractère zoologique des iles de l'Archipel indien." In *Le moniteur des indes orientales et occidentales, recueil de mémoires et de notices scientifiques et industriels, de nouvelles et de faits importants concernant les possessions néerlandais d'Asie et d'Amérique*, edited by Philipp Franz von Siebold and Pieter Melvill van Carnbee, vol. 2, 129–136. La Haye: Belinfante Frères, 1847–1849.

Müller, Salomon. "Een woord over den koperrijkdom van Timor." *Nieuwe Rotterdamsche Courant*, no. 266 (November 7, 1849): 2.

Müller, Salomon. “Ingezonden stuk.” *Nieuwe Rotterdamsche Courant*, no. 194 (August 15, 1850): 2.

Horner, Ludwig, Salomon Müller, and H. L. Osthoff. “Barometer-waarnemingen in de jaren 1834 tot 1839, verrigt op Sumatra door Dr. Horner, S. Müller en H. L. Osthoff, verzameld en berekend door H. L. Osthoff.” *Natuurkundig tijdschrift voor Nederlandsch Indië* 2 (1851): 153–156.

Müller, Salomon, and Ludwig Horner. “Fragmenten uit de reizen en onderzoekingen in Sumatra, gedaan op last der Nederlandsch-Indische regering, in de jaren 1833 tot 1838: I.” *Bijdragen tot de taal-, land- en volkenkunde* 2 (1854): 212–242.

Müller, Salomon, and Ludwig Horner. “Fragmenten uit de reizen en onderzoekingen in Sumatra, gedaan op last der Nederlandsch-Indische regering, in de jaren 1833 tot 1838: II.” *Bijdragen tot de taal-, land- en volkenkunde* 2 (1854): 243–271.

Müller, Salomon, and Ludwig Horner. “Fragmenten uit de reizen en onderzoekingen in Sumatra, gedaan op last der Nederlandsch-Indische regering, in de jaren 1833 tot 1838: III.” *Bijdragen tot de taal-, land- en volkenkunde* 3 (1855): 65–72.

Müller, Salomon, and Ludwig Horner. “Fragmenten uit de reizen en onderzoekingen in Sumatra, gedaan op last der Nederlandsch-Indische regering, in de jaren 1833 tot 1838: IV.” *Bijdragen tot de taal-, land- en volkenkunde* 3 (1855): 193–228.

Müller, Salomon, and Ludwig Horner. “Fragmenten uit de reizen en onderzoekingen in Sumatra, gedaan op last der Nederlandsch-Indische regering, in de jaren 1833 tot 1838: V.” *Bijdragen tot de taal-, land- en volkenkunde* 3 (1855): 229–249.

Müller, Salomon, and Ludwig Horner. “Fragmenten uit de reizen en onderzoekingen in Sumatra, gedaan op last der Nederlandsch-Indische regering, in de jaren 1833 tot 1838: VI.” *Bijdragen tot de taal-, land- en volkenkunde* 3 (1855): 313–346.

Müller, Salomon, and Ludwig Horner. “Fragmenten uit de reizen en onderzoekingen in Sumatra, gedaan op last der Nederlandsch-Indische regering, in de jaren 1833 tot 1838: VII.” *Bijdragen tot de taal-, land- en volkenkunde* 3 (1855): 347–373.

Müller, Salomon, and Ludwig Horner. “Fragmenten uit de reizen en onderzoekingen in Sumatra, gedaan op last der Nederlandsch-Indische regering, in de jaren 1833 tot 1838: VIII.” *Bijdragen tot de taal-, land- en volkenkunde* 3 (1855): 374–388.

Müller, Salomon, and Ludwig Horner. *Reizen en onderzoekingen in Sumatra, gedaan op last der Nederlandsch Indische regering, tusschen de jaren 1833 en 1838*. ‘s-Gravenhage: K. Fuhri, 1855.

Müller, Salomon. “Over eenige oudheden van Java en Sumatra.” *Bijdragen tot de taal-, land- en volkenkunde* 4 (1856): 98–122.

Müller, Salomon. *Reizen en onderzoekingen in den Indischen Archipel, gedaan op last der Nederlandsche Indische regering, tusschen de jaren 1828 en 1836*. Werken van het Koninklijk Instituut voor Taal-, Land- en Volkenkunde van Nederlandsch-Indië. Tweede afdeeling: afzonderlijke werken. 2 vols. Amsterdam: Frederik Muller, 1857.

Müller, Salomon. “Contributions to the Knowledge of New Guinea.” *Proceedings of the Royal Geographical Society of London* 2, no. 3 (1857–1858): 181–185.

Müller, Salomon. “Contributions to the Knowledge of New Guinea.” *The Journal of the Royal Geographical Society of London* 28 (1858): 264–272.

Müller, Salomon, and John Yeats. “Geognostic Sketch of the Western Portion of Timor.” *Proceedings of the Royal Geographical Society of London* 3, no. 6 (June 27 1859): 368–369.

Oort, Pieter van

Oort, Pieter van, and Salomon Müller. “Aanteekeningen gehouden op eene reize over een gedeelte van het eiland Java, in den aanvang van het jaar 1833.” *Verhandelingen van het Bataviaasch Genootschap van Kunsten en Wetenschappen* 16 (1836): 81–156.

Oort, Pieter van, and Pieter Willem Korthals. “Verslag van eenen togt van Padang naar de Boengas- en Setans- of houtbaaijen.” *Tijdschrift voor Neêrland’s Indië* 4, 1 (1842): 451–461.

Weber, Andreas, and Sylvia van Zanen. *Kunstenaar op Java: De reisdagboeken en natuurtekeningen van Pieter van Oort (1825–1833)*. Werken uitgegeven door de Linschoten-Vereeniging, 119. Walburg, 2021.

Weber, Andreas, and Sylvia van Zanen. *Kunstenaar op Sumatra: De reisdagboeken en natuurtekeningen van Pieter van Oort (1833–1834)*. Werken uitgegeven door de Linschoten-Vereeniging, 120. Walburg, 2021.

Pers, Auguste van

Pers, Auguste van, and J. C. Hageman J. Cz. *Nederlandsch Oost-Indische typen = Types indiens neerlandais: Verzameling van groote gelithografieerde platen in kleurdruk: Naar de natuur geteekend door A. van Pers: Met een verklarenden tekst in ‘t Hollandsch en Fransch: Opgedragen aan zijne excellentie den gouverneur-generaal van Nederlandsch-Indië den heer Mr. A. J. Duijmaer van Twist.* ‘s Gravenhage: C. W. Mieling, 1856.

Schwaner, Carl Anton Ludwig Maria

Schwaner, Carl Anton Ludwig Maria. “De dood van George Muller.” *Tijdschrift voor Nederlandsch Indie* 11, 1, no. 2 (1849): 139–148.

Schwaner, Carl Anton Ludwig Maria. "Resultaten van een onderzoek naar den Barito-stroom, ten opzigte zijner bevaarbaarheid voor grootere vaartuigen." *Indisch archief* 2 (1849): 394–410.

Schwaner, Carl Anton Ludwig Maria. "Bijdrage tot de kronijk der aardbevingen en vulkanische verschijnselen in Nederlandsch Indië." *Natuurkundig tijdschrift voor Nederlandsch Indië* 1 (1850): 295–302.

Brouwer, A. G. "Het schrijven van Dr. S. Müller, over de koperrijkdom van Timor, en tegen Mr. A. G. Brouwer." *Nieuwe Rotterdamsche Courant*, no. 101 (April 16, 1851): 2–3. The article contains a transcript of a letter from Schwaner to Brouwer.

Schwaner, Carl Anton Ludwig Maria, and J. H. Croockewit Hz. "Reis naar en aanteekeningen betreffende de steenkolen van de Batoe Belian (zuidoostkust van Borneo)." *Natuurkundig tijdschrift voor Nederlandsch Indië* 3 (1852): 673–688.

Schwaner, Carl Anton Ludwig Maria, and J. H. Croockewit Hz. "Aanteekeningen betreffende eenige maatschappelijke instellingen en gebruiken der Dajaks van Doesson, Moeroeng en Siang, aangetroffen onder de bij het Gouvernement van N. Indië berustende papieren van C. M. Schwaner, in leven lid der Natuurkundige commissie in N. Indië." *Tijdschrift voor Indische taal-, land- en volkenkunde* 1 (1853): 201–234.

Schwaner, Carl Anton Ludwig Maria, E. Netscher, and H. von Dewall. "Historische, geographische en statistieke aanteekeningen betreffende Tanah Boemboe; aangetroffen onder de bij het gouvernement van Nederlandsch-Indië berustende papieren van C. M. Schwaner, in leven Lid der Natuurkundige Commissie in N. Indië." *Tijdschrift voor Indische taal-, land- en volkenkunde* 1 (1853): 335–371.

Schwaner, Carl Anton Ludwig Maria. "Geologische uitstapjes in de omstreken van Buitenzorg, voornamelijk op den Salak." *Natuurkundig tijdschrift voor Nederlandsch Indië* 5 (1853): 369–426.

Schwaner, Carl Anton Ludwig Maria, and J. Pijnappel Gz. *Borneo: Beschrijving van het stroomgebied van den Barito en reizen langs eenige voorname rivieren van het zuid-oostelijk gedeelte van dat eiland, door Dr. C. A. L. M. Schwaner, op last van het Gouvernement van Nederl. Indië gedaan in de jaren 1843–1847.* 2 vols. Amsterdam: P. N. van Kampen, 1853–1854.

Schwaner, Carl Anton Ludwig Maria. "De steenkolen in het rijk van Bandjermasin." *Tijdschrift voor Nederlandsch Indie* 19, 2 (1857): 129–156.

Spanoghe, Johan Baptist

Spanoghe, Johan Baptist. "Extract of a Letter from M. Spanoghe, the Dutch resident at Coupang, Relating to the Natural History of the Island of Timor; with Some Account of the Upas Tree, Discovered There by That Gentleman." *Companion to the Botanical Magazine* 1 (1835): 308–317.

Spanoghe, Johan Baptist. "Catalogue of the Plants Found on Timor and the Neighbouring Islands." *Companion to the Botanical Magazine* 1 (1835): 344–351.

Spanoghe, Johan Baptist. "De heidenen of Badoewienen van Bantam." *Tijdschrift voor Neêrland's Indië* 1, 2 (1838): 295–305.

Spanoghe, Johan Baptist. "Prodromus florae timorensis." *Linnaea* 15 (1841): 161–208, 314–350, 476–480.

Zipelius, Alexander

Zipelius, Alexander. "Uittreksel uit eenen brief van den Heer Zippelius, natuuronderzoeker in dienst der Nederlanden, aan en medegedeeld door doct. C. L. Blume, ridder der Orde van den Nederlandschen Leeuw, hoogleraar en directeur van 's Rijks Herbarium te Brussel." *Algemeene konst- en letter-bode*, no. 19 (May 8, 1829): 294–298.

Appendix 4: Table of Contents of the *Verhandelingen*

This section contains a list of all the individual articles published as part of all three volumes of Susanna, J. A., ed., Verhandelingen over de natuurlijke geschiedenis der Nederlandsche overzeesche bezittingen door de Leden der Natuurkundige commissie in Indië en andere schrijvers, uitgegeven op last van den koning door C. J. Temminck. *3 vols. Leiden: S. & J. Luchtmans en C. C. Van der Hoek, 1839–1847. The volumes are not numbered, but* Zoologie *is generally considered to be volume 1,* Botanie *volume 2 and* Land- en Volkenkunde *volume 3.*

Zoologie, Section 1

Müller, Salomon. "Over de zoogdieren van den Indischen Archipel." 1–57.

Zoologie, Section 2: Mammals

Schlegel, Hermann, and Salomon Müller. "Bijdragen tot de natuurlijke historie van den orang-oetan (Simia satyrus)." 1–28, 54–56.

Sandifort, G. "Ontleedkundige beschouwing van een' volwassen orang-oetan (Simia satyrus, Linn.) van het mannelijk geslacht." 29–53.

Müller, Salomon, and Hermann Schlegel. "Monografisch overzicht van het geslacht Semnopithecus." 57–84.

Müller, Salomon, and Hermann Schlegel. "Over de tot heden bekende eekhorens (Sciurus) van den Indischen Archipel." 85–102.

Schlegel, Hermann, and Salomon Müller. "Bijdragen tot de natuurlijke geschiedenis der vliegende eekhorens (Pteromys)." 103–114.

Schlegel, Hermann, and Salomon Müller. "Beschrijving van een vleeschetend zoogdier uit de familie der civetkatten, Potamophilus barbatus." 115–120.

Müller, Salomon, and Hermann Schlegel. "Over eene nieuwe soort van civetkat van Borneo, Viverra Boiei." 121–128.

Schlegel, Hermann, and Salomon Müller. "Over drie buideldieren uit de familie der kengoeroe's." 129–148.

Müller, Salomon, and Hermann Schlegel. "Beschrijving eener nieuwe soort van vleeschetende buideldieren Phascogalea melas." 149–152.

 | DOI:10.1163/9789004696877_015

Müller, Salomon, and Hermann Schlegel. "Beschrijving van een merkwaardig insektenetend zoogdier, Hylomys suillus." 153–158.
Schlegel, Hermann, and Salomon Müller. "Over de, op de Oost-Indische eilanden levende soorten van het geslacht, Hylogalea." 159–168.
Müller, Salomon, and Hermann Schlegel. "Over de wilde zwijnen van den Indischen Archipel." 169–182.
Müller, Salomon, and Hermann Schlegel. "Bijdragen tot de natuurlijke historie der rhinocerossen van den Indischen Archipel." 183–194.
Schlegel, Hermann, and Salomon Müller. "Over de ossen van den Indischen Archipel." 195–208.
Müller, Salomon, and Hermann Schlegel. "Over de herten van den Indischen Archipel." 209–228.

Zoologie, Section 3: Birds

Müller, Salomon, and Hermann Schlegel. "Overzigt der in den Indischen Archipel levende soorten van het geslacht Pitta." 1–20.
Schlegel, Hermann, and Salomon Müller. "Over de neushoornvogels (Buceros) van den Indischen Archipel." 21–34.
Schlegel, Hermann, and Salomon Müller. "Over eenen nieuwen roofvogel van het ondergeslacht Lophotes, (Falco (Lophotes) Reindwardtii)." 35–40.
Müller, Salomon, and Hermann Schlegel. "Over de vischarenden (Pandion) van den Indischen Archipel." 41–48.
Schlegel, Hermann, and Salomon Müller. "Over den wespenbuizert van Oost-Indië, Falco (Pernis) ptilorhynchus." 49–52.
Müller, Salomon, and Hermann Schlegel. "Bijdragen tot de natuurlijke geschiedenis der honigvogels (Nectariniae) van den Indischen Archipel." 53–72.

Zoologie, Section 4: Reptiles

Müller, Salomon, and Hermann Schlegel. "Over de krokodillen van den Indischen Archipel." 1–28.
Schlegel, Hermann, and Salomon Müller. "Over de schildpadden van den Indischen Archipel en beschrijving eener nieuwe soort van Sumatra." 29–36.
Müller, Salomon, and Hermann Schlegel. "Over de in den Indischen Archipel levende hagedisachtige dieren van het geslacht Monitor." 37–48.

Müller, Salomon, and Hermann Schlegel. "Over de, in den Indischen Archipel voorkomende soorten van het slangengeslacht Trigonocephalus." 49–58.
Schlegel, Hermann, and Salomon Müller. "Over de slangensoorten van het geslacht Homalopsis, uit den Indischen Archipel." 59–62.
Schlegel, Hermann, and Salomon Müller. "Over de slangen van het geslacht Elaps, welke den Indischen archipel bewonen." 63–68.
Müller, Salomon, and Hermann Schlegel. "Over de brilslangen van den Indischen Archipel." 69–72.

Zoologie, Section 5: Fishes

Müller, Salomon, and Hermann Schlegel. "Beschrijving van een' nieuwen zoetwatervisch van Borneo, Osteoglossum formosum." 1–8.
Schlegel, Hermann, and Salomon Müller. "Beschrijving van vier Oost-Indische zeevisschen, behoorende tot het geslacht Amphacanthus." 9–16.
Schlegel, Hermann, and Salomon Müller. "Overzigt der uit de Sunda- en Moluksche zeeën bekende visschen van de geslachten Amphiprion, Premnas, Pomacentrus, Glyphisodon, Dascyllus en Heliases." 17–26.

Zoologie, Section 6: Insects

Haan, Wilhem de. "Bijdragen tot de kennis der Papilionidea." 1–44.
Haan, Wilhem de. "Bijdragen tot de kennis der Orthoptera." 45–248.

Botanie

Korthals, Pieter Willem. "Over het geslacht Nepenthes." 1–44.
Korthals, Pieter Willem. "Over eenige soorten van de familie der Dipterocarpeae." 45–76.
Korthals, Pieter Willem. "Over eenige soorten van het geslacht Bauhinia." 77–92.
Korthals, Pieter Willem. "Bijdrage tot de kennis der Ternstroemiaceae." 93–174.
Korthals, Pieter Willem. "Bijdragen tot de geschlachten Cratoxylon en Tridesmis." 175–180.
Korthals, Pieter Willem. "Bijdragen tot de geslachten Salacia en Hippocratea." 181–188.
Korthals, Pieter Willem. "Over de geslachten Praravinia en Omphacarpus." 189–194.
Korthals, Pieter Willem. "Bijdragen tot de kennis der eiken van den Indischen Archipel." 195–217.

Korthals, Pieter Willem. "Bijdrage tot de kennis der Indische Melastomaceae." 218–255.
Korthals, Pieter Willem. "Bijdragen tot de geslachten Cleisocratera, Boschia en Maranthes." 256–259.

Land- en volkenkunde

Müller, Salomon. "Bijdragen tot de kennis van Nieuw-Guinea." 1–80.
Müller, Salomon. "Schetsen uit de oostelijke streken des Indischen Archipels." 81–128.
Müller, Salomon. "Bijdragen tot de kennis van Timor en eenige andere naburige eilanden." 129–320.
Müller, Salomon. "Reis in het zuidelijk gedeelte van Borneo." 321–446.
Müller, Salomon. "Gezigten van bergen, kraters, kusten en eilanden van Java, Sumatra en de Straat Sunda." 447–469.

Bibliography

Works about and by members of the Natuurkundige Commissie, and papers from the Verhandelingen *(as listed in Appendix 2, 3 and 4) are only included in this bibliography if they are cited in the main text.*

Aa, A. J. van der. *Biographisch woordenboek der Nederlanden, bevattende levensbeschrijvingen van zoodanige personen, die zich op eenigerlei wijze in ons vaderland hebben vermaard gemaakt*. 22 vols. Haarlem: J. J. van Brederode, 1852–1878.

Abendanon, E. C. *Geologische en geographische doorkruisingen van Midden-Celebes (1909–1910)*. 4 vols. E. J. Brill, 1915–1918.

Akerman, James R., ed. *Cartographies of Travel and Navigation*, The Kenneth Nebenzahl, Jr., Lectures in the History of Cartography. University of Chicago Press, 2006.

Akerman, James R., ed. *The Imperial Map: Cartography and the Mastery of Empire*, The Kenneth Nebenzahl, Jr., Lectures in the History of Cartography. University of Chicago Press, 2009.

Akerman, James R. "Introduction." In *The Imperial Map: Cartography and the Mastery of Empire*, edited by James R. Akerman. The Kenneth Nebenzahl, Jr., Lectures in the History of Cartography. University of Chicago Press, 2009.

Akerman, James R., ed. *Decolonizing the Map: Cartography from Colony to Nation*, The Kenneth Nebenzahl, Jr., Lectures in the History of Cartography. University of Chicago Press, 2017.

Aldrich, Robert. *Greater France: A History of French Overseas Expansion*. European Studies. Macmillan, 1996.

Allen, David Elliston. *The Naturalist in Britain: A Social History*. Penguin Books, 1976.

Anderson, Clare. *Convicts: A Global History*. Cambridge University Press, 2022.

Anderson, Clare. *Subaltern Lives: Biographies of Colonialism in the Indian Ocean World, 1890–1920*. Cambridge University Press, 2012.

Anderson, Warwick. "Remembering the Spread of Western Science." *Historical Records of Australian Science* 29, no. 2 (2018): 73–81.

Anonymus. "Notice sur les voyages de M. Duvaucel." *Le musée des variétés litteraires* 4, no. 24 (1824): 201–206.

Anonymus. "Korte schets van het eiland Banka." *Tijdschrift voor Neêrland's Indië* 8, 4 (1846): 125–155.

Anonymus. "Levensberigt van Dr. C. A. L. M. Schwaner." *Natuurkundig tijdschrift voor Nederlandsch Indië* 2 (1851): 189–191.

Anonymus. "Untitled." *Siebzehntes Neujahrsblatt zum Besten des Waisenhauses auf das Neujahr 1854 von einer Gesellschaft herausgegeben*, no. 76 (1854).

Anonymus. "Johan Baptist Spanoghe, resident van Timor." *Handelingen en geschriften van het Indisch Genootschap* 2 (1855): 393–400.

Anonymus. "Gerrit van Raalten." *Handelingen en geschriften van het Indisch Genootschap* 5 (1858): 88–92.

Anonymus, *Gedenkboek der Nederlandsch Indische theecultuur, 1824–1924: Uitgegeven door het Proefstation voor Thee bij gelegenheid van het Theecongres met Tentoonstelling, Bandoeng 1924*. G. Kolff, 1924.

Antunes, Anderson Pereira, Luisa Massarani, and Ildeu de Castro Moreira. "'Practical Botanists and Zoologists': Contributions of Amazonian Natives to Natural History Expeditions (1846–1865)." *Historia Crítica*, no. 73 (2019): 137–160.

Antunes, Cátia, and Jos Gommans, eds. *Exploring the Dutch Empire: Agents, Networks and Institutions, 1600–2000*. Bloomsbury Academic, 2015.

Attenborough, David. *Life On Air*. Updated and revised edition. BBC Books, 2014.

Badenhuizen, Nicolaas. "The Botanical Garden at Bogor, Java." *Canadian Journal of Netherlandic Studies* 16, no. 2 (1995): 1–4.

Bakels, Jet. *Het verbond met de tijger: Visies op mensenetende dieren in Kerinci, Sumatra.* CNWS Publications, 93. Research School of Asian, African, and Amerindian Studies, Universiteit Leiden, 2000.

Balen, Bas van. "Gevleugeld eremetaal: De natuuronderzoekers van Indonesië en hun vogelnamen." In *Gevleugelde geschiedenis van Nederland: De Nederlanders en hun vogels*, edited by Jan Luiten van Zanden, Helena Čordašev and Erik de Bruin. Ambo|Anthos, 2022.

Balen, Bas van. "Ornithological History." In *Birds of the Indonesian Archipelago: Greater Sundas and Wallacea*, edited by James A. Easton, Bas van Balen, Nick W. Brickle and Frank E. Rheindt. Lynx, 2016.

Barnes, Ruth. "Recording Cultures: Collecting in Eastern Indonesia." In *Colonial Collections Revisited*, edited by Pieter ter Keurs. CNWS Publications, 2007.

Bartholemew, J. G., ed. *The Century Atlas and Gazetteer of the World.* John Walker, 1890.

Basalla, George. "The Spread of Western Science." *Science. New Series* 156, no. 3775 (May 5, 1967): 611–622.

Bastin, John. "The Natural History Researches of Dr Thomas Horsfield (1773–1859): First American Naturalist of Indonesia." In *Zoological Researches in Java, and the Neighbouring Islands: With a Memoir by John Bastin*. Oxford University Press, 1990.

Bastin, John. "Sir Stamford Raffles and the Study of Natural History in Penang, Singapore and Indonesia." *Journal of the Malaysian Branch of the Royal Asiatic Society* 63, no. 2 (no. 259) (1990): 1–25.

Bastin, John Sturgus. *Sir Stamford Raffles and Some of His Friends and Contemporaries: A Memoir of the Founder of Singapore*. World Scientific, 2019.

Bastin, John, and Bea Brommer. *Nineteenth Century Prints and Illustrated Books of Indonesia: With Particular Reference to the Print Collections of the Tropenmuseum, Amsterdam: A Descriptive Bibliography*. Spectrum, 1979.

Bastin, John, and D. T. Moore. "The Geological Researches of Dr Thomas Horsfield in Indonesia 1801–1819." *Bulletin of the British Museum (Natural History). Historical Series* 10, no. 3 (1982): 75–115.

Bennett, Edward T. "Evidences in Proof of Certain Statements Contained in the 'Gardens and Menagerie of the Zoological Society Delineated'." *The Magazine of Natural History, and Journal of Zoology, Botany, Mineralogy, Geology, and Meteorology* 4 (1831): 199–206.

Bennett, Edward Turner. *The Gardens and Menagerie of the Zoological Society Delineated*. 2 vols. London: Thomas Tegg; N. Hailes, 1830.

Berkel, Klaas van. "Een onwillige mecenas? De rol van de VOC bij het natuurwetenschappelijk onderzoek in de zeventiende eeuw." In *VOC en cultuur: Wetenschappelijke en culturele relaties tussen Europa en Azië ten tijde van de Verenigde Oostindische Compagnie*, edited by J. Bethlehem and A. C. Meijer. Schiphouwer en Brinkman, 1993.

Berkel, Klaas van. "Empire without Science? The Dutch Scholarly World and Colonial Science around 1800." In *Empire and Science in the Making: Dutch Colonial Scholarship in Comparative Global Perspective*, edited by Peter Boomgaard. Palgrave Studies in the History of Science and Technology. Palgrave Macmillan, 2013.

Berkel, Klaas van. *Universiteit van het Noorden: Vier eeuwen academisch leven in Groningen: Deel 1 De oude universiteit, 1614–1876*. Studies over de geschiedenis van de Groningse universiteit, 8. Verloren, 2014.

Berkel, Klaas van, Albert van Helden, and Lodewijk Palm, eds. *A History of Science in the Netherlands: Survey, Themes and Reference*. Brill, 1999.

Bertens, Laura M. F., and Ann Marie Wilson. "Wonder, Empire, Science: The Quagga and Other Extinctions on Display at Naturalis." *Museum & Society* 20, no. 1 (2022): 33–49.

Bethlehem, J., and A. C. Meijer, eds. *VOC en cultuur: Wetenschappelijke en culturele relaties tussen Europa en Azië ten tijde van de Verenigde Oostindische Compagnie*, Thesaurus, 7. Schiphouwer en Brinkman, 1993.

Bik, Johannes Theodorus. "Aanteekeningen op eene reis door het westelijk gedeelte van de residentie Bantam, met de Heeren Van Hasselt en Meaurevert in het jaar 1823." *Tijdschrift voor Indische taal-, land- en volkenkunde* 16 (1867): 260–276.

Blais, Hélène. L'empire de la nature: Une histoire des jardins botaniques coloniaux (Fin XVIIIe siècle–années 1930). Champ Vallon, 2023.

Bleeker, Pieter. "Algemeen verslag der werkzaamheden van de Natuurkundige Vereeniging in Nederlandsch Indië." *Natuurkundig tijdschrift voor Nederlandsch Indië* 2 (1851): 1–24.

Bleeker, Pieter. *Reis door de Minahassa en den Molukschen Archipel gedaan in de maanden september en oktober 1855 in het gevolg van den gouverneur generaal Mr. A. J. Duymaer van Twist: Eerste deel.* Batavia: Lange, 1856.

Bleeker, Pieter "De invoering der volkszijdeteelt in Nederlandsch Indië." *Tijdschrift voor Nederlandsch Indië. Nieuwe serie* 1, 1 (1863): 11–29, 82–98, 127–136.

Bloembergen, Marieke, and Martijn Eickhoff. *The Politics of Heritage in Indonesia: A Cultural History*. Asian Connections. Cambridge University Press, 2020.

Blume, Carl Ludwig. *Bijdragen tot de flora van Nederlandsch Indië*. Batavia: Lands Drukkerij, 1825.

Blussé, Leonard, and Ilonka Ooms, eds. *Kennis en Compagnie: De Verenigde Oost-Indische Compagnie en de moderne wetenschap*. Balans, 2002.

Boie, Heinrich. "Manuel d'ornithologie, ou tableau systèmatique des oiseaux qui se trouvent en Europe par C. J. Temminck." *Heidelbergische Jahrbücher der Literatur* 25–26 (1816): 387–416.

Boomgaard, Peter. "Hunting and Trapping in the Indonesian Archipelago, 1500–1950." In *Paper Landscapes: Explorations in the Environmental History of Indonesia*, edited by Peter Boomgaard, Freek Colombijn and David Henley. Verhandelingen van het Koninklijk Instituut voor Taal-, Land- en Volkenkunde, 178. KITLV, 1997.

Boomgaard, Peter. "The Making and Unmaking of Tropical Science: Dutch Research on Indonesia, 1600–2000." *Bijdragen tot de taal-, land- en volkenkunde* (*BKI*) 162, no. 2/3 (2008): 191–217.

Boomgaard, Peter, ed. *Empire and Science in the Making: Dutch Colonial Scholarship in Comparative Global Perspective, 1760–1830*, Palgrave Studies in the History of Science and Technology. Palgrave Macmillan, 2013.

Bosma, Ulbe. *The Sugar Plantation in India and Indonesia: Industrial Production, 1770–2010*. Studies in Comparative World History. Cambridge University Press, 2013.

Bosma, Ulbe, Juan Giusti-Cordero, and G. Roger Knight, eds. *Sugarlandia Revisited: Sugar and Colonialism in Asia and the Americas, 1800–1940*, International Studies in Social History, 9. Berghahn Books, 2007.

Bossenbroek, Martin. *Van Holland naar Indië: Het transport van koloniale troepen voor het Oost-Indische leger, 1815–1809*. Amsterdam: Bataafsche Leeuw, 1986.

Bossenbroek, Martin. *De wraak van Diponegoro: Begin en einde van Nederlands-Indië*. Athenaeum-Polak & Van Gennep, 2020.

Braam, J. J. van. *Vues de Java: Desinées d'après natures et lithographiées sous la direction de monsieur J. J. van Braam*. Amsterdam: Frans Buffa, 1842.

Brébion, Antoine. "Diard, naturaliste français dans l'Extrême-Orient." *T'oung Pao. Second Series* 15, no. 2 (1914): 203–213.

Breman, Jan. *Koelies, planters en koloniale politiek: Het arbeidsregime op de grootlandbouwondernemingen aan Sumatra's Oostkust in het begin van de twintigste eeuw.*

Verhandelingen van het Koninklijk Instituut voor Taal-, Land- en Volkenkunde, 123. Derde herziene druk. KITLV, 1992.

Breman, Jan. "Colonialism and Its Racial Imprint." *Sojourn: Journal of Social Issues in Southeast Asia* 35, no. 3 (2020): 463–492.

Brendonck, A. A. *Mijne reize naar Nederlandsch-Indië: Militaire loopbaan op Java en Bali en mijn terugtogt naar het vaderland.* Lagerweij, 1859.

Brescius, Moritz von. *German Science in the Age of Empire: Enterprise, Opportunity and the Schlagintweit Brothers.* Cambridge University Press, 2019.

Brockway, Lucille H. *Science and Colonial Expansion: The Role of the British Royal Botanic Gardens.* Studies in Social Discontinuity. Academic Press, 1979.

Bürger, Heinrich. "Aanmerkingen gehouden op eene reize door eenige districten der Padangsche bovenlanden." *Verhandelingen van het Bataviaasch Genootschap van Kunsten en Wetenschappen* 16 (1836): 157–226.

Burnett, D. Graham. "'It Is Impossible to Make a Step without the Indians': Nineteenth-Century Geographical Exploration and the Amerindians of British Guiana." *Ethnohistory* 49, no. 1 (2002): 3–40.

Camerini, Jane R. "Wallace in the Field." *Osiris* 11 (1996): 44–65.

Champoiseau, Noel. "Voyages aux Indes Orientales: Notice sur les voyages de M. Diard, naturaliste français." *Journal des voyages, découvertes et navigations modernes ou archives géographiques du XIX*[e] *siècle* 43 (1829): 222–239.

Chandler, David Porter. "Cambodia before the French: Politics in a Tributary Kingdom, 1794–1848." PhD, University of Michigan, 1977.

Chansigaud, Valérie. *The History of Ornithology.* New Holland, 2009.

Chansigaud, Valérie. *All About Birds: A Short Illustrated History of Ornithology.* Princeton University Press, 2010.

Chevalier, C. "Compte-rendu annuel du secrétaire perpétuel." *Annales de la Société d'Agriculture, Sciences, Arts, et Belles-Lettres du Département d'Indre-et-Loire. Deuxième série* 57 (1878): 39–46.

Chijs, J. A. van der. *Geschiedenis van de gouvernements thee-cultuur op Java.* Landsdrukkerij, 1903.

Choi, Byung Wook. *Southern Vietnam under the Reign of Minh Mạng (1820–1841): Central Policies and Local Response.* Southeast Asia Program Series, 20. Southeast Asia Program Publications, Southeast Asia Program, Cornell University, 2004.

Claeys, Jean-Yves. "Deux Français à la conquête de Singapour: Les naturalistes Alfred Duvaucel et Pierre-Médard Diard." *France-Asie* 9, no. 10 (1954): 1001–1007.

Claver, Alexander. *Dutch Commerce and Chinese Merchants in Java: Colonial Relationships in Trade and Finance, 1800–1942.* Verhandelingen van het Koninklijk Instituut voor Taal-, Land- en Volkenkunde, 291. Brill, 2014.

Colombijn, Freek. "Hoe de Nederlanders op Sumatra aan de weg timmerden: De ontwikkeling van het wegennet op Midden-Sumatra, 1600–1870." In *Macht en majesteit:*

Opstellen voor Cees Fasseur bij zijn afscheid als hoogleraar in de geschiedenis van Indonesië aan de Universiteit Leiden, edited by J. Thomas Lindblad and Willem van der Molen. Semaian. Universiteit Leiden: Opleiding Talen en Culturen van Zuidoost-Azië en Oceanië, 2002.

Cook, Harold J. *Matters of Exchange: Commerce, Medicine, and Science in the Dutch Golden Age*. Yale University Press, 2007.

Cooper, Frederick, and Ann Laura Stoler, eds. *Tensions of Empire: Colonial Cultures in a Bourgeois World*. University of California Press, 1997.

Cordier, Henri. *Le Consulat de France à Hué sous la Restoration: Documents inédits tirés des archives des départements des affaires étrangères de la marine et des colonies*. Paris: Ernest Leroux, 1884.

Cornets de Groot van Kraaijenburg, Johan Pieter. *Overzigt der reis in Nederlandsch Indië, door Zijne Koninklijke Hoogheid Prins Willem Frederik Hendrik der Nederlanden, gedurende het jaar 1837*. Batavia: Lands-Drukkerij, 1838.

Crais, Clifton C., and Pamela Scully. *Sara Baartman and the Hottentot Venus: A Ghost Story and a Biography*. Princeton University Press, 2009.

Crawfurd, John. *History of the Indian Archipelago: Containing an Account of the Manners, Arts, Languages, Religions, Institutions, and Commerce of its Inhabitants*. Edinburgh: Archibald Constable, 1820.

Crawfurd, John. *Journal of an Embassy from the Governor-General of India to the Courts of Siam and Cochin China: Exhibiting a View of the Actual State of Those Kingdoms*. Second edition. 2 vols. London: Henry Colburn, 1830.

Curry, H. A., N. Jardine, J. A. Secord, and E. C. Spary, eds. *Worlds of Natural History*. Cambridge University Press, 2018.

Cuvier, Georges. "Notice sur les voyages de MM. Diard et Duvaucel, naturalistes français, dans les Indes orientales et dans les îles de la Sonde; extraite de leur correspondence, et lue à l académie des sciences." *Revue encyclopédique, ou analyses et annonces raisonnées des productions les plus remarquables dans la Littérature, les Sciences et les Arts* 10, no. 30 (1821): 473–482.

Cuvier, Georges. *Discours sur les révolutions de la surface du globe, et sur les changemens qu'elles ont produits dans le règne animal*. Troisième édition française. Paris: G. Dufour et Edmond d'Ocagne, 1825.

Cuvier, Georges. *Histoire des progrès des sciences naturelles, depuis 1789 jusqu'a ce jour*. 5 vols. Paris: Librairie encyclopédique de Roret; Pourrat Frères, 1834.

Cuvier, Georges, and Achille Valenciennes. *Histoire naturelle des poissons*. 22 vols. Paris: F. G. Levrault, 1828–1849.

Das, Subhadra, and Miranda Lowe. "Nature Read in Black and White: Decolonial Approaches to Interpreting Natural History Collections." *Journal of Natural Science Collections* 6 (2018): 4–14.

De Prins, Bart. *Voor keizer en koning: Leonard du Bus de Gisignies, 1780–1849, commissaris-generaal van Nederlands-Indië*. Balans, 2002.

Deerr, Noël. *Cane Sugar: A Text-Book on the Agriculture of the Sugar Cane, the Manufacture of Cane Sugar, and the Analysis of Sugar House Products*. Norman Rodger, 1911.

Delden, E. E. van. *Indisch Genootschap, 1854–1940*. Klein Repertorium, 4. Koninklijk Instituut voor de Tropen, 1988.

Delden, E. E. van. *Tijdschrift voor Nederlandsch Indië, 1838–1866*. Klein Repertorium, 5. Koninklijk Instituut voor de Tropen, 1990.

Delden, E. E. van. *Acht algemene tijdschriften, 1834–1864*. Klein Repertorium, 7. Koninklijk Instituut voor de Tropen, 1993.

Deleuze, Joseph Philippe François. *Histoire et description du Muséum royal d'histoire naturelle*. 2 vols. Paris: M. A. Royer, 1823.

Derfelden van Hinderstein, Gijsbert Franco von. *Algemeene kaart van Nederlandsch Oostindië, met hoogstdeszelfs toestemming opgedragen aan Zijne Majesteit den Koning der Nederlanden, volgens echte bescheiden, grootendeels in handschrift berustende bij de archieven van het Koloniaal Bestuur te 's Gravenhage en te Batavia, mitsgaders uit mededeelingen van geleerden, beambten, zee- en landofficieren in Nederlandsch Indië te zamengesteld*. La Haye: Imprimerie de l'État, 1842.

Desmarest, A. G. *Mammalogie ou Description des espèces de mammifères*. 2 vols. Paris: Veuve Agasse, 1820.

Desmond, Adrian. "The Making of Institutional Zoology in London 1822–1836: Part 1." *History of Science* 23, no. 2 (1985): 153–185.

Desmond, Adrian. "The Making of Institutional Zoology in London 1822–1836: Part 2." *History of Science* 23, no. 3 (1985): 223–250.

Diard, Hippolyte. "Pierre-Médard Diard, naturaliste." *Annales de la Société d'Agriculture, Sciences, Arts, et Belles-Lettres du Département d'Indre-et-Loire. Deuxième série* 56 (1877): 198–210.

Diard, Pierre. *Catalogue raisonné des plantes qui croissent naturellement a St.-Calais et dans ses environs*. Edited by Ed. Guéranger. Saint-Calais: Peltier-Voisin, 1852.

Diard, Pierre-Médard. "Chine et Indo-Chine: Faits commerciaux, No. 34: Cochinchine." *Annales du commerce extérieur* troisième série, no. 1466 (1863).

Dick, Howard, Vincent J. H. Houben, J. Thomas Lindblad, and Thee Kian Wie. *The Emergence of a National Economy: An Economic History of Indonesia, 1800–2000*. Asian Studies Association of Australia: Southeast Asia Publications Series. Allen & Unwin, 2002.

Diessen, J. R. van, ed. *Grote atlas van Nederlands Oost-Indië = Comprehensive Atlas of the Netherlands East Indies*. Tweede editie. Asia Maior, 2004.

Diessen, J. R. van, ed. *Grote atlas van de Verenigde Oost-Indische Compagnie = Comprehensive Atlas of the Dutch United East India Company*. 7 vols. Asia Maior/Atlas Maior, 2006–2010.

Dorai, Francis, and Martyn E. Y. Low, eds. *Diard & Duvaucel: French Natural History Drawings of Singapore and Southeast Asia (1818–1820) = Dessins d'histoire naturelle français de Singapour et d'Asie du Sud-Est (1818–1820)*. Epigram Books, 2021.

Dozier, Laura, ed. *Natural History Drawings: The Complete William Farquhar Collection, Malay Peninsula 1803–1818*. Didier Millet, 2010.

Drayton, Richard. *Nature's Government: Science, Imperial Britain, and the 'Improvement' of the World*. Yale University Press, 2000.

Drieënhuizen, Caroline, and Fenneke Sysling. "Java Man and the Politics of Natural History: An Object Biography." *Bijdragen tot de taal-, land- en volkenkunde* 177 (2021): 290–311.

Driver, Felix. *Geography Militant: Cultures of Exploration and Empire*. Blackwell, 2001.

Driver, Felix. "Hidden Histories Made Visible? Reflections on a Geographical Exhibition." *Transactions of the Institute of British Geographers* 38, no. 3 (2013): 420–435.

Driver, Felix. "Exploration as Knowledge Transfer: Exhibiting Hidden Histories." In *Mobilities of Knowledge*, edited by Heike Jöns, Peter Meusburger and Michael Heffernan. Knowledge and Space, 10. Springer, 2017.

Driver, Felix, and Lowri Jones. *Hidden Histories of Exploration: Researching the RGS-IBG Collections*. Royal Holloway, 2009.

Drouin, Jean-Marc. "De Linné à Darwin: Les voyageurs naturalistes." In *Éléments d'histoire des sciences*, edited by Michel Serres. Bordas Cultures, 1989.

Duvaucel, Alfred, Victor Jacquemont, and Jérôme Petit. *Impressions de voyages en Inde, 1818–1832*. Agora, 385. Pocket, 2015.

Edney, Matthew H. *Mapping an Empire: The Geographical Construction of British India, 1765–1843*. University of Chicago Press, 1990.

Eijnden, Roeland van, and Adriaan van der Willigen. *Geschiedenis der vaderlandsche schilderkunst, sedert de helft der XVIII eeuw: Deel 3*. Haarlem: A. Loosjes, 1820.

Fan, Fa-Ti. *British Naturalists in Qing China: Science, Empire, and Cultural Encounter*. Harvard University Press, 2004.

Farquhar, William, A. Seton, Pierre-Médard Diard, and G. J. Siddons. "Account of a New Species of Tapir Found in the Peninsula of Malacca." *Asiatick Researches* 13 (1820): 417–427.

Fasseur, C. *Kultuurstelsel en koloniale baten: De Nederlandse exploitatie van Java 1840–1860*. Universitaire Pers, 1975.

Fasseur, C. *The Politics of Colonial Exploitation: Java, the Dutch, and the Cultivation System*. Translated by R. E. Elson. Studies on Southeast Asia, 8. Southeast Asia Program, Cornell University, 1992.

Fasseur, C. *Indischgasten*. Bert Bakker, 1997.

Finlayson, George, and Thomas Stamford Sir Raffles. *The Mission to Siam, and Hué, the Capital of Cochin China, in the Years 1821–2*. London: John Murray, 1826.

Finnegan, Diarmid A. "The Spatial Turn: Geographical Approaches in the History of Science." *Journal of the History of Biology* 41, no. 2 (2008): 369–388.

Finnegan, Diarmid A., and Jonathan Jeffrey Wright, eds. *Spaces of Global Knowledge: Exhibition, Encounter and Exchange in an Age of Empire*, Studies in Historical Geography. Routledge, 2015.

Fisher, David R. "Ricketts, Charles Milner (1776–1867)." In *The History of Parliament: The House of Commons, 1790 1820: V Members Q – Y*, edited by R. G. Thorne. Secker & Warburg, 1986.

Förschler, Silke, and Anne Mariss, eds. *Akteure, Tiere, Dinge: Verfahrensweisen der Naturgeschichte in der Frühen Neuzeit*. Böhlau, 2017.

Forsten, Eltio Alegondus. *Responsio de aethere et naphthis*. Leiden: S. & J. Luchtmans, 1832.

Forsten, Eltio Alegondus. *Dissertatio botanico-pharmaceutico-medica inauguralis de cedrela febrifuga*. Leiden: C. C. van der Hoek, 1836.

Fransen, C. H. J. M., L. B. Holthuis, and J. P. H. M. Adema. "Type-catalogue of the Decapod Crustacea in the Collections of the Nationaal Natuurhistorisch Museum, with Appendices of Pre-1900 Collectors and Material." *Zoologische verhandelingen* 311 (1997): i–xvi, 1–344.

Frenken, Koen. "Geography of Scientific Knowledge: A Proximity Approach." *Quantitative Science Studies* 1, no. 3 (2020): 1007–1016.

Friedrich, Susanne. "The Importance of Being a Good Employee: Georg Everhard Rumphius, the Dutch East India Company, and Knowledge in the Late Seventeenth Century." *Early Modern Low Countries* 3, no. 2 (2019): 183–207.

Friedrich, Susanne, Arndt Brendecke, and Stefan Ehrenpreis, eds. *Transformations of Knowledge in Dutch Expansion*, Pluralisierung & Autorität, 44. De Gruyter, 2015.

Fritze, E. A. "Voorberigt." *Verhandelingen van het Bataviaasch Genootschap van Kunsten en Wetenschappen* 16 (1836): i–xxvi.

G. M. [Guérin-Méneville, Félix Édouard]. "Mémoires pour servir à l'histoire naturelle des possessions Néerlandaises dans l'Inde, par les membres de la commission d'Histoire Naturelle de l'Inde, et quelques autres auteurs." *Revue zoologique* 5, no. 5 (1842): 150–154.

Gaastra, Femme S. *The Dutch East India Company: Expansion and Decline*. Walburg, 2003.

Gaastra, Femme S. *Geschiedenis van de VOC: Opkomst, bloei en ondergang*. Elfde druk. Walburg, 2012.

Gaffron, Heinrich von, and J. Pijnappel Gz. "Beschrijving van het westelijke gedeelte van de Zuid- en Oosterafdeling van Borneo (de afdeeling Sampit en de Zuidkust)." *Bijdragen tot de taal-, land- en volkenkunde* 7 (1860): 243–346.

Gage, A. T., I. H. Burkill, and William Jack. "William Jack's Letters to Nathaniel Wallich, 1819–1821." *Journal of the Straits Branch of the Royal Asiatic Society*, no. 73 (1916): 147–268.

Gassó Miracle, M. Eulàlia. "The Significance of Temminck's Work on Biogeography: Early Nineteenth Century Natural History in Leiden, the Netherlands." *Journal of the History of Biology* 41, no. 4 (2008): 677–716.

Gassó Miracle, M. Eulàlia. "On Whose Authority? Temminck's Debates on Zoological Classification and Nomenclature: 1820–1850." *Journal of the History of Biology* 44, no. 3 (2011): 445–481.

Gassó Miracle, M. Eulàlia. "Temminck's Order: Debates on Zoological Classification, 1800–1850." PhD, Leiden University, 2019.

Gassó Miracle, M. Eulàlia. *Coenraad Jacob Temminck and the Emergence of Systematics (1800–1850)*. Emergence of Natural History, 4. Brill, 2021.

Gassó Miracle, Eulàlia, L. Stork, Andreas Weber, M. Ameryan, and K. Wolstencroft. "Natuurkundige Commissie Archives Online." Brill, 2020.

Gassó Miracle, Eulàlia, and Tiny Monquil-Broersen, eds. *Van onschatbare waarde: 200 jaar Naturalis*. Walburg, 2020.

Gebhardt, Ludwig. *Die Ornithologen Mitteleuropas: Ein Nachschlagewerk*. Brühlscher Verlag, 1964.

Gebroeders Dag en Nacht. *Licht- en schaduwbeelden uit de binnenlanden van Java: Over het karakter, de mate van beschaving, de zeden en gebruiken der Javanen; over de invoering van het Christendom op Java, het bezigen van vrijen arbeid en andere vragen van den dag: Verhalen en gesprekken, verzameld op reizen door gebergten en bosschen, in de woningen van armen en rijken*. Leiden: Jacs. Hazenberg Corns. zoon, 1854.

Gentil, Ambroise. "Notice sur Pierre Diard, botaniste manceau." *Bulletin de la Societé d'Agriculture, Sciences et Arts de la Sarthe* 38, no. 3 (1901–1902): 229–252.

Geoffroy Saint Hilaire, Étienne, and Frédéric Cuvier. *Histoire naturelle des mammifères, avec des figures originales, coloriées, dessinées d'après des animaux vivans*. 7 vols. Paris: A. Belin, 1824.

Gibson, Susannah. "The Careering Naturalists: Creating Career Paths in Natural History, 1790–1830." *Archives of Natural History* 44, no. 2 (2017): 195–214.

Gijzen, Agatha. *'s Rijks Museum van Natuurlijke Historie, 1820–1915*. W. L. & J. Brusse, 1938.

Glas, Elly. "Een Oostenrijker op Borneo: Leven en werk van H. A. Henrici, 1783–1838." *Caert-Thresoor* 1, no. 2 (1982): 18–20.

Glaubrecht, Matthias. "The 'Experience' of Nature: From Salomon Müller to Ernst Mayr, or the Insights of Travelling Naturalists Toward a Zoological Geography and Evolutionary Biology." In *Die Entstehung biologischer Disziplinen II: Beiträge zur 10. Jahrestagung der DGGTB in Berlin 2001*, edited by Uwe Hoßfeld and Thomas Junker. Verhandlungen zur Geschichte und Theorie der Biologie, 9, 245–282. VWB, 2001.

Glendinning, Victoria. *Raffles and the Golden Opportunity, 1781–1826*. Revised paperback edition. Profile Books, 2018.

Gols, Paul. "Rijke oud-Indië gangers kopen Wateren (versie 4, januari 2018)." (2018). https://www.deruyter.org/uploads/media/606ddeed9f4b6.docx.

Gorsel, J. T. van. *Pioneers and Milestones of Indonesian Geology*. 4 vols. ITB Press, 2022.

Goss, Andrew. *The Floracrats: State-Sponsored Science and the Failure of the Enlightenment in Indonesia*. New Perspectives in Southeast Asian Studies. University of Wisconsin Press, 2011.

Goss, Andrew. "Reinventing the Kebun Raya in the New Republic: Scientific Research at the Bogor Botanical Gardens in the Age of Decolonization." *Studium* 11, no. 3 (2018).

Goss, Andrew. "Introduction: An Imperial Turn in the History of Science." In *The Routledge Handbook of Science and Empire*, edited by Andrew Goss. Routledge Handbooks. Routledge, 2021.

Graaf, Ton de. *Voor handel en maatschappij: Geschiedenis van de Nederlandsche Handel-Maatschappij, 1824–1964*. Boom, 2012.

Greshoff, M. *Rumphius gedenkboek: 1702–1902*. Koloniaal Museum, 1902.

Groen, Petra, Anita van Dissel, Mark Loderichs, Rémy Limpach, and Thijs Brocades Zaalberg. *Krijgsgeweld en kolonie: Opkomst en ondergang van Nederland als koloniale mogendheid, 1816–2010*. Militaire geschiedenis van Nederland, 6. Boom, 2021.

Groot, Hans. *Van Batavia naar Weltevreden: Het Bataviaasch Genootschap van Kunsten en Wetenschappen, 1778–1867*. Verhandelingen van het Koninklijk Instituut voor Taal-, Land- en Volkenkunde, 243. KITLV, 2009.

Grouw, Hein van, Wim Dekkers, and Kees Rookmaaker. "Third Time Lucky for Forsten's Pigeon; *taeniura, forsterii, forsteni*." *Bulletin of the British Ornithologists' Club* 144, no. 1 (2024): 29–38.

Guan, Kwa Chong. "Drawing Nature in the East Indies: Framing Farquhar's Natural History Drawings." In *Natural History Drawings: The Complete William Farquhar Collection, Malay Peninsula 1803–1818*, edited by Laura Dozier. Didier Millet, 2010.

Hagen, Piet. *Koloniale oorlogen in Indonesië: Vijf eeuwen verzet tegen vreemde overheersing*. Arbeiderspers, 2018.

Harris, Moses, and John O. Westwood. *The Aurelian: A Natural History of English Moths and Butterflies, Together with the Figures of Their Transformation and of the Plants on Which They Feed*. New edition. London: Henry G. Bohn, 1840.

Hedinger, Daniel, and Nadine Hée. "Transimperial History: Connectivity, Cooperation and Competition." *Journal of Modern European History = Zeitschrift für moderne europäische Geschichte = Revue d'histoire européenne contemporaine* 16, no. 4 (2018): 429–452.

Heiningen, Teunis Willem van, ed. *The Correspondence of Caspar Georg Carl Reinwardt (1773–1854)*. 2 vols., Tools and Sources for the History of Science in the Netherlands, 2. Dutch – History of Science – Web Centre, 2011.

Henley, David. "Minahasa Mapped: Illustrated Notes on Cartography and History in Minahasa, 1512–1942." In *Minahasa Past and Present: Tradition and Transition in an Outer Island Region of Indonesia*, edited by R. Schefold. CNWS Publications, 28. Research School CNWS, 1995.

Henley, David. *Fertility, Food and Fever: Population, Economy and Environment in North and Central Sulawesi, 1600–1930*. Verhandelingen van het Koninklijk Instituut voor Taal-, Land- en Volkenkunde, 201. KITLV, 2005.

Historische Kommission bei der Bayerischen Akademie der Wissenschaften, ed. *Neue Deutsche Biographie*. 28 vols. Duncker & Humblot, 1953–2024.

Holthuis, L. B. *1820–1958: Rijksmuseum van Natuurlijke Historie*. Nationaal Natuurhistorisch Museum, 1995.

Honig, Pieter, and Frans Verdoorn, eds. *Science and Scientists in the Netherlands Indies*, Natuurwetenschappelijk tijdschrift voor Nederlandsch Indië, 102 – Special Supplement. Board for the Netherlands Indies, Surinam and Curaçao, 1945.

Honings, Rick. "Naar het "Verlokkend Oost": De onderzoeksreizen van Carl Ludwig Blume in Indië." *Jaarboek Bilderijk* 3 (2022): 87–103.

Honings, Rick. *De ontdekking van Insulinde: Op reis in Nederlands-Indië in de negentiende eeuw*. Prometheus, 2023.

Honings, Rick, Coen van 't Veer, and Jacqueline Bel, eds. *De postkoloniale spiegel: De Nederlands-Indische letteren herlezen*. Tweede, gewijzigde druk. Leiden University Press, 2022.

Hoogmoed, M. S., "Nomenclatural Problems Relating to *Atractus trilineatus* Wagler, 1828." *Zoologische mededelingen* 56, no. 10 (1982): 131–138.

Horner, Friedrich. "Briefe und Tagebuchskizzen des Dr. med. Ludwig Horner (1811–1838) aus Niederländisch-Indien." *Zurcher Taschenbuch* (1919): 183–208.

Horsfield, Thomas. *Zoological Researches in Java, and the Neighbouring Islands*. London: Printed for Kingsbury, Parbury, & Allen, 1824.

Horsfield, Thomas, and John Bastin. *Zoological Researches in Java, and the Neighbouring Islands: With a Memoir by John Bastin*. Oxford University Press, 1990.

Houben, Vincent J. H., J. Thomas Lindblad, Ellen Leenarts, Trudi Nierop, Carin van Empel, and Jurrien van den Berg. *Coolie Labour in Colonial Indonesia: A Study of Labour Relations in the Outer Islands, c.1900–1940*. Harrassowitz, 1999.

Howe, Stephen, ed. *The New Imperial Histories Reader*. Routledge, 2010.

Hugenholtz, W. R. *Het geheim van Paleis Kneuterdijk: De wekelijkse gesprekken van koning Willem II met zijn minister J.C. Baud over het koloniale beleid en de herziening van de grondwet, 1841–1848*. Verhandelingen van het Koninklijk Instituut voor Taal-, Land- en Volkenkunde, 239. KITLV, 2008.

Husson, A. M., and L. B. Holthuis. "The Dates of Publication of 'Verhandelingen over de natuurlijke geschiedenis der Nederlandsche overzeesche bezittingen' edited by C. J. Temminck." *Zoologische mededelingen* 34, no. 2 (1955): 17–24.

Huyssen van Kattendijke, W. J. C. ridder. *Met Prins Hendrik naar de Oost: De reis van W. J. C. Huyssen van Kattendijke naar Nederlands-Indië, 1836–1838*. Werken uitgegeven door de Linschoten-Vereeniging, 102. Walburg, 2004.

Impelen, Corine van, ed. *Het nieuwe Naturalis = The New Naturalis*. Naturalis Biodiversity Center, 2019.

Israel, Jonathan. *The Dutch Republic: Its Rise, Greatness, and Fall, 1477–1806*. Oxford History of Early Modern Europe. Clarendon, 1995.

Jacobs, Hans J. "Frühe Beschreibungen und Darstellungen des Dunklen Tigerpythons, *Python bivittatus* Kuhl 1820." *Sekretär* 21 (2021): 3–36.

Jacobs, Hans J. "Salomon Müller – ein unterschätzter deutscher Naturforscher." *Sekretär* 22 (2022): 21–56.

Jacobs, Hans J. "Heinrich Christian Macklot (20.10.1799–12.5.1832) und die Natuurkundige Commissie in Niederländisch Ostindien 1825–32." *Sekretär* 23, no. 2 (2023): 18–51.

Jacobs, Hans J., and André Koch. "On the Discovery and Scientific Description of the Emerald Tree Monitor, *Varanus prasinus* (Schlegel, 1839)." *Bibliotheca Herpetologica* 15, no. 7 (2021): 61–76.

Jacobs, Hans J., and Glenn M. Shea. "Eight Weeks in Lobo Bay: The Natuurkundige Commissie on New Guinea in 1828: I. *Scincus* and *Centroplites* (Scincidae)." *Bibliotheca herpetologica* 16, no. 6 (2022): 48–81.

Jardine, N., J. A. Secord, and E. C. Spary, eds. *Cultures of Natural History*. Cambridge University Press, 1996.

Jeurgens, Charles. "Op zoek naar betrouwbare informatie: De comissarissen-generaal en de stichting van de kolonial staat, 1816–1819." In *Het verre gezicht: Politieke en culturele relaties tussen Nederland en Azië, Afrika en Amerika: Opstellen aangeboden aan prof. dr. Leonard Blussé*, edited by J. Thomas Lindblad and Alicia Schrikker. Van Wijnen, 2011.

Jones, Lowri Madeleine. "Local Knowledge and Indigenous Agency in the History of Exploration: Studies from the RGS-IBG Collections." PhD, Royal Holloway University of London, 2010.

Junghuhn, Franz Wilhelm. "Bijdragen tot de geschiedenis der vulkanen in den Indischen archipel, tot 1842." *Tijdschrift voor Neêrland's Indië* 5 (1843): 97–133, 185–227, 257–280, 614–626, 745–763.

Junghuhn, Franz Wilhelm. *Java, deszelfs gedaante, bekleeding en inwendige structuur*. 3 vols. Amsterdam: P. N. van Kampen, 1850–1854.

Juynboll, H. H. "Wall (Hermann Theodor Friedrich Karl Emil Wilhelm August Casimir von de)." In *Nieuw Nederlandsch biografisch woordenboek: Derde deel*, edited by P. C. Molhuysen and P. J. Blok. A. W. Sijthoff's, 1914.

Kaiser, Katja. "The Coloniality of Natural History Collections." In *Provenance Research on Collections from Colonial Contexts: Principles, Approaches, Challenges*, edited by Claudia Andratschke, Lars Müller and Katja Lembke. Publications of the Network for Provenance Research in Lower Saxony. Arthistoricum.net – Fachinformationsdienst Kunst, Fotografie, Design, 2023.

Karstens, W. K. H., and Herman Kleibrink. *De Leidse Hortus: Een botanische erfenis*. Waanders, 1982.

Kennedy, Dane. *The Last Blank Spaces: Exploring Africa and Australia*. Harvard University Press, 2013.

Kennedy, Dane, ed. *Reinterpreting Exploration: The West in the World*, Reinterpreting History. Oxford University Press, 2014.

Kernkamp, P. M. "Heinrich Bürger (Hamelen 1804/1806-Indramajoe 1858) en zijn Duitse en Nederlands-Indische familie." *De Nederlandsche leeuw* 132, no. 3 (2015): 108–132.

Keurs, Pieter ter. "Introduction: Theory and Practice of Colonial Collecting." In *Colonial Collections Revisited*, edited by Pieter ter Keurs. Mededelingen van het Rijksmuseum voor Volkenkunde Leiden, 36. CNWS, 2007.

Keurs, Pieter ter. "Collecting in the Colony: Hybridity, Power and Prestige in the Netherlands East Indies." *Indonesia and the Malay World* 37, no. 108 (2009): 147–161.

Kießling, Charlotte. *Wissenspoetik und koloniale Naturgeschichte: G. E. Rumphius' D'Amboinsche Rariteitkamer (1705)*. Schriften aus dem Haus der Niederlande, 11. Universitäts- und Landesbibliothek Münster, 2021.

Kitzan, Laurence. "Lord Amherst and the Declaration of War on Burma, 1824." *Journal of Asian History* 9, no. 2 (1975): 101–127.

Klaver, Charles. *Inseparable Friends in Life and Death: The Life and Work of Heinrich Kuhl (1797–1821) and Johan Conrad van Hasselt (1797–1823), Students of Prof. Theodorus van Swinderen*. Barkhuis, 2007.

Klein Douwel, Mon. "Dus je bent er nog wel, zwartbrauwmuistimalia: Wat het betekent als uitgestorven gewaande soorten worden teruggevonden." *De Volkskrant*, August 19, 2021. https://www.volkskrant.nl/wetenschap/dus-je-bent-er-nog-wel-zwartbrauwmuistimalia-wat-het-betekent-als-uitgestorven-gewaande-soorten-worden-teruggevonden~b77c9bb9/.

Knight, G. Roger. *Colonial Production in Provincial Java: The Sugar Industry in Pekalongan-Tegal, 1800–1942*. Comparative Asian Studies, 13. VU University Press, 1993.

Knight, G. Roger. *Sugar, Steam and Steel: The Industrial Project in Colonial Java, 1830–1885*. University of Adelaide Press, 2004.

Koch, Jeroen. *Koning Willem I, 1772–1843*. Boom, 2013.

Kohler, Robert E. *All Creatures: Naturalists, Collectors, and Biodiversity, 1850–1950*. Princeton University Press, 2006.

Kohler, Robert E. "History of Field Science: Trends and Prospects." In *Knowing Global Environments: New Historical Perspectives on the Field Sciences*, edited by Jeremy Vetter. Studies in Modern Science, Technology, and the Environment. Rutgers University Press, 2011.

Konishi, Shino, Maria Nugent, and Tiffany Shellam, eds. *Indigenous Intermediaries: New Perspectives on Exploration Archives*. Australian National University Press, 2015.

Kuhl, Heinrich. *Die deutschen Fledermäuse*. Hanau, 1817.

Kuhl, Heinrich, and Johan Conrad van Hasselt. "Uittreksels uit brieven van de heeren Kuhl en Van Hasselt, aan de heeren C. J. Temminck, Th. van Swinderen en W. de Haan, geschreven in de straat Sunda, den 17den December 1820." *Algemeene konst- en letter-bode*, no. 6 (February 8, 1822): 82–88.

Kuipers, Matthijs. *A Metropolitan History of the Dutch Empire: Popular Imperialism in the Netherlands, 1850–1940*. Heritage and Memory Studies, 17. Amsterdam University Press, 2022.

Kuitert, Lisa. *Met een drukpers de oceaan over: Koloniale boekcultuur in Nederlands-Indië, 1816–1920*. Prometheus, 2020.

Kuklick, Henrika, and Robert E. Kohler, eds. *Science in the Field*, Osiris: A Research Journal Devoted to the History of Science and its Cultural Influences, Second Series, 11. University of Chicago Press, 1996.

Kuster, Megan. "Global Commodity Chains and Local Use-Value: William Colenso, Natural History Collecting and Indigenous Labour." *Journal of Colonialism and Colonial History* 22, no. 2 (2021): 19–46.

Lambert, David, and Alan Lester, eds. *Colonial Lives Across the British Empire: Imperial Careering in the Long Nineteenth Century*. Cambridge University Press, 2006.

Land, J. van der, ed. *The History of Natural History in Leiden*. National Museum of Natural History Naturalis, 2001.

Latour, Bruno. *Science in Action: How to Follow Scientists and Engineers Through Society*. Harvard University Press, 1987.

Legêne, Susan. *De bagage van Blomhoff en Van Breugel: Japan, Java, Tripoli en Suriname in de negentiende-eeuwse Nederlandse cultuur van het imperialisme*. KIT, 1998.

Leidelmeijer, Margaret. *Van suikermolen tot grootbedrijf: Technische vernieuwingen in de Java-suikerindustrie in de negentiende eeuw*. NEHA-Series III, 25. NEHA, 1997.

Leonhard, K. C. von. *Fremdenbuch für Heidelberg und die Umgegend*. Heidelberg: Karl Groos, 1834.

Lesson, René-Primevère. *Voyage autour du monde entrepris par ordre du gouvernement sur la corvette La Coquile*. 4 vols. Brussels: N.-J. Gregoir, V. Wouters, 1839.

Liliencron, Rochus Freiherr von, ed. *Allgemeine Deutsche Biographie*. 56 vols. Leipzig: Duncker & Humblot, 1875–1912.

Lindblad, J. Thomas. "The Outer Islands in the 19th Century: Contest for the Periphery." In *The Emergence of a National Economy: An Economic History of Indonesia, 1800–2000*, edited by Howard Dick, Vincent J. H. Houben, J. Thomas Lindblad and Thee Kian Wie. Asian Studies Association of Australia. Southeast Asia Publications Series. Allen & Unwin, 2002.

Livingstone, David. *Putting Science in Its Place: Geographies of Scientific Knowledge*. science.culture. University of Chicago Press, 2003.

Livingstone, David N. "Text, Talk and Testimony: Geographical Reflections on Scientific Habits: An Afterword." *British Journal for the History of Science* 38, no. 1 (2005): 93–100.

Livingstone, David N., and Charles W. J. Withers, eds. *Geographies of Nineteenth-Century Science*. University of Chicago Press, 2011.

Locher-Scholten, Elsbeth. *Sumatran Sultanate and Colonial State: Jambi and the Rise of Dutch Imperialism, 1830–1907*. Translated by Beverley Jackson. Studies on Southeast Asia, 37. Southeast Asia Program, Cornell University, 2004.

MacGregor, Arthur, ed. *Naturalists in the Field: Collecting, Recording and Preserving the Natural World from the Fifteenth to the Twenty-First Century*, Emergence of Natural History, 2. Brill, 2018.

Mackay, David. *In the Wake of Cook: Exploration, Science & Empire, 1780–1801*. St. Martin's, 1985.

Macklot, Heinrich Christian. "Uittreksels uit brieven van Dr. H. C. Macklot, natuuronderzoeker in Oost-Indië, aan Dr. C. J. Temminck, directeur, en J. A. Susanna, administrateur van 's Rijks Museum te Leiden." *Bijdragen tot de natuurkundige wetenschappen* 4 (1829): 298–309.

Macklot, Heinrich Christian. "Verslag van het Land, de Bewoners en Voortbrengselen van eenige plaatsen op de kust van Nieuw Guinea, welke in den loop van het jaar 1828, door de Natuurkundige kommissie in Oost-Indië, aan boord van Z. M. korvet Triton zijn bezocht, alsmede van de voorwerpen van Natuurlijke Historie, welke gedurende de reis, van den 20 Mei tot den 30 Augustus, op en langs die kust, door gemelde kommissie zijn verzameld." *Bijdragen tot de natuurkundige wetenschappen* 5 (1830): 142–182.

MacLeod, Roy. "On Visiting the 'Moving Metropolis': Reflections on the Architecture of Imperial Science." *Historical Records of Australian Science* 5 (1982): 1–16.

MacLeod, Roy, ed. *Nature and Empire: Science and the Colonial Enterprise*, Osiris: A Research Journal Devoted to the History of Science and its Cultural Influences, Second Series, 15. University of Chicago Press, 2000.

Magee, Judith. *The Art of India*. Images of Nature. Natural History Museum, 2013.

Magee, Judith. *Chinese Art and the Reeves Collection*. Images of Nature. Natural History Museum, 2011.

Manse, Maarten. "Kennis is macht: De veelzijdige expedities van botanicus Pieter Willem Korthals (1807–1892)." *Studium* 6, no. 1 (2013): 42–50.

Maronier, J. H. *Pictures of the Tropics: A Catalogue of Drawings, Water-Colours, Paintings, and Sculptures in the Collection of the Royal Institute of Linguistics and Anthropology in Leiden*. Martinus Nijhoff, 1967.

Marsden, William. *The History of Sumatra, Containing an Account of the Government, Laws, Customs, and Manners of the Native Inhabitants, with a Description of the Natural Productions, and a Relation of the Ancient Political State of that Island*. London, 1783. Printed for the author.

Martin, K. "Neue Fundpunkte von Tertiaer-gesteinen im indischen Archipel: Nach Sammlungen von Horner, Korthals, Macklot, Müller und Reinwardt." *Sammlungen des Geologischen Reichs-Museums in Leiden: Serie 1, Beiträge zur Geologie Ost-Asiens und Australiens* 1, no. 1 (1881): 131–193.

Mathijsen, Marita. *L: De lezer van de 19de eeuw*. Balans, 2021.

McClellan III, James E. *Colonialism and Science: Saint Domingue in the Old Regime*. University of Chicago Press Edition 2010. University of Chicago Press, 2010.

Mees, G. F. "A List of the Birds Recorded from Bangka Island, Indonesia." *Zoologische verhandelingen* 232, no. 1 (1986): 1–176.

Mees, G. F. "Vogelkundig onderzoek op Nieuw Guinea in 1828: Terugblik op de ornithologische resultaten van de reis van Zr. Ms. Korvet *Triton* naar de zuid-west kust van Nieuw-Guinea." *Zoologische bijdragen* 40 (1994): 1–64.

Melvill van Carnbee, Pieter baron. "Kaart van de Minahassa in de residentie Menado." Batavia: Van Haren Noman & Kolff, 1853.

Melvill van Carnbee, Pieter baron, and W. F. Versteeg. *Algemeene atlas van Nederlandsch Indie: Uit officieele bronnen en met goedkeuring van het gouvernement zamengesteld*. Batavia: Van Haren Noman & Kolff, 1853–1862.

Meusburger, Peter, David N. Livingstone, and Heike Jöns, eds. *Geographies of Science*, Knowledge and Space, 3. Springer, 2010.

Mikaberidze, Alexander. *The Napoleonic Wars: A Global History*. Oxford University Press, 2020.

Miller, David Philip, and Peter Hanns Reill. *Visions of Empire: Voyages, Botany, and Representations of Nature*. Cambridge University Press, 1996.

Minteer, Ben A., James P. Collins, Karen E. Love, and Robert Puschendorf. "(Re)extinction." *Science. New Series* 344, no. 6161 (18 April 2014): 260–261.

Molhuysen, P. C., and P. J. Blok. *Nieuw Nederlandsch biografisch woordenboek*. 10 vols. A. W. Sijthoff, 1911–1937.

Moor, J. A. de. "Warmakers in the Archipelago: Dutch Expeditions in Nineteenth Century Indonesia." In *Imperialism and War: Essays on Colonial Wars in Asia and Africa*, edited by J. A. de Moor and H. L. Wesseling. Comparative Studies in Overseas History: Publications of the Leiden Centre for the History of European Expansion, 8. Brill, 1989.

Moore, Bob, and Henk van Nierop, eds. *Colonial Empires Compared: Britain and the Netherlands, 1750–1850: Papers Delivered at the Fourteenth Anglo-Dutch Historical Conference, 2000*. Ashgate, 2003.

Morris, Jennifer R. "Non-western Collectors and Their Contributions to Natural History, c. 1750–1940." In *The Routledge Handbook of Science and Empire*, edited by Andrew Goss. Routledge Handbooks. Routledge, 2021.

Mörzer Bruyns, Willem F. J. *Met de Triton en Iris naar Nieuw-Guinea: De reisverhalen van Justin Modera en Arnoldus Johannes van Delden uit 1828*. Werken uitgegeven door de Linschoten-Vereeniging, 117. Walburg, 2018.

Mörzer Bruyns, Willem F. J. "The Taking Possession of Part of New Guinea by the Dutch in 1828, and Their Contribution to the Knowledge of the Arafura Sea." *The Great Circle* 41, no. 1 (2019): 38–59.

Mörzer Bruyns, Willem F. J. "De Nederlandse hydrografische opnemingen aan de zuidwestkust van Nieuw-Guinea en in de Arafurazee, 1826–1835." *Caert-Thresoor* 38, no. 3 (2019): 9–13.

Müller, Salomon. *Berigten over Sumatra, met eene kaart van een gedeelte van hetzelve, voornamelijk aantoonende de wegen en rivieren, welke uit de Padangsche binnenlanden naar de oostkust afloopen, getrokken uit het dagboek van den Heer S. Muller, Lid van de Nederlandsche Natuurkundige Commissie in Oost-Indie*. Amsterdam: G. J. A. Beijerinck, 1837.

Müller, Salomon. "Over de zoogdieren van den Indischen Archipel." In *Verhandelingen over de natuurlijke geschiedenis der Nederlandsche overzeesche bezittingen door de Leden der Natuurkundige commissie in Indië en andere schrijvers, uitgegeven op last van den koning door C. J. Temminck: Zoologie*, edited by J. A. Susanna. Leiden: S. & J. Luchtmans en C. C. van der Hoek, 1839–1845.

Müller, Salomon. "Bijdragen tot de kennis van Timor en eenige andere naburige eilanden." In *Verhandelingen over de natuurlijke geschiedenis der Nederlandsche overzeesche bezittingen door de Leden der Natuurkundige commissie in Indië en andere schrijvers, uitgegeven op last van den koning door C. J. Temminck: Land- en volkenkunde*, edited by J. A. Susanna. Leiden: S. & J. Luchtmans en C. C. van der Hoek, 1840–1847.

Müller, Salomon. *Bijdragen tot de kennis van Sumatra, bijzonder in geschiedkundig en ethnographisch opzigt*. Leiden: S. & J. Luchtmans, 1846.

Müller, Salomon. *Reizen en onderzoekingen in den Indischen Archipel, gedaan op last der Nederlandsche Indische regering, tusschen de jaren 1828 en 1836*. Werken van het Koninklijk Instituut voor Taal-, Land- en Volkenkunde van Nederlandsch-Indië. Tweede afdeeling: afzonderlijke werken. 2 vols. Amsterdam: Frederik Muller, 1857.

Müller, Salomon, and Hermann Schlegel. "Over de krokodillen van den Indischen Archipel." In *Verhandelingen over de natuurlijke geschiedenis der Nederlandsche overzeesche bezittingen door de Leden der Natuurkundige commissie in Indië en andere schrijvers, uitgegeven op last van den koning door C. J. Temminck: Zoologie*, edited by J. A. Susanna. Leiden: S. & J. Luchtmans en C. C. van der Hoek, 1839–1845.

Müller, Salomon, and Hermann Schlegel. “Beschrijving van een merkwaardig insektenetend zoogdier, hylomys suillus.” In *Verhandelingen over de natuurlijke geschiedenis der Nederlandsche overzeesche bezittingen door de Leden der Natuurkundige commissie in Indië en andere schrijvers, uitgegeven op last van den koning door C. J. Temminck: Zoologie*, edited by J. A. Susanna. Leiden: S. & J. Luchtmans en C. C. van der Hoek, 1839–1845.

Müller, Salomon, and Hermann Schlegel. “Bijdragen tot de natuurlijke historie der rhinocerossen van den Indischen Archipel.” In *Verhandelingen over de natuurlijke geschiedenis der Nederlandsche overzeesche bezittingen door de Leden der Natuurkundige commissie in Indië en andere schrijvers, uitgegeven op last van den koning door C. J. Temminck: Zoologie*, edited by J. A. Susanna. Leiden: S. & J. Luchtmans en C. C. van der Hoek, 1839–1845.

Muller, W. C. “Junghuhn-bibliographie.” In *Gedenkboek Franz Junghuhn, 1809–1909*, edited by Junghuhn-commissie. Martinus Nijhoff, 1910.

Muller, W. C. “Hageman (Johannes).” In *Nieuw Nederlandsch biografisch woordenboek: Eerste deel*, edited by P. C. Molhuysen and P. J. Blok. A. W. Sijthoff, 1911.

Naylor, Simon. “Introduction: Historical Geographies of Science – Places, Contexts, Cartographies.” *British Journal for the History of Science* 38, no. 1 (2005): 1–12.

Nieuhof, Joan. *Zee en lantreize, door verscheide gewesten van Oostindien, behelzende veel zeltzaame en wonderlijke voorvallen en geschiedenissen, beneffens een beschrijving van lantschappen, steden, dieren, gewassen, draghten, zeden en godsdienst der inwoonders: En inzonderheit een wijtloopig verhael der stad Batavia, versiert doorgaens met verscheide koopere platen*. Amsterdam, 1682. Voor de weduwe van Jacob van Meurs, op de Keizers-gracht.

Nieuwenhuys, Rob, Frits Jacquet, and Franz Wilhelm Junghuhn. *Java's onuitputtelijke natuur: Reisverhalen, tekeningen en fotografieën van Franz Wilhelm Junghuhn*. A. W. Sijthoff, 1980.

Noltie, H. J. *Raffles' Ark Redrawn: Natural History Drawings from the Collection of Sir Thomas Stamford Raffles*. British Library, 2009.

Ormeling, Ferjan. “De Topografische Dienst van Nederlands-Indië, Batavia 1864–1950.” *Caert-Thresoor* 36, no. 3 (2017): 102–113.

Osborne, Milton E. *The French Presence in Cochinchina and Cambodia: Rule and Response (1859–1905)*. Cornell University Press, 1969.

Otterspeer, Willem. *Groepsportret met Dame III: De werken van de wetenschap: De Leidse universiteit, 1776–1876*. Bert Bakker, 2005.

Ouden, A. den. “C. L. Blume, periode 1826–1832.” 1979.

Oury, Clément. “Diard, Duvaucel and the Muséum National d'Histoire Naturelle = Diard, Duvaucel et le Muséum national d'histoire naturelle.” In *Diard & Duvaucel: French Natural History Drawings of Singapore and Southeast Asia (1818–1820) = Dessins d'histoire naturelle français de Singapour et d'Asie du Sud-Est (1818–1820)*, edited by Francis Dorai and Martyn E. Y. Low. Epigram Books, 2021.

Outram, Dorinda. *Georges Cuvier: Vocation, Science, and Authority in Post-Revolutionary France*. Manchester University Press, 1984.

Outram, Dorinda. "New Spaces in Natural History." In *Cultures of Natural History*, edited by N. Jardine, J. A. Secord and E. C. Spary. Cambridge University Press, 1996.

Overweel, Jeroen. "Keep Them Out!: Early Nineteenth Century English/Dutch Rivalry in Eastern Indonesia and Australia, and the Founding of Merkus-Oord." *The Great Circle* 43, no. 1 (2021): 19–44.

Paulus, J., ed. *Encyclopaedie van Nederlandsch-Indië: Tweede druk*. 8 vols. Martinus Nijhoff, 1917–1939.

Pers, Auguste van, and J. C. Hageman J. Cz. *Nederlandsch Oost-Indische typen = Types indiens neerlandais: Verzameling van groote gelithografieerde platen in kleurdruk: Naar de natuur geteekend door A. van Pers: Met een verklarenden tekst in 't Hollandsch en Fransch: Opgedragen aan zijne excellentie den gouverneur-generaal van Nederlandsch-Indië den heer Mr. A. J. Duijmaer van Twist*. 's Gravenhage: C. W. Mieling, 1856.

Peyssonnaux, J. H. "Vie, voyages et travaux de Pierre Médard Diard: Naturaliste français aux Indes-Orientales, 1794–1863: Voyage dans l'Indochine, 1821–1824." *Bulletin des amis du vieux Hué* 22, no. 1 (1935): 3–115.

Phillips, Denise. *Acolytes of Nature: Defining Natural Science in Germany, 1770–1850*. University of Chicago Press, 2012.

Pilon, Maxime, and Danièle Weiler. *The French in Singapore: An Illustrated History (1819-Today)*. Didier Millet, 2011.

Pols, Hans. "Notes from Batavia, the Europeans' Graveyard: The Nineteenth-Century Debate on Acclimatization in the Dutch East Indies." *Journal of the History of Medicine and Allied Sciences* 67, no. 1 (2012): 120–148.

Pols, Hans. *Nurturing Indonesia: Medicine and Decolonisation in the Dutch East Indies*. Global Health Histories. Cambridge University Press, 2018.

Posewitz, Theodor. *Borneo: Entdeckungsreisen und Untersuchungen: Gegenwärtiger Stand der geologischen Kenntnisse: Verbreitung der nutzbaren Mineralien*. Berlin: R. Friedländer, 1889.

Pyenson, Lewis. *Empire of Reason: Exact Sciences in Indonesia 1840–1940*. E. J. Brill, 1989.

Qureshi, Sadiah. "Displaying Sara Baartman, the 'Hottentot Venus'." *History of Science* 42, no. 2 (2004): 233–257.

Raat, Alexis J. P. "Het Rijksmuseum van Natuurlijke Historie in Leiden in de 19e eeuw." *De negentiende eeuw* 1, no. 1 (1977): 31–35.

Raat, Alexis J. P. "Coenraad Jacob Temminck (1778–1858): Een biografische schets." *De negentiende eeuw* 2, no. 2 (1978): 89–104.

Raben, Remco. "A New Dutch Imperial History? Perambulations in a Prospective Field." BMGN – *Low Countries Historical Review* 128, no. 1 (2013): 5–30.

Raffles, Sophia. *Memoir of the Life and Public Services of Sir Thomas Stamford Raffles, F. R. S. &c., Particularly in the Government of Java, 1811–1816, and of Bencoolen and its Dependencies, 1817–1824: With Details of the Commerce and Resources of the Eastern Archipelago and Selections from his Correspondence*. London: John Murray, 1830.

Raffles, Thomas Stamford. *The History of Java*. 2 vols. London: Black, Parbury, and Allen, 1817.

Raffles, Thomas Stamford. "Descriptive Catalogue of a Zoological Collection, Made on Account of the Honorable East India Company, in the Island of Sumatra and its Vicinity, under the Direction of Sir Thomas Stamford Raffles, Lieutenant-Governor of Fort Marlborough: With Additional Notices Illustrative of the Natural History of those Countries." *The Transactions of the Linnean Society of London* 13 (1822): 239–274, 277–340.

Raffles, Thomas Stamford. *A History of Java*. Cambridge Library Collection. 2 vols. Cambridge University Press, 2010.

Raina, Dhruv. "From West to Non-West? Basalla's Three-Stage Model Revisited." *Science as Culture* 8, no. 4 (1999): 497–516.

Raj, Kapil. *Relocating Modern Science: Circulation and the Construction of Knowledge in South Asia and Europe, 1650–1900*. Palgrave Macmillan, 2007.

Reeuwijk, Alexander, and Hay Wijnhoven. *De duiven van Temminck en Knip*. Sterck & De Vreese, 2023.

Reveil, O., and A. Dupuis. *Flore médicale, usuelle et industrielle du XIXe siècle: Donnant la description, la culture, la composition chimique, les propriétés curatives our dangereuses, les usages économiques et industrielle des plantes: Atlas iconografique du tome troisième*. Paris: L. Guérin, 1872.

Rice, Tony. *Voyages of Discovery: A Visual Celebration of Ten of the Greatest Natural History Expeditions*. Natural History Museum, 2000.

Riffenburgh, Beau. *Journeys of Discovery: The Heroic Exploits of the World's Greatest Explorers*. SevenOaks, 2017.

Rijnberg, Theo F. *'s Lands plantentuin, Buitenzorg 1817–1992: Kebun Raya Indonesia, Bogor*. Johanna Oskamp, 1992.

Roller, Heather Flynn. "Colonial Collecting Expeditions and the Pursuit of Opportunities in the Amazonian Sertão, c. 1750–1800." *The Americas* 66, no. 4 (2010): 435–467.

Ronkel, Ph. S. van. "Maleisch." In *Encyclopaedie van Nederlandsch-Indië: Tweede druk: Tweede deel H-M*, edited by S. de Graaff and D. G. Stibbe. Martinus Nijhoff, 1918.

Rookmaaker, K. "Mauled by a Rhinoceros: The Final Years of Alfred Duvaucel (1793–1824) in India." *Zoosystema* 41, no. 1 (2019): 259–267.

Rookmaaker, L. C. (Kees). *The Rhinoceros of South Asia*. Emergence of Natural History, 4. Brill, 2024.

Ruiter, Fred de. *Natuurschatten: Natuurhistorische collecties in Nederlandse musea.* Matrijs, 2020.

Ruiter, Peter de. "Het Mijnwezen in Nederlands-Oost-Indië, 1850–1950." PhD, Universiteit Utrecht, 2016.

Rutten, L. M. R., ed. *Science in the Netherlands East Indies.* Koninklijke Akademie van Wetenschappen, 1929.

Salles, André. *Un mandarin breton au service du roi de Cochinchine: Jean-Baptiste Chaigneau et sa famille: Suivi du mémoire sur la Cochinchine de Jean-Baptiste Chaigneau.* Bretons à travers le monde. Portes du large, 2006.

Savage, Jay M., Marcus Sparreboom, and Gregory K. Pregill, *The Dutch Indies Commission and the Herpetological Legacy of Heinrich Boie (1794–1827).* Contributions to Herpetology. Society for the Study of Amphibians and Reptiles, forthcoming.

Scalliet, Marie-Odette. *Antoine Payen, peintre des Indes Orientales: Vie et écrits d'un artiste du XIXe siècle (1792–1853).* CNWS Publications, 34. Research School CNWS, 1995.

Scalliet, Marie-Odette. "Beelden van Oost-Indië: De collectie Bik in het Rijksprentenkabinet." *Bulletin van het Rijksmuseum* 49, no. 4 (2001): 342–379.

Schaffer, Simon, Lissa Roberts, Kapil Raj, and James Delbourgo, eds. *The Brokered World: Go-betweens and Global Intelligence, 1770–1820*, Uppsala Studies in History of Science, 35. Science History, 2009.

Schama, Simon. *Patriots and Liberators: Revolution in the Netherlands, 1780–1813.* Second edition. Harper, 1992.

Schär, Bernhard C. *Tropenliebe: Schweizer Naturforscher und niederländischer Imperialismus in Südostasien um 1900.* Globalgeschichte, 20. Campus, 2015.

Schär, Bernhard C. "Earth Scientists as Time Travelers and Agents of Colonial Conquest: Swiss Naturalists in the Dutch East Indies." *Historical Social Research* 40, no. 2 (2015): 67–80.

Schär, Bernhard C. "From Batticaloa via Basel to Berlin: Transimperial Science in Ceylon and Beyond around 1900." *The Journal of Imperial and Commonwealth History* 48, no. 2 (2020): 230–262.

Schär, Bernhard. "Switzerland, Borneo and the Dutch Indies: Towards a New Imperial History of Europe, c .1770–1850." *Past & Present* 257, no. 1 (2022): 134–167.

Schefold, Reimar, ed. *Minahasa Past and Present: Tradition and Transition in an Outer Island Region of Indonesia*, CNWS Publications, 28. Research School CNWS, 1995.

Schiebinger, Londa, and Claudia Swan, eds. *Colonial Botany: Science, Commerce, and Politics in the Early Modern World.* University of Pennsylvania Press, 2005.

Schilthuizen, Menno. "Terabytes aan tropenjaren." *Bionieuws* 27, no. 4 (2017): 8–9.

Schilthuizen, Menno, and Freek Vonk. *Wie wat bewaart: Twee eeuwen Nederlandse natuurhistorie.* Spectrum, 2020.

Schmidt, Max C. P. "Biographische Skizze: 12. Seine Schriften." In *Franz Junghuhn: Biographische Beiträge zur 100. Wiederkehr seines Geburtstages*. Verlag der Dürr'schen Buchhandlung, 1909.

Schwaner, Carl Anton Ludwig Maria, and J. Pijnappel Gz. *Borneo: Beschrijving van het stroomgebied van den Barito en reizen langs eenige voorname rivieren van het zuid-oostelijk gedeelte van dat eiland, door Dr. C. A. L. M. Schwaner, op last van het Gouvernement van Nederl. Indië gedaan in de jaren 1843–1847*. 2 vols. Amsterdam: P. N. van Kampen, 1853–1854.

Secord, James A. "Knowledge in Transit." *Isis* 95, no. 4 (2004): 654–672.

Sens, Angelie. *De kolonieman: Johannes van den Bosch (1780–1844), volksverheffer in naam van de koning*. Balans, 2019.

Shellam, Tiffany, Maria Nugent, Shino Konishi, and Allison Cadzow, eds. *Brokers and Boundaries: Colonial Exploration in Indigenous Territory*. Australian National University Press, 2016.

Sirks, M. J. *Indisch natuuronderzoek: Een beknopte geschiedenis van de beoefening der natuurwetenschappen in de Nederlandsche koloniën*. Koloniaal Instituut, 1915.

Sivasundaram, Sujit. "Sciences and the Global: On Methods, Questions, and Theory." *Isis* 101 (2010): 146–158.

Sivasundaram, Sujit. "Islanded: Natural History in the British Colonization of Ceylon." In *Geographies of Nineteenth-Century Science*, edited by David N. Livingstone and Charles W. J. Withers. University of Chicago Press, 2011.

Smit, P. "International Influences on the Development of Natural History in the Netherlands and its East Indian Colonies between 1750 and 1850." *Janus* 65 (1978): 45–65.

Spreeuwenberg, A. F. van. "Een blik op de Minahassa." *Tijdschrift voor Neêrland's Indië* 7, no. 4 (1845): 161–214, 301–333.

Spreeuwenberg, A. F. van. "Een blik op de Minahassa." *Tijdschrift voor Neêrland's Indië* 8, no. 1 (1846): 23–49.

Steenis, C. G. G. J. van. "Dedication." *Flora Malesiana. Series 1, Spermatophyta* 10, no. 1 (1984): 6–43.

Steenis-Kruseman, M. J. van. "Malaysian Plant Collectors and Collections Being a Cyclopaedia of Botanical Exploration in Malaysia and a Guide to the Concerned Literature up to the Year 1950." *Flora Malesiana. Series 1, Spermatophyta* 1, no. 1 (1950).

Steenis-Kruseman, M. J. van. "Malaysian Plant Collectors and Collections Supplement I." *Flora Malesiana. Series 1, Spermatophyta* 5, no. 4 (1955): 1–108.

Steenis-Kruseman, M. J. van. "Malaysian Plant Collectors and Collections Supplement II." *Flora Malesiana. Series 1, Spermatophyta* 8, no. 1 (1974): 1–115.

Stemfoort, J. W., and J. J. ter Siethoff. *Atlas der Nederlandsche bezittingen in Oost-Indië naar de nieuwste bronnen samengesteld en aan de regeering opgedragen*. 's-Gravenhage: Topographische Inrichting, 1883–1885.

Sternagel, Renate. *Der Humboldt von Java: Leben und Werk des Naturforschers Franz Wilhelm Junghuhn 1809–1864*. 2., durchgesehene Auflage. Mitteldeutscher Verlag, 2018.

Steur, J. "Staatssecretarie en Kabinet des Konings onder Willem I." *Bijdragen en mededelingen van het Historisch Genootschap* 84 (1969): 88–138.

Storey, William Kelleher. *Science and Power in Colonial Mauritius*. Rochester Studies in African History and the Diaspora. University of Rochester Press, 1997.

Storms, Martijn, Mario Cams, Imre Josef Demhardt, and Ferjan Ormeling, eds. *Mapping Asia: Cartographic Encounters Between East and West: Regional Symposium of the ICA Commission on the History of Cartography, 2017*, Lecture Notes in Geoinformation and Cartography: Publications of the International Cartographic Association (ICA). Springer, 2019.

Stresemann, Erwin. *Die Entwicklung der Ornithologie: Von Aristoteles bis zur Gegenwart*. Hans Limberg, 1951.

Stresemann, Erwin. *Ornithology: From Aristotle to the Present*. Harvard University Press, 1975.

Stuchtey, Benedikt, ed. *Science Across the European Empires, 1800–1950*, Studies of the German Historical Institute London. Oxford University Press, 2005.

Suarez, Andrew V., and Neil D. Tsutsui. "The Value of Museum Collections for Research and Society." *BioScience* 54, no. 1 (2014): 66–74.

Susanna, J. A. *Lebens-Abriss und Würdigung der Verdienste Heinrich Boies*. Schleswig: K. Taubstummen-institut, 1830.

Susanna, J. A. *Levensschets van Hendrik Boie, en hulde aan zijne deugden en verdiensten, benevens eenige door hem geschrevene brieven, gedurende zijne reis en verblijf in Oost-Indië*. Amsterdam: P. Meijer Warnars, 1834.

Susanna, J. A., ed. *Verhandelingen over de natuurlijke geschiedenis der Nederlandsche overzeesche bezittingen door de Leden der Natuurkundige commissie in Indië en andere schrijvers, uitgegeven op last van den koning door C. J. Temminck*. 3 vols. Leiden: S. & J. Luchtmans en C. C. van der Hoek, 1839–1847.

Susanna, J. A. "Levensschets van Coenraad Jacob Temminck." *Handelingen der jaarlijksche algemeene vergadering van de Maatschappij der Nederlandsche Letterkunde te Leiden* (1858): 47–78.

Sutton, Elizabeth A. *Capitalism and Cartography in the Dutch Golden Age*. University of Chicago Press, 2015.

Swainson, William. "A Defence of 'Certain French Naturalists'." *The Magazine of Natural History, and Journal of Zoology, Botany, Mineralogy, Geology, and Meteorology* 4 (1831): 97–108.

Sysling, Fenneke. *Racial Science and Human Diversity in Colonial Indonesia*. NUS Press, 2016.

Tarling, Nicholas. *Anglo-Dutch Rivalry in the Malay World, 1780–1824*. Cambridge University Press, 1962.

Taylor, Jean Gelman. *The Social World of Batavia: European and Eurasian in Dutch Asia*. University of Wisconsin Press, 1983.

Teichfischer, Philipp. "Transnational Entanglements in Colonial Medicine: German Medical Practitioners as Members of the Health Service in the Dutch East Indies (1816–1884)." *Histoire, médecine et santé* 10 (2016): 63–78.

Temminck, Coenraad Jacob. *Monographies de mammalogie, ou description de quelques genres de mammifères, dont les espèces ont été observées dans les différens musées de l'Europe*. 2 vols. Paris: G. Dufour et Edmond d'Ocagne, 1827–1841.

Temminck, Coenraad Jacob. *Coup-d'oeil général sur les possessions néerlandaises dans l'Inde archipélagique*. 3 vols. Leiden: A. Arnz, 1846–1849.

Temminck, Coenraad Jacob, and Guillaume Michel Jerôme Meiffren Laugier de Chartrouse. *Nouveau recueil de planches coloriées d'oiseaux: Pour servir de suite et de complément aux planches enluminées de Buffon*. 5 vols. Paris: F. G. Levrault, 1820–1838.

Terwen-De Loos, J. *Nederlandse schilders en tekenaars in de Oost: 17de-20ste eeuw*. Rijksmuseum Amsterdam, 1972.

The, Lian, and Paul van der Veur. *The Verhandelingen van het Bataviaasch Genootschap: An annotated Content Analysis*. Papers in International Studies: Southeast Asia Series, 26. Ohio University, Center for International Studies, 1973.

Thomas, Martin, ed. *Expedition into Empire: Exploratory Journeys and the Making of the Modern World*. Routledge, 2015.

Tjiook-Liem, Patricia. *Chinezen uit Indonesië: De geschiedenis van een minderheid*. Walburg, 2022.

Touw, Dries. "Wat bezielde onze voorgangers uit de negentiende eeuw? Nog meer passie voor mossen." *Buxbaumiella* 101 (2014): 1–44.

Treub, Melchior, ed. *'s Lands Plantentuin te Buitenzorg: 18 Mei 1817–18 Mei 1892*. Batavia: Landsdrukkerij, 1892.

Troelstra, A. S. *Bibliography of Natural History Travel Narratives*. KNNV, 2016.

Troelstra, A. S. *Supplement to Bibliography of Natural History Travel Narratives*. KNNV, 2018.

Uéno, Masuzo. "A Japanese Portrait of Heinrich Bürger." *Zoologische mededelingen* 49, no. 10 (1975): 91–94.

Ukers, William H. *All about Tea*. 2 vols. Tea and Coffee Trade Journal Company, 1935.

Velde, C. W. M. van de. *Gezigten uit Neêrlands Indië, naar de natuur getekend en beschreven*. Amsterdam: Frans Buffa, 1845.

Veth, H. J. *Overzicht van hetgeen, in het bijzonder door Nederland, gedaan is voor de kennis der fauna van Nederlandsch Indië: Academisch proefschrift*. Leiden: S. C. van Doesburgh, 1879.

Vigors, N. A. "A Reply to some Observations in the "Dictionnaire des Sciences Naturelles," upon the Newly Characterized Groups of the Psittacidae." *The Zoological Journal* 3 (1828): 91–123.

Vigors, N. A., and Thomas Horsfield. "Observations on Some of the Mammalia Contained in the Museum of the Zoological Society." *The Zoological Journal* 4 (1829): 105–113.

Visser, Rob. "De ontwikkeling van het natuurhistorische museum sedert 1750." In *Kabinetten, galerijen en musea: Het verzamelen en presenteren van naturalia en kunst van 1500 tot heden*, edited by Ellinoor Bergvelt, Debora J. Meijers and Mieke Rijnders. Open Universiteit, 2013.

Visser, R. W. P. "Het Rijksmuseum van Natuurlijke Historie in de 19e eeuw." In *Het verdwenen museum: Natuurhistorische verzamelingen 1750–1850*, edited by B. C. Sliggers and M. H. Besselink. V+K, 2002.

Vriese, W. H. de, and J. Pijnappel Gz, eds. *Reis naar het oostelijk gedeelte van den Indischen Archipel in het jaar 1821; door C. G. C. Reinwardt: Uit zijne nagelaten aanteekeningen opgesteld met een levensberigt en bijlagen vermeerderd door W. H. de Vriese*, Werken van het Koninklijk Instituut voor Taal-, Land- en Volkenkunde van Nederlandsch-Indië. Tweede afdeeling: afzonderlijke werken. Amsterdam: Frederik Muller, 1858.

Walters, Michael. *A Concise History of Ornithology: The Lives and Works of its Founding Figures*. Christopher Helm, 2003.

Weber, Andreas. "Encountering the Netherlands Indies: Caspar G. C. Reinwardt's Field Trip to the East (1816–1822)." *Itinerario* 23, no. 1 (2009): 45–60.

Weber, Andreas. *Hybrid Ambitions: Science, Governance, and Empire in the Career of Caspar G. C. Reinwardt (1773–1854)*. Leiden University Press, 2012.

Weber, Andreas. "Bitter Fruits of Accumulation: The Case of Caspar Georg Carl Reinwardt (1773–1854)." *History of Science* 52, no. 3 (2014): 297–318.

Weber, Andreas. "A Garden as a Niche: Botany and Imperial Politics in the Early Nineteenth Century." *Studium* 11, no. 3 (2018): 178–190.

Weber, Andreas. "Collecting Colonial Nature: European Naturalists and the Netherlands Indies in the Early Nineteenth Century." *BMGN – Low Countries Historical Review* 134, no. 3 (2019): 72–95.

Weber, Andreas. "Over krokodillenvangers en schedeltentoonstellingen." *Wonderkamer* 1 (2020): 40–41.

Weber, Andreas, Mahya Ameryan, Katherine Wolstencroft, Lise Stork, Maarten Heerlien, and Lambert Schomaker. "Towards a Digital Infrastructure for Illustrated Handwritten Archives." In *Digital Cultural Heritage: Final Conference of the Marie Skłodowska-Curie Initial Training Network for Digital Cultural Heritage, ITN-DCH 2017, Olimje, Slovenia, May 23–25, 2017, Revised Selected Papers*, edited by Marinos Ioannides. Springer, 2018.

Weber, Andreas, and Robert-Jan Wille. "Laborious Transformations: Plants and Politics at the Bogor Botanical Gardens." *Studium* 11, no. 3 (2018): 169–177.

Weber, Andreas, and Sylvia van Zanen. *Kunstenaar op Java: De reisdagboeken en natuurtekeningen van Pieter van Oort (1825–1833)*. Werken uitgegeven door de Linschoten-Vereeniging, 119. Walburg, 2021.

Weber, Andreas, and Sylvia van Zanen. *Kunstenaar op Sumatra: De reisdagboeken en natuurtekeningen van Pieter van Oort (1833–1834)*. Werken uitgegeven door de Linschoten-Vereeniging, 120. Walburg, 2021.

Weber, Andreas, and Sylvia van Zanen. "Matters of Perspective: Local Visual Expertise and Natural History Drawings in Java, 1820–1850." *Wacana* 23, no. 1 (2022): 38–63.

Weiler, Danièle. "Pierre Médard Diard and Alfred Duvaucel: Two French Naturalists in the Service of Sir Stamford Raffles (December 7, 1818–March 27, 1820)." In *Voyageurs, explorateurs et scientifiques: The French and Natural History in Singapore*, edited by Martyn E. Y. Low, Kate Pocklington and Wan F. A. Jusoh. Lee Kong Chian Natural History Museum, 2019.

Wesseling, H. L. "The Giant That Was a Dwarf, or the Strange History of Dutch Imperialism." *The Journal of Imperial and Commonwealth History* 16, no. 3 (1988): 58–70.

Wesseling, H. L. *The European Colonial Empires, 1815–1919*. Pearson Longman, 2004.

Wheeler, Terry. "Vital Bodies of Evidence." *NewScientist*, no. 2967 (May 3, 2014) (2014).

Wille, Robert-Jan. "From Laboratory Lichens to Colonial Symbiosis: Melchior Treub Bringing German Evolutionary Plant Embryology to Dutch Indonesia, 1880–1909." *Studium* 18, no. 3 (2018): 191–205.

Wille, Robert-Jan. *Mannen van de microscoop: De laboratiumbiologie op veldtocht in Nederland en Indië, 1840–1910*. Vantilt, 2019.

Williams, Glyndwr. *Naturalists at Sea: Scientific Travellers from Dampier to Darwin*. Yale University Press, 2013.

Wingerden, Pieter van. "Goudkoorts in de gordel van smaragd." *Wonderkamer*, no. 1 (2020): 46–47.

Wingerden, Pieter van. "Science on the Edge of Empire: E. A. Forsten (1811–1843) and the Natural History Committee (1820–1850) in the Netherlands Indies." *Centaurus* 62, no. 4 (2020): 797–821.

Wingerden, Pieter van. "Op reis vanuit een drijvende stad: De expeditie van de Natuurkundige Commissie op Borneo in 1836." *Indische letteren* 38, no. 3 (2023): 179–197.

Wolters, Willem. "The 'Doit Infestation in Java': Exchange Rates between Silver and Copper Coins in Netherlands India in the Period 1816–1854." In *Money in Asia (1200–1900): Small Currencies in Social and Political Contexts*, edited by Jane Kate Leonard and Ulrich Theobald. Monies, Markets, and Finance in East Asia, 1600–1900, 6. Brill, 2015.

Wulf, Andrea. *Chasing Venus: The Race to Measure the Heavens*. Windmill Books, 2013.

Wulf, Andrea. *The Invention of Nature: The Adventures of Alexander von Humboldt, the Lost Hero of Science*. John Murray, 2015.

Yoo, Genie. "Wars and Wonders: The Inter-Island Information Networks of Georg Everhard Rumphius." *British Journal for the History of Science* 51, no. 4 (2018): 559–584.

Zanen, Sylvia van, and Andreas Weber. "Schreeuwende apen, zingende vogels en fluitende kogels: Natuuronderzoek op Sumatra in oorlogstijd (1833–1834)." *Indische letteren* 36, no. 4 (2021): 279–300.

Zangger, Andreas. *Koloniale Schweiz: Ein Stück Globalgeschichte zwischen Europa und Südostasien (1860–1930)*. 1800/2000 Kulturgeschichten der Moderne, 8. Transcript, 2011.

Zipelius, Alexander. "Uittreksel uit eenen brief van den Heer Zippelius, natuuronderzoeker in dienst der Nederlanden, aan en medegedeeld door doct. C. L. Blume, ridder der Orde van den Nederlandschen Leeuw, hoogleraar en directeur van 's Rijks Herbarium te Brussel." *Algemeene konst- en letter-bode*, no. 19 (May 8, 1829): 294–298.

Zobel, Maurice. "Les naturalistes voyageurs français et les grands voyages maritimes du XVIIIe et XIXe siècle." Doctorat en Médecine (Diplôme d'État), Faculté de Médecine de Paris, 1961.

Index of Names

This name index includes personal names of both historical characters and modern authors as they are mentioned in the main text of the introduction, chapters 1 to 7, and the conclusion. Please note that footnotes (except for discussions), figure captions, tables and appendices are not covered in this index.

Index of Places

This place index includes geographical names as they are mentioned in the main text of the introduction, chapters 1 to 7, and the conclusion. Please note that footnotes (except for discussions), tables and appendices are not covered in this index.

www.ingramcontent.com/pod-product-compliance
Lightning Source LLC
Chambersburg PA
CBHW070253260926
51849CB00002BA/405
9789004696860